ELECTROMAGNETIC NONDESTRUCTIVE EVALUATION (VIII)

Studies in Applied Electromagnetics and Mechanics

Volume 24

Previously published in this series:

Vol. 23. F. Kojima, T. Takagi, S.S. Udpa and J. Pávó (Eds.), Electromagnetic Nondestructive Evaluation (VI)
Vol. 22. A. Krawczyk and S. Wiak (Eds.), Electromagnetic Fields in Electrical Engineering
Vol. 21. J. Pávó, G. Vértesy, T. Takagi and S.S. Udpa (Eds.), Electromagnetic Nondestructive Evaluation (V)
Vol. 20. Z. Haznadar and Ž. Štih, Electromagnetic Fields, Waves and Numerical Methods
Vol. 19. J.S. Yang and G.A. Maugin (Eds.), Mechanics of Electromagnetic Materials and Structures
Vol. 18. P. Di Barba and A. Savini (Eds.), Non-Linear Electromagnetic Systems
Vol. 17. S.S. Udpa, T. Takagi, J. Pavo and R. Albanese (Eds.), Electromagnetic Nondestructive Evaluation (IV)
Vol. 16. H. Tsuboi and I. Vajda (Eds.), Applied Electromagnetics and Computational Technology II
Vol. 15. D. Lesselier and A. Razek (Eds.), Electromagnetic Nondestructive Evaluation (III)
Vol. 14. R. Albanese, G. Rubinacci, T. Takagi and S.S. Udpa (Eds.), Electromagnetic Nondestructive Evaluation (II)
Vol. 13. V. Kose and J. Sievert (Eds.), Non-Linear Electromagnetic Systems
Vol. 12. T. Takagi, J.R. Bowler and Y. Yoshida (Eds.), Electromagnetic Nondestructive Evaluation
Vol. 11. H. Tsuboi and I. Sebestyen (Eds.), Applied Electromagnetics and Computational Technology
Vol. 10. A.J. Moses and A. Basak (Eds.), Nonlinear Electromagnetic Systems
Vol. 9. T. Honma (Ed.), Advanced Computational Electromagnetics
Vol. 8. R. Collins, W.D. Dover, J.R. Bowler and K. Miya (Eds.), Nondestructive Testing of Materials
Vol. 7. C. Baumgartner, L. Deecke, G. Stroink and S.J. Williamson (Eds.), Biomagnetism: Fundamental Research and Clinical Applications

Volumes 1-6 were published by Elsevier Science under the series title "Elsevier Studies in Applied Electromagnetics in Materials".

ISSN 1383-7281

Electromagnetic Nondestructive Evaluation (VIII)

Edited by

Thierry Sollier

CEA-LIST, Gif-sur-Yvette, France

Denis Prémel

CEA-LIST, Gif-sur-Yvette, France

and

Dominique Lesselier

L2S-DRE, CNRS-SUPELEC-UPS, Gif-sur-Yvette, France

IOS Press

Amsterdam • Berlin • Oxford • Tokyo • Washington, DC

ISBN 1 58603 407 3
Library of Congress Control Number: 2003116195

Publisher
IOS Press
Nieuwe Hemweg 6B
1013 BG Amsterdam
The Netherlands
fax: +31 20 620 3419
e-mail: order@iospress.nl

Distributor in the UK and Ireland
IOS Press/Lavis Marketing
73 Lime Walk
Headington
Oxford OX3 7AD
England
fax: +44 1865 75 0079

Distributor in the USA and Canada
IOS Press, Inc.
5795-G Burke Centre Parkway
Burke, VA 22015
USA
fax: +1 703 323 3668
e-mail: iosbooks@iospress.com

PRINTED IN THE NETHERLANDS

Foreword

The E'NDE 2003 Workshop in Saclay, May 15-16, 2003, was the ninth of its kind, and is inscribed as such within a long and successful series of annual Workshops, which are intended to enable a fruitful and synergistic investigation of Electromagnetic Non-Destructive Evaluation. Indeed, this topic is crucial to an increasing number of engineering activities and environmental evaluation and protection issues, and it requires theoretical insight, efficient models and simulations, robust data interpretation, and proper instrumentation combined accordingly.

This Workshop has been organized – under the auspices of the ENDE International Standing Committee chaired by Professor Rubinacci, University di Cassino, Italy – by the Laboratoire de Contrôle par Méthodes Electromagnétiques (LCME), Service d'Instrumentation et Simulation pour le Contrôle non destructif (SISC), CEA, in cooperation with the Japan Society of Applied Electromagnetics and Mechanics (JSAEM), and with the contribution of the Laboratoire des Signaux et Systèmes (L2S) in Gif-sur-Yvette. It was held at the excellent facilities of the Institut National pour les Sciences et Techniques Nucléaires (INSTN) nearby.

68 participants were officially registered, from Europe, the United States, Korea and Japan, and 40 papers were presented in all.

After brief welcoming remarks by the two co-chairs, T. Sollier and D. Prémel, a 20-minute keynote address on "E-NDE in the Nuclear Industry: Viewpoint of the User" was given in introduction by R. Levy from Framatome ANP Intercontrôle, which, in summarized form, opens these edited proceedings.

Seven oral sessions were organized: "EC Modeling (I)" chaired by D. Lesselier and J. Pávó, "EC Modeling (II)" chaired by J. Bowler and A. Tamburrino, "Inverse Problem" chaired by G. Rubinacci and D. Prémel, "Inverse Problem and Signal Processing" chaired by E. Matsumoto and M. Lambert, on the first day; "Material Characterization" chaired by G. Dobmann and T. Théodoulidis, "Advances in Electromagnetic Non-Destructive Techniques (I)" chaired by T. Takagi and S. Udpa, and "Advances in Electromagnetic Non-Destructive Techniques (II)" chaired by F. Kojima and G. Berthiau, on the second day. On each day one poster session allowed also very interesting discussions in more informal fashion.

Short versions of all contributions have been published in the Workshop Digest, and 36 reviewed and accordingly revised full papers have been accepted and are now found in these edited proceedings, "Electromagnetic Non-Destructive Evaluation (VIII)" published by IOS Press in the series "Studies in Applied Electromagnetics and Mechanics".

To conclude, the organizers would like to thank all Committees members who have tirelessly contributed to the Workshop, the 29 reviewers of the extended papers, listed below, who have played a key role in the making of this book.

T. Sollier, CEA, D. Prémel, CEA, on leave from SATIE, D. Lesselier, L2S
Co-Editors

ENDE (2003) – Saclay, Paris

List of Referees and Scientific Committee

P. Aknin – INRETS, France
R. Albanese – University of Reggio Calabria, Italy
G. Berthiau – Université de Nantes, France
A. Baussard – Ecole Normale Supérieure de Cachan, France
J.R. Bowler – Iowa State University, Center for Nondestructive Evaluation, USA
G. Dobmann – IZFP Germany
B. Duchêne – Département de Recherche en Electromagnétisme, Laboratoire des Signaux et Systèmes, France
S. Gautier – EDF/Direction Recherche et Développement, France
C. Gilles-Pascaud – CEA Saclay, Direction de la Recherche Technologique, France
P.Y. Joubert – Ecole Normale Supérieure de Cachan, France
F. Kojima – Kobe University, Japan
M. Lambert – Département de Recherche en Electromagnétisme, Laboratoire des Signaux et Systèmes, France
Y. Le Bihan – Laboratoire de Génie Electrique de Paris, France
D. Lesselier – Département de Recherche en Electromagnétisme, Laboratoire des Signaux et Systèmes, France
C. Marchand – Laboratoire de Génie Electrique de Paris, France
M. Mayos – EDF-Groupe des Laboratoires, France
G. Micolau – Institut Fresnel, Marseillle, France
J. Pávó – Budapest University of Technology and Economics, Hungary
G. Pichenot – CEA Saclay, Direction de la Recherche Technologique, France
D. Prémel – CEA Saclay, Direction de la Recherche Technologique, France
A. Razek – Laboratoire de Génie Electrique de Paris, France
G. Rubinacci – University of Cassino, Italy
T. Sollier – CEA Saclay, Direction de la Recherche Technologique, France
T. Takagi – Tohoku University, Japan
A. Tamburrino – University of Cassino, Italy
T. Theodoulidis – Ariostotle University, Greece
L. Udpa – Michigan State University, U.S.A.
S.S. Udpa – Michigan State University, U.S.A.
R. Zorgati – EDF R&D / Département de Recherche en Electromagnétisme, Laboratoire des Signaux et Systèmes, France

Standing Committee

G. Rubinacci – Universita Cassino, Italy, Chairman
J.R. Bowler – Iowa State University, U.S.A.
R. Albanese – Universita Reggio Calabria, Italy
G. Dobmann – Fraunhofer Institute for NDT, Germany
R. Grimberg – National Institute of R&D for Technical Physics, Romania
H.K. Jung – Seoul National University, South Korea
F. Kojima – Kobe University, Japan
D. Lesselier – DRE-LSS CNRS-SUPELEC-UPS, France
V. Lunin – Moscow Power Engineering Institute, Russia
K. Miya – Keio University, Japan
G.Z. Ni – Zhejing University, China
J. Pávó – Budapest University of Technology and Economics, Hungary
A. Razek – LGEP CNRS-SUPELEC-UPS-UPMC, France
J.N. Sheng – X'ian Jiatong University, China
T. Sollier – CEA Paris, France
T. Takagi – Tohoku University, Japan
H. Takamatsu – KEPCO, Japan
S.S. Udpa – Iowa State University, U.S.A.
V. Vengrinovich – Institute of Applied Physics, Belarus

Organizing Committee

D. Prémel and T. Sollier, Co-chairs – CEA Saclay, France,
D. Lesselier – DRE-LSS CNRS-SUPELEC-UPS, France,
G. Berthiau – GE44, Université de Nantes, France.

ENDE (2003) – Saclay, Paris, List of Participants

Mr. Patrice AKNIN
INRETS
2 avenue du Général Malleret-Joinville
94114 Arcueil
France
Phone : 33 1 47 40 73 37
Fax : 33 1 45 47 56 06
aknin@inrets.fr

Mr. Raffaele ALBANESE
DIMET-University of Mediterranea RC
Via Graziella, Loc. Feo di Vito
89060 Reggio Calabria
Italy
Phone : 3909 6587 5242
Fax : 3909 658 75247
albanese@unirc.it

Ms. Iris ALTPETER
Fraunhofer-Institute für Zerstörungsfreie Prüfverfahren, Universität, Geb. 37
66123 Saarbrücken
Germany
Phone : 4968 19302 3827
Fax : 4968 1930 25920
iris.altpeter@izfp.fraunhofer.de

Mr. Katsuyuki ARA
NDE & SR Center, Faculty of Engineering, Iwate University, UEDA 4-3-5
020-8551 Morioka-shi
Japan
Phone : 81 19 621 6891
Fax : 81 19 621 6891
kara@iwate-u.ac.jp

Mr. Philippe BENOIST
CEA SACLAY
Bâtiment 611, Service SISC
91191 Gif-sur-Yvette
France
Phone : 33 1 69 08 66 08
Fax : 33 1 69 08 75 97
philippe.benoist@cea.fr

Mr. Mohamed BENTOUMI
CRAN – ESSTIN
2 rue Jean Lamour
54500 Vandoeuvre-Les-Nancy
France
Phone : 33 3 83 68 51 39
bentoumi@esstin.uhp-nancy.fr

Mr. Gérard BERTHIAU
CRTT
Bd de l'Université BP 406
44602 Saint Nazaire
France
Phone : 33 2 40 17 26 34
Fax : 33 2 40 17 26 18
gerard.berthiau@ge44.univ-nantes.fr

Mr. John R. BOWLER
CNDE, Iowa State University
1915 Scholl Road
50011 Ames, IA
USA
Phone : +1 515-294-2093
Fax : +1 515-294-7771
jbowler@iastate.edu

Mr. Fabrice BUVAT
CEA SACLAY
Bâtiment 611, Service SISC
91191 Gif-sur-Yvette
France
Phone : 33 1 69 08 65 24
Fax : 33 1 69 08 75 97
fabrice.buvat@cea.fr

Mr. Sébastien CADEAU-BELLIARD
CEDRAT
10 Chemin de pré carré, ZIRST
38246 Meylan Cedex
France
Phone : 334 76 90 50 45
Fax : 33 1 76 90 16 09
cedrat@cedrat.com

Mr. Tomasz CHADY
Technical University of Szczecin
ul. Sikorskiego 37
70-313 Szczecin
Poland
Phone : (+48 91) 4 49 42 83
Fax : (+48 91) 4 34 09 26
tchady@ps.pl

Mr. Zhenmao CHEN
International Institute of Universality
Bldg. 8F, 1-4-6 Nezu, Bunkyo-ku
113-0031 Tokyo
Japan
Phone : 81-3-5814-5350
Fax : 81-3-3827-0682
chen@iiu.co.jp

Ms. Weiying CHENG
Tsurumi R&D center, JAPEIC
Bentenn-cho 14-1, Tsurumi-ku
230-0044 Yokohama-Shi
Japan
Phone : 86-45-511-2751
Fax : 86-45-511-2750
cheng-weiying@japeic.or.jp

Mr. Ho-Yong CHOI
Seoul National University,ENG420-040,
301-614, School of Electrical Engineering,
Seoul National University, Shilim-Dong,
Kwanak-Gu
151-744 Seoul
South Korea
Phone : 82 2 880 7262
Fax : 82 2 878 1452
plate@elecmech.snu.ac.kr

Mr. Marco CIOFFI
Second University of Napoli,
Dipartimento di Ingegneria delli
Informazione
Via Roma 29
81031 Aversa
Italy
Phone : 39 81 5010342
Fax : 3981 5037042
marco.cioffi@unina2.it

Mr. Jean-Marc DECITRE
SATIE, ENS Cachan
61 av. du président Wilson
94230 Cachan
France
Phone : 33 1 47 50 55 84
Fax : 33 1 47 40 21 99
decitre@satie.ens-cachan.fr

Mr. Gerd DOBMANN
Fraunhofer-Institute für Zerstörungsfreie
Prüfverfahren, Universität, Geb. 37
66123 Saarbrücken
Germany
Phone : 4968 1930 23855
Fax : 4968 1930 2593
gerd.dobmann@izfp.fraunhofer.de

Ms. Stephanie DUBOST
EDF
6 quai Watier
78401 Chatou Cedex
France
Phone : 33 1 30 87 85 92
Fax : 33 1 30 87 75 89
stephanie.dubost@edf.fr

Mr. Bernard DUCHENE
L2S-DRE, SUPELEC
Plateau de Moulon, 3 rue Joliot Curie
91192 Gif-sur-Yvette
France
Phone : 33 1 69 85 15 57
Fax : 33 1 69 85 17 65
duchene@lss.supelec.fr

Mr. Gilles DURIEUX
Technocentre RENAULT
API : TCR LAB 0 35, Sce 64130
1 avenue du Golf
78288 Guyancourt
France
Phone : 33 1 34 95 49 95
Fax : 33 1 34 95 10 29
gilles.durieux@renault.com

Mr. Hisashi ENDO
Institute of Fluid Science, Tohoku University
2-1-1 Katahira, Aoba
980-8577 Sendai
Japan
Phone : 81 22 217 5298
Fax : 81 22 217 5298
endo@ifs.tohoku.ac.jp

Mr. Antonio FABA
Via Pentima Bassa
21 Faculty of Engineering
5100 Terni
Italy
Phone : 390 7444 92812
Fax : 3907 444 92925
faba@unipg.it

Mr. Fabrizio FERRAIOLI
Second University of Napoli, Dipartimento di Ingegneria delli Informazione
Via Roma 29
81031 Aversa
Italy
Phone : 39 81 5010342
Fax : 3981 503742
ferraioli.fabrizio.ari@ansaldobreda.it

Mr. Hervé GAC
RDTECH
3 avenue de la Baltique, ZA de Courtaboeuf
91953 Courtaboeuf
France
Phone : 33 1 69 18 99 20
Fax : 33 1 69 18 99 30
hgac@rd-tech.fr

Mr. Antal GASPARICS
Research Institute for Technical Physics and Materials Science
Konkoly T. M. 29-33
1121 Budapest
Hungary
Phone : 36 1 39 22222
Fax : 36 1 3922226
gaspari@mfa.kfki.hu

Ms. Catherine GILLES-PASCAUD
CEA SACLAY
Bâtiment 611, Service SISC
91191 Gif-sur-Yvette
France
Phone : 33 1 69 08 64 29
Fax : 33 1 69 08 75 97
catherine.gilles-pascaud@cea.fr

Mr. Fedor GOMORY
Inst. Electrical Engineering, Slovak Acad. Sciences
Dubravska 9
84101 Bratislava
Slovakia
fedor.gomory@savba.sk

Mr. Hidatoshi HASHIZUME
Tohoku University
01 Amaki Aza Aota, Aoba
980-8579 Sendai
Japan
Phone : 81 22 217 7904
Fax : 81 22 217 7904
hidatoshi.hashizume@qse.tohoku.ac.jp

Mr. Satoshi ITO
Tohoku University
01 Amaki Aza Aota, Aoba
980-8579 Sendai
Japan
Phone : 81 22 217 7906
Fax : 81 22 217 7906
sito@karma.qse.tohoku.ac.jp

Mr. Ladislav JANOUSEK
International Institute of Universality
Bldg. 8F, 1-4-6 Nezu, Bunkyo-ku
113-0031 Tokyo
Japan
Phone : 81-3-5814-5354
Fax : 81-3-3827-0682
janousek@jsaem.gr.jp

Mr. Hiroaki KIKUCHI
NDE & SR Center, Faculty of Engineering, Iwate University, UEDA 4-3-5
020-8551 Morioka-shi
Japan
Phone : 81 19 621 6890
Fax : 81 19 621 66890
hkiku@iwate-u.ac.jp

Mr. Jae-Kwang KIM
Seoul National University,ENG420-040, 301-614, School of Electrical Engineering, Seoul National University, Shilim-Dong, Kwanak-Gu
151-744 Seoul
South Korea
Phone : 82 2 880 7262
Fax : 82 2 878 1452
light7@elecmech.snu.ac.kr

Mr. Fumio KOJIMA
Graduate School of Science and Technology, Kobe University
1-1, Rokkodai, Nada-ku
657-8501 Kobe, Hyougo
Japan
Phone : 81 78 803 8493
Fax : 81 78 803 6493
kojima@cs.kobe-u.ac.jp

Mr. Marc LAMBERT
L2S-DRE, SUPELEC
Plateau de Moulon, 3 rue Joliot Curie
91192 Gif-sur-Yvette
France
Phone : 33 1 69 85 15 67
Fax : 33 1 69 85 17 65
lambert@lss.supelec.fr

Mr. Yann LE BIHAN
SUPELEC/LGEP Plateau de Moulon
91192 Gif-sur-Yvette
France
Phone : 33 1 69 85 16 61
Fax : 33 1 69 41 83 18
yann.le-bihan@lgep.supelec.fr

Mr. Dominique LESSELIER
L2S-DRE, SUPELEC
Plateau de Moulon, 3 rue Joliot Curie
91192 Gif-sur-Yvette
France
Phone : 33 1 69 85 15 61
Fax : 33 1 69 85 17 65
lesselier@lss.supelec.fr

Mr. Robert LEVY
Intercontrôle
13, rue du Capricorne, SILIC 433
94583 Rungis Cedex
France
Phone : 33 1 49 78 40 44
Fax : 33 1 49 78 41 66
robert.levy@framatome-anp.com

Mr. Eiji MATSUMOTO
Kyoto University, Department of Energy Conversion Science
606-8501 Kyoto
Japan
Phone : 81 75 753 5247
Fax : 81 75 753 5247
matumoto@energy.kyoto-u.ac.jp

Mr. Michel MAYOS
EDF/GDL/SIMA
2 rue Ampère
93206 Saint Denis
France
Phone : 33 1 43 69 54 40
Fax : 33 1 43 69 91 04
michel.mayos@edf.fr

Mr. Philip MEILLAND
IRSID
Voie romaine BP 30320
57283 Maizieres les Metz
France
Phone : 33 3 87 70 48 59
Fax : 33 3 87 70 41 05
philip.meilland@irsid.arcelor.com

Mr. Ovidiu MIHALACHE
Japan Nuclear Cycle Development
Institute, ISI System Development Group
1 Shiraki, TSURUGA-SHI, FUKUI-KEN
919-1279 Tsuruga-Shi
Japan
Phone : 81 770 39 1031
Fax : 81 770 39 9227
omih@t-hq.jnc.go.jp

Mr. Kenzo MIYA
International Institute of Universality
Bldg. 8F, 1-4-6 Nezu, Bunkyo-ku
113-0031 Tokyo
Japan
Phone : 81-3-5814-5350
Fax : 81-3-3827-0682
miya@iiu.co.jp

Mr. Maxim MOROZOV
DAEIMI
via G. di Biasio, 43
03043 Cassino
Italy
Phone : +39-0776-299677
Fax : 39 0776 299707
morozov@unicas.it

Mr. Subhas MUKHOPADHYAY
Massey University
PVR.BAG 11 222 Palmerston North
New Zealand
Phone : 64 6 3505799
Fax : 64 6 3502257
S.C.Mukhopadhyay@massey.ac.nz

Mr. Shoichiro NAGATA
Miyazaki University
1-1 Gakuenkibanadai-nishi, Miyazaki
889-2192 Miayazaki pref
Japan
Phone : 81 985 58 7355
Fax : 81 985 58 7327
nagata@ee.miyazaki-u.ac.jp

Mr. Pavel NOVOTNY
Institute of Chemical Technology
Technicka 5
166 28 Prague 6
Czech Republic
Phone : 4202 2435 5163
Fax : 4202 2431 0337
pavel.novotny@vscht.cz

Mr. Jozsef PAVO
Department of Electromagnetic Theory,
Budapest Univ. Of Tech. And Econ.
Egry J. u. 18
H-1521 Budapest
Hungary
Phone : 36 1 4632913
Fax : 36 1 4633189
pavo@evtsz1.evt.bme.hu

Mr. Grégoire PICHENOT
CEA SACLAY
Bâtiment 611, Service SISC
91191 Gif-sur-Yvette
France
Phone : 33 1 69 08 91 50
Fax : 33 1 69 08 75 97
gregoire.pichenot@cea.fr

Mr. Jean-Paul POULET
DELI
16bis rue Guejet
71100 Chalon sur Saone
France
Phone : 33 3 85 98 96 00
Fax : 33 3 85 98 95 95
jp.poulet@deli.fr

Mr. Denis PREMEL
CEA SACLAY
Bâtiment 611, Service SISC
91191 Gif-sur-Yvette
France
Phone : 33 1 69 08 59 63
Fax : 33 1 69 08 75 97
denis.premel@cea.fr

Mr. Mihai REBICAN
International Institute of Universality
Bldg. 8F, 1-4-6 Nezu, Bunkyo-ku
113-0031 Tokyo
Japan
Phone : 81-3-5814-5350
Fax : 81-3-3827-0682
mihai-r@jsaem.gr.jp

Ms. Belén RIBES SAEZ
TECNATOM, SA.
Avda. Montes de Oca 1
28709 San Sebastian de los Reyes
Spain
Phone : 9165 986 64
Fax : 91 6586 77
bribes@tecnatom.es

Mr. Benoit RICHARD
EDF
6 quai Watier
78401 Chatou Cedex
France
Phone : 33 1 30 87 81 83
Fax : 33 1 30 87 75 89
benoit-r.richard@edf.fr

Mr. Guglielmo RUBINACCI
DAEIMI
via G. di Biasio, 43
03043 Cassino
Italy
Phone : +39-0776-299629
Fax : 3907 7629 9707
rubinacci@unicas.it

Mr. Jean-Marc SARTEEL
EXCELDEF
24, Parc d'activités La Tuilerie
71640 Dracy Le Fort
France
Phone : 33 3 85 98 15 14
Fax : 33 3 85 87 68 47
direct.luziesa@wanadoo.fr

Mr. Takuya SHIBATA
Tohoku University
01 Amaki Aza Aota, Aoba
980-8579 Sendai
Japan
Phone : 81 22 217 7906
Fax : 81 22 217 7906
tshiba@karma.qse.tohoku.ac.jp

Mr. Ryszard SIKORA
Technical University of Szczecin
ul. Sikorskiego 37
70-313 Szczecin
Poland
Phone : (+48 91) 4 49 49 67
Fax : (+48 91) 4 34 09 26
rs@ps.pl

Mr. Thierry SOLLIER
CEA SACLAY
Bâtiment 611, Service SISC
91191 Gif-sur-Yvette
France
Phone : 33 1 69 08 35 14
Fax : 33 1 69 08 75 97
thierry.sollier@cea.fr

Mr. Oleksandr STUPAKOV
NA Slovance 2, PRAHA 8
18221 Prague
Czech Republic
Phone : 42026 605 2879
Fax : 4202 868 90527
stupak@fzu.cz

Mr. Toshiyuki TAKAGI
Tohoku University, Institute of Fluid Science
Katahira 2-1-1, Aoba-ku
980-8577 Sendai
Japan
Phone : 81 22 217 5248
Fax : 81 22 217 5248
takagi@ifs.tohoku.ac.jp

Mr. Seiki TAKAHASHI
NDE & SR Center, Faculty of Engineering
Iwate University, UEDA 4-3-5
020-8551 Morioka-shi
Japan
Phone : 81 19 621 6348
Fax : 81 19 621 6348
seiki.t@iwate-u.ac.jp

Mr. Antonello TAMBURRINO
DAEIMI
via G. di Biasio, 43
03043 Cassino
Italy
Phone : +39-0776-299675
Fax : 39077 62 99707
tamburrino@unicas.it

Mr. Theodoros THEODOULIDIS
Aristotle University of Thessaloniki
Kastorias & Fleming
50100 Kozani
Greece
Phone : 3023 1099 1307
Fax : 3023 1099 1304
theodoul@hellasnet.gr

Ms. Françoise THEVENOT
Dassault Aviation
1 avenue du Parc
95100 Argenteuil
France
Phone : 33 1 34 11 86 99
Fax : 33 1 34 11 88 24
cde@dassault-aviation.fr

Mr. Satish UDPA
Michigan State University, Electrical and
Computer Engineering
2120 Eng. Bldg., E. Lansing,
MI 48824 East Lansing
USA
Phone : +1 517-432-4787
Fax : +1 (517) -353-1980
udpa@egr.msu.edu

Ms. Lalita UDPA
Michigan State University, Electrical and
Computer Engineering
2120 Eng. Bldg., E. Lansing,
MI 48824 East Lansing
USA
Phone : (517) 355-9261
Fax : +1 (517) -353-1980
udpal@msu.edu

Mr. Noritaya YUSA
International Institute of Universality
Bldg. 8F, 1-4-6 Nezu, Bunkyo-ku
113-0031 Tokyo
Japan
Phone : 81-3-5814-5350
yusa@iiu.co.jp

Mr. Riadh ZORGATI
EDF R & D
L2S-DRE, SUPELEC
Plateau de Moulon, 3 rue Joliot Curie
91192 Gif-sur-Yvette
France
Phone : 33 1 69 85 15 60
Fax : 33 1 69 85 17 65
zorgati@lss.supelec.fr

Contents

Keynote Address

E-NDE IN THE NUCLEAR INDUSTRY: VIEWPOINT OF A USER

Robert LEVY
Framatome ANP/Intercontrôle – 13, rue du Capricorne 94583 Rungis Cedex, France

1. INTRODUCTION

The French nuclear industry has made extensive use of electromagnetic non destructive evaluation for more than twenty five years now. Initially, eddy current testing of steam generator tubes was performed in 1975 at a level of technology that did not exist in other industrial sectors. The performances of the NDE method fostered new applications and today, eddy current inspections also occur, for example, on the reactor pressure vessel in nozzles where ultrasonic testing and radiography were the only methods used before, or on the vessel head penetrations.

When considering the applications over a quarter of century, the main improvements have come from the development of new probes and from signal processing.

2. VESSEL HEAD PENETRATIONS

2.1 The role of E-NDE

Since the detection of a leak during hydraulic pressure testing and acoustic monitoring of the vessel in 1994, the inspection of vessel head penetrations has become a requirement of the Regulatory Body.

As can be seen from the drawing in Figure 1, the area is particularly difficult to reach and to scan. In many cases, the thermal sleeve cannot be removed and the space left free for a probe is thus no larger than 3mm.

The requirements of the Regulator covering both the detection and the characterisation of cracks, it was decided to use eddy currents (ET) for the ability of this method to detect anomalies with fast scanning, and to apply ultrasound (UT) as a complementary method only when sizing is needed.

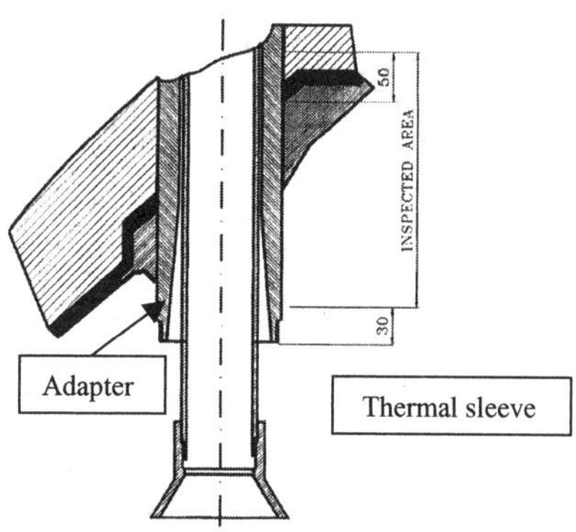

Figure 1 : Vessel closure head adapter

Blade probes (Fig. 2) mounted on a robotic arm are remotely operated:
- the ET probe (Fig. 3b) makes it possible to detect and position cracks at ± 3mm in height and ± 1° in angular location.
- wherever cracks are longer than 2 mm, their depth is measured using a specific UT probe, (Fig. 3a) also mounted on a blade which enables a measurement accuracy of ± 1mm in depth, as was demonstrated on real cracks.

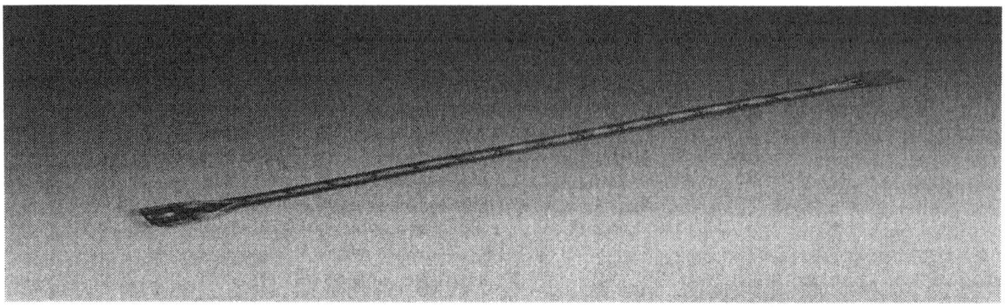

Figure 2: Blade probe

(a) 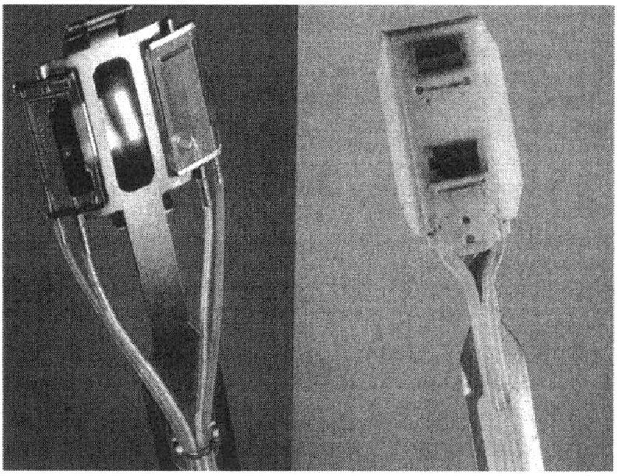 (b)

Figure 3 : Sensors mounted on the blade probes
a) UT transducer b) ET probe

2.2 The inspections

Today, the number of inspections carried out in France has considerably decreased. Nevertheless, the experience gained in the inspections is applicable to other regions where this degradation is now emerging; this is the case for nuclear plants in the U.S.A where defects found on head penetrations have led to the qualification of the techniques described and their implementation on site by Framatome ANP Inc.

3. REACTOR VESSEL NOZZLES

3.1 A historical result

In 1982, a specific eddy current inspection was performed on steam generator tubesheets at Koeberg NPP (RSA). The objective was the detection of cracks at the interface of the tubesheet with the cladding and for that purpose a specific probe had to be developed and qualified.

On a rather large number of recordings, the eddy current signal unambiguously showed a larger thickness of the cladding. When reporting the corresponding positions on the map of the tubesheet, the analysts came out with the map of a repair.

3.2 A new configuration

A certain number of well-identified outlet (H) nozzles of the reactor pressure vessel were repaired in the shop during the manufacturing process. The inner surface was ground over areas were small defects had been detected and the removed metal was replaced by a layer of Inconel.

The development of cracks at the interface being possible, a specific inspection was required starting from 1993.

The objectives of the inspection were initially:

- to draw a map of the repaired zones
- to provide information on any existing crack within these zones.

In order not to dramatically increase the duration of nozzle inspections, it was decided to adopt a multi-technique approach.

Eddy currents are very sensitive to changes in conductivity and metallurgic structure; therefore, they were selected for detecting the boundary between repaired zones and sound material. The operating conditions were fine-tuned in order to use the same scan of the probe for both drawing maps and detecting cracks, if any.

The performance of the electromagnetic method did not meet the requirements as far as sizing of defects was concerned. The inspection tool was then equipped with four UT transducers enabling a sizing accuracy of ± 1mm in depth which was demonstrated on real defects.

Between 1993 and 1995, all the vessels concerned were inspected: the repaired areas were successfully mapped (Fig. 4) and no cracks were detected.

Last May 2003, in the framework of a ten-year in-service inspection, the vessel nozzles were examined with an upgraded tool incorporating improvements as result of the experience acquired.

The inspection was limited to the zones identified during the previous inspections. The eddy current probe scans the surface at a speed of 300 mm/s, then an angular movement of 0.2° of the tool places it on a new area to be scanned; if a crack is detected, the examination is resumed with the UT probe at the location provided by the eddy current results and the UT transducers scan the limited portion at a speed of 15 mm/s, with the same angular increment of the tool after each linear scan.

In both ET and UT examinations, data is taken every 1 mm.

Figure 4 : Map of a repair in a nozzle.

4. STEAM GENERATOR TUBES

4.1 French experience

In 2002, the French nuclear industry inspected 156 steam generators of the total fleet, using eddy current techniques.

In these generators, approximately 305 000 tubes were inspected with an "axial" probe that travels inside the tube at a linear speed of 50 cm/s (when being pulled); the total length inspected is close to 700 kilometres. 228 000 portions on as many tubes were scanned by rotating probes in the roll transition area; the probe rotates at 300 rpm, representing a linear speed of 30 cm/s; the total surface of scanning is close to 84 400 m^2 .

In terms of amount of data, 460 Gbytes from axial probes were fed to dedicated software (Surcouf, Aida…) while 90 Gbytes from rotating probes were processed by another programme (Estelle).

For axial probe data, it takes less than *five seconds* to have 1.5 Mbyte processed; the processing includes a calibration on a reference signal, a data quality check with 20 verifications, a detection of all "events"; from these events, anomalies are identified and characterised by measurement of the amplitude and phase of the signal at three frequencies and two multifrequency mixes in order to propose a diagnosis.

In the case of rotating probe data, 400 Kbytes are processed in *two seconds* and the result includes the number of cracks, the length of each crack and its position with respect to the last contact point between the tube and the tubesheet.

4.2 Developments

The performances of the rotating probes (Fig 5) in terms of accuracy and resolution being somewhat hindered by the rather slow scanning speed, this situation fostered the development of a probe that could travel as fast as the axial probe and could deliver as precise an amount of information as the rotating probe.

Several attempts were made in the world to produce array probes on an industrial basis. Some encouraging results had first been obtained by Cecco in Canada and around 1996, in France, the French Atomic Energy Commission (CEA) and Framatome were studying in the lab their own approach. Figure 6 clearly illustrates the complexity of the problem.
The only industrial array probes nowadays are produced by RDTech (Canada) and MHI (Japan).

Having to conciliate *performance requirements and reduced costs*, no request for implementation of such probes was formulated by the French nuclear industry, thus leaving the "user" with a feeling of frustration! The only innovation today consists in the use of a new rotating probe which is in fact the assembly in a single body of two sets of sensors (Fig. 7).

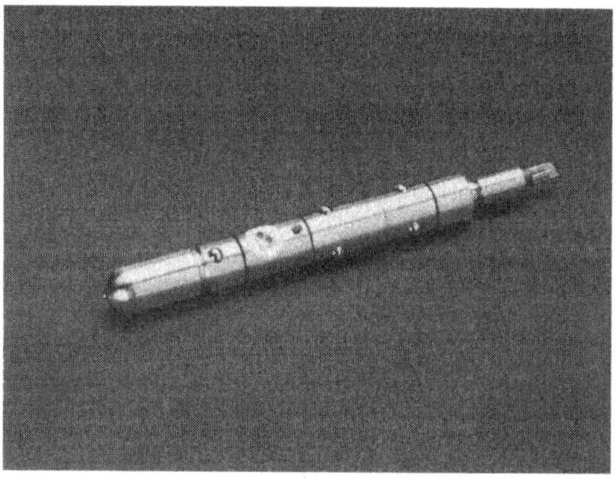

Figure 5 : ET rotating probe

Figure 6 : Prototype of an axial array probe

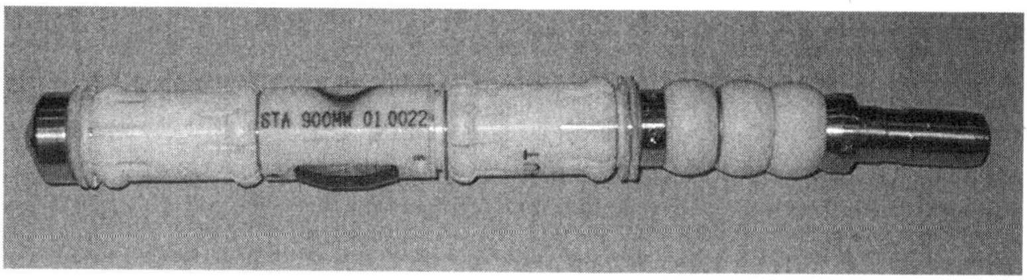

Figure 7 : The improved rotating probe

Since the analytical model developed by Dodd and Deeds, modelling in electromagnetic non-destructive evaluation has been the subject of numerous works and publications. The papers following this keynote are a small evidence, if need be!

As in other places, analytical and numerical models as well as computational codes have been developed by members of the French nuclear industry or university laboratories: Mascotte (analytical model), Trifou, Jasmine (EDF), Messine (CEA). Still, in almost forty years now, there has been little (or no) "penetration" of these works in industrial applications of the nuclear sector.

Nevertheless, the increasing demand for "productivity" opens a larger space to the use of software programmes for automatic data analysis and this could very well be the appropriate field where models and inversion should play an important role.

5. CONCLUSION

After the excitement linked to the novelty and the feeling of pioneering, e-NDE in the nuclear industry has entered an era of apparent stability. Typically, the request for rapidity and economy should mainly encourage the optimisation of the existing tools and their refinement where relevant.

Provided that the R&D works, which have been going on for several years now, produce results compatible with an industrial implementation, the nuclear industry however remains an open and large field for e-NDE applications.

EC Modeling

Electromagnetic Nondestructive Evaluation (VIII)
T. Sollier et al. (Eds.)
IOS Press, 2004

Theoretical and Experimental Study of Eddy Current Induction in a Conducting Cylinder

Theodoros P. THEODOULIDIS
Department of Engineering and Management of Energy Resources, Aristotle University of Thessaloniki, Kastorias & Fleming, 50100 Kozani, Greece. Email: theodoul@hellasnet.gr

Stephen K. BURKE
Defence Science and Technology Organisation, 506 Lorimer St, Fishermans Bend, Victoria 3207, Australia. Email: steve.burke@dsto.defence.gov.au.

Abstract. Eddy-current induction in a conducting cylinder is studied using the second order vector potential formulation. A general expression for the impedance change of the excitation coil is derived using the reciprocity theorem and the solution to the electromagnetic field problem. It is shown that the only information needed about the coil is a Fourier transform of the coil's magnetic field in free-space. Theoretical results obtained from a numerical treatment of this transform as well as experimental results are presented for an air-cored cylindrical coil positioned with its axis perpendicular to the cylinder axis. The theoretical results are in excellent agreement with experiment.

1. Introduction

The analysis of eddy-current induction in a conducting cylinder excited by a coil of arbitrary shape is a challenging problem for existing analytical modelling methods. Previous studies have been concerned either with filamentary coils or concentrated on the calculation of the induced eddy current distribution [1]–[5]. Following a similar study on eddy current induction in boltholes [6], we examine the electromagnetic field in the case of a coil outside a conducting cylinder and focus on the derivation of a general expression for the coil impedance change ΔZ due to eddy-current induction. The analysis is then applied to the specific case of a horizontal coil, i.e. a cylindrical air-cored coil with its axis perpendicular to the cylinder axis and coincident with the cylinder radius, as shown in Fig.1. To verify the theory, a series of experiments is performed to measure ΔZ as a function of frequency for a cylindrical air-cored coil and a range of conducting cylinders.

2. Analysis

2.1 The SOVP formulation

As early as 1973, it was shown by Hannakam [1] that the time-harmonic 3D eddy current induction problem in a conducting cylinder can be solved using the second order vector potential (SOVP) **W,** which is connected to the magnetic vector potential **A** and the magnetic flux density **B** through:

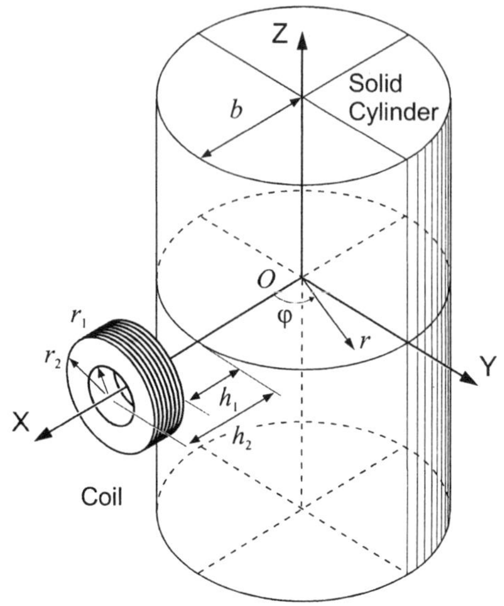

Figure 1. Horizontal coil outside a conducting cylinder. The cylindrical air cored coil has a rectangular cross section and the coil axis coincides with the x axis.

$$\mathbf{B} = \nabla \times \mathbf{A} = \nabla \times \nabla \times \mathbf{W} = \nabla \times \nabla \times (W_a \hat{\mathbf{z}} + \hat{\mathbf{z}} \times \nabla W_b) = \nabla \left(\frac{\partial W_a}{\partial z} \right) - k^2 (W_a \hat{\mathbf{z}} + \hat{\mathbf{z}} \times \nabla W_b), \qquad (1)$$

where $k^2 = j\omega\mu_r\mu_0\sigma$, σ is the electrical conductivity, μ_r is the relative permeability, μ_0 is the permeability of free space and ω is the angular frequency of the excitation. In Eq.(1), W_a and W_b are second order potential functions (similar to the electric and magnetic Hertz potentials) that satisfy the scalar Helmholtz or scalar Laplace equations in conducting and non-conducting regions respectively.

The electromagnetic field will be calculated in two regions: the annular region R_0 which lies between the minimum radial coordinate of the coil windings and the wall of the cylinder ($b < r < b+h_1$) and the volume of the conducting cylinder R_1 ($0 < r < b$).

From Eq.(1) it is deduced that in the non-conducting region R_0 ($k^2 = 0$), the magnetic flux density can be described using W_a only and knowledge of W_b is not required. The W_a potential in R_0 can be regarded as the superposition

$$W_{a0} = W_{a0}^{(s)} + W_{a0}^{(ec)} \qquad (2)$$

of a (known) incident or source potential $W_{a0}^{(s)}$ due to excitation of the coil in free space and the potential $W_{a0}^{(ec)}$ due to eddy-current induction in the conducting material of the cylinder. As described previously [1], closed form integral representations for the potentials in regions R_0 and R_1 can be obtained in cylindrical coordinates using the method of separation of variables and applying physical restrictions for the potentials at $r \to \infty$ and $r \to 0$:

$$W_{a0}^{(s)} = \int_{-\infty}^{\infty} \sum_{m=-\infty}^{\infty} C_s(a,m)\, I_m(|\alpha| r) e^{jm\varphi} e^{j\alpha z} d\alpha\,, \tag{3}$$

$$W_{a0}^{(ec)} = \int_{-\infty}^{\infty} \sum_{m=-\infty}^{\infty} D_{ec}(a,m)\, K_m(|\alpha| r) e^{jm\varphi} e^{j\alpha z} d\alpha\,, \tag{4}$$

$$W_{a1} = \int_{-\infty}^{\infty} \sum_{m=-\infty}^{\infty} C_{a1}(a,m)\, I_m(\alpha_1 r) e^{jm\varphi} e^{j\alpha z} d\alpha\,, \tag{5}$$

$$W_{b1} = \int_{-\infty}^{\infty} \sum_{m=-\infty}^{\infty} C_{b1}(a,m)\, I_m(\alpha_1 r) e^{jm\varphi} e^{j\alpha z} d\alpha\,, \tag{6}$$

where C_s, D_{ec}, C_{a1}, C_{b1} are unknown coefficients, I_m and K_m are modified Bessel functions of the first and second kind respectively and $\alpha_1^{\,2}=\alpha^2+k^2$. The coefficient C_s which enters the representation for the source potential $W_{a0}^{(s)}$ in (3) depends only on the coil geometry and is assumed to be a known quantity. The expressions for the remaining coefficients are obtained in terms of C_s by applying the appropriate interface conditions at the wall of the cylinder:

$$C_{a1} = \frac{C_s(\alpha,m)\Lambda(\alpha_1 b)}{K_m(|\alpha| b) I_m(\alpha_1 b)} \times \frac{\alpha^2 b \mu_r \alpha_1}{k^2 m^2 \mu_r + \alpha_1^2 b^2 \Lambda(\alpha_1 b)[\alpha^2 \mu_r \Lambda(\alpha_1 b) - |\alpha| \alpha_1 \mathrm{M}(|\alpha| b)]}\,, \tag{7}$$

$$C_{b1} = \frac{C_s(\alpha,m)}{K_m(|\alpha| b) I_m(\alpha_1 b)} \times \frac{m \alpha \mu_r}{k^2 m^2 \mu_r + \alpha_1^2 b^2 \Lambda(\alpha_1 b)[\alpha^2 \mu_r \Lambda(\alpha_1 b) - |\alpha| \alpha_1 \mathrm{M}(|\alpha| b)]}\,, \tag{8}$$

$$D_{ec} = -C_s(\alpha,m) R(\alpha,m)\,, \tag{9}$$

where

$$R(\alpha,m) = \frac{I_m(|\alpha| b)}{K_m(|\alpha| b)} \times \frac{k^2 m^2 \mu_r + \alpha_1^2 b^2 \Lambda(\alpha_1 b)[\alpha^2 \mu_r \Lambda(\alpha_1 b) - |\alpha| \alpha_1 \Lambda(|\alpha| b)]}{k^2 m^2 \mu_r + \alpha_1^2 b^2 \Lambda(\alpha_1 b)[\alpha^2 \mu_r \Lambda(\alpha_1 b) - |\alpha| \alpha_1 \mathrm{M}(|\alpha| b)]}\,, \tag{10}$$

is an even function of α and m, and plays the role of a reflection coefficient, and

$$\Lambda(x) = \frac{I'_m(x)}{I_m(x)} = \frac{m}{x} + \frac{I_{m+1}(x)}{I_m(x)} \quad \text{and} \quad \mathrm{M}(x) = \frac{K'_m(x)}{K_m(x)} = \frac{m}{x} - \frac{K_{m+1}(x)}{K_m(x)}\,.$$

The coefficients in (7 – 9) can now be substituted into (4 – 6) to give closed form expressions for the second order potentials in terms of the known source coefficient C_s.

2.2 Calculation of coil impedance

We now focus on the calculation of the impedance change ΔZ, which is the quantity of interest in eddy-current testing. Following the derivation of the impedance change for a coil in a tube [7] and in a borehole [6], ΔZ can be written in the following form:

$$\Delta Z = \frac{j\omega}{\mu_0 I^2} \int_{-\infty}^{\infty}\int_{-\pi}^{\pi} \left(\frac{\partial^2 W_{a0}^{(s)}}{\partial r \partial z} \frac{\partial W_{a0}^{(ec)}}{\partial z} - \frac{\partial^2 W_{a0}^{(ec)}}{\partial r \partial z} \frac{\partial W_{a0}^{(s)}}{\partial z} \right)_{r=b} b \, d\varphi \, dz \, . \tag{11}$$

Upon substituting for $W_a^{(s)}$ and $W_a^{(ec)}$ in (11) using (3) and (4) and applying the Parseval theorem for Fourier transforms and Fourier series, the impedance change can be expressed as

$$\begin{aligned} \Delta Z &= -\frac{j\omega 4\pi^2}{\mu_0 I^2} \int_{-\infty}^{\infty} \sum_{m=-\infty}^{\infty} \alpha^2 C_s(\alpha,m)\, D_{ec}(-\alpha,-m)\, d\alpha \\ &= \frac{j\omega 4\pi^2}{\mu_0 I^2} \int_{-\infty}^{\infty} \sum_{m=-\infty}^{\infty} \alpha^2 C_s(\alpha,m)\, C_s(-\alpha,-m)\, R(\alpha,m)\, d\alpha \end{aligned} \tag{12}$$

where D_{ec} from (9) was used to obtain the final expression. Eq. (12) is valid for an arbitrary air-cored coil where the coil characteristics enter through the source coefficient $C_s(\alpha,m)$ and the properties of the cylinder through the reflection coefficient $R(\alpha,m)$.

The coefficient C_s can be computed from the knowledge of the source (or free-space) magnetic flux density $\mathbf{B}^{(s)}$. From (3) and (1) the z-component of $\mathbf{B}^{(s)}$ on the surface $r = b$ is given by:

$$B_z^{(s)}(r=b,\varphi,z) = -\int_{-\infty}^{\infty} \sum_{m=-\infty}^{\infty} \alpha^2 C_s I_m(|\alpha| b) e^{jm\varphi} e^{j\alpha z} d\alpha \, . \tag{13}$$

Inverting (13) by taking the appropriate Fourier transforms with respect to z and φ, it follows that for an arbitrary air-cored coil C_s can be computed from the double Fourier transform of $B_z^{(s)}$ on the surface $r=b$.

$$C_s(\alpha,m) = -\hat{B}_z^{(s)}(\alpha,m) / [\alpha^2 I_m(|\alpha| b)] \tag{14}$$

where

$$\hat{B}_z^{(s)}(\alpha,m) = \frac{1}{(2\pi)^2} \int_{-\infty}^{\infty}\int_{-\pi}^{\pi} B_z^{(s)}(r=b,\varphi,z) \mathrm{e}^{-j\alpha z} e^{-jm\varphi} d\varphi \, dz \, . \tag{15}$$

Upon substituting (14) into (12) the final expression for the impedance change becomes:

$$\Delta Z = \frac{j\omega 4\pi^2}{\mu_0 I^2} \int_{-\infty}^{\infty} \sum_{m=-\infty}^{\infty} \frac{\hat{B}_z^{(s)}(-\alpha,-m)\ \hat{B}_z^{(s)}(\alpha,m)}{\left[\alpha\, I_m(|\alpha| b)\right]^2} R(\alpha,m)\ d\alpha \, . \tag{16}$$

Using the same approach, it can be shown that the source coefficient is related to the radial component and the azimuthal component of $\mathbf{B}^{(s)}$. The impedance change can then be expressed accordingly.

3. Cylindrical air-cored coil

3.1. Numerical treatment of the ΔZ expression

In the previous section, it was shown that C_s can be calculated by evaluating a Fourier integral involving the free-space magnetic field on the surface of the cylinder. While there are specific coil shapes where the double integration in (15) can be performed analytically, in the general case of a complicated coil shape or a non-axisymmetric coil position, C_s has to be computed numerically.

Such is the case of the cylindrical air-cored coil positioned as shown in Fig.1. While a closed-form solution was derived in [6] for the related case of a "horizontal" coil placed in a borehole, in the present study C_s is computed numerically. Two independent methods were used for the numerical calculations of the double Fourier transform given by (15). The first one focused on accuracy and the second one on speed of calculation. In the first method, adaptive quadrature routines from the IMSL FORTRAN library were used for the accurate calculation of the Fourier transform of $B_z^{(s)}$. In the second method, a 2D-Fast Fourier Transform routine, also from the IMSL library, was used to compute $\hat{B}_z^{(s)}(\alpha,m)$. Since the FFT provided $\hat{B}_z^{(s)}(\alpha,m)$ at discrete "α" values, a spline-interpolation routine was also used to interpolate for intermediate values. The impedance change was then computed from (16) using adaptive quadrature to evaluate the integral and a simple summation over "m". It must be emphasized that the aforementioned integration routines are automatic routines, i.e. an upper limit in the infinite integration range is not needed in the computations. The only parameter that must be supplied by the user is the relative accuracy. This was set in every case to 0.1%.

All these calculations must be accompanied by an efficient algorithm for computing $B_z^{(s)}$. There are a number of methods for computing the magnetic flux density from a cylindrical coil of rectangular cross-section: the most familiar rely either on numerical integration [8] or evaluation of algebraic expressions involving elliptic integrals [9]. In the present work, $B_z^{(s)}$ is obtained numerically by appropriate coordinate transformations of the formulae in [9] because they require less computation time.

The theoretical results for the impedance change using the two methods for the calculation of C_s agreed to within 1%. However, in terms of computation time, the method based on the use of FFT was significantly faster. For example, on a Pentium 1.8GHz PC, the FFT approach required about 10sec of CPU time in contrast to the method based on adaptive quadratures that required about 10 min.

3.2 Experiment

The validity of the method described above was tested by measuring ΔZ for a cylindrical air-cored coil and a series of Al alloy and Nickel Aluminium Bronze (NAB) cylinders of varying radius, shown in Fig.2. The NAB alloy is interesting because it is weakly magnetic and thus provides a further test of the theoretical model. The same coil was used in all experiments and the nominal coil lift-off was held constant. Measurements were also made of the corresponding ΔZ due to a conducting half-space (ie a thick plate) for the same coil and lift-off.

The change in coil impedance ΔZ, defined as the difference in the impedance for the coil outside the cylinder and the coil in air, was determined using a Hewlett-Packard HP-4192A impedance analyzer. Measurements were made in equal log steps over the nominal frequency range 500Hz-200kHz: the upper frequency dictated by the increased influence of stray capacitance and the lower frequency limited by the decreased sensitivity in the

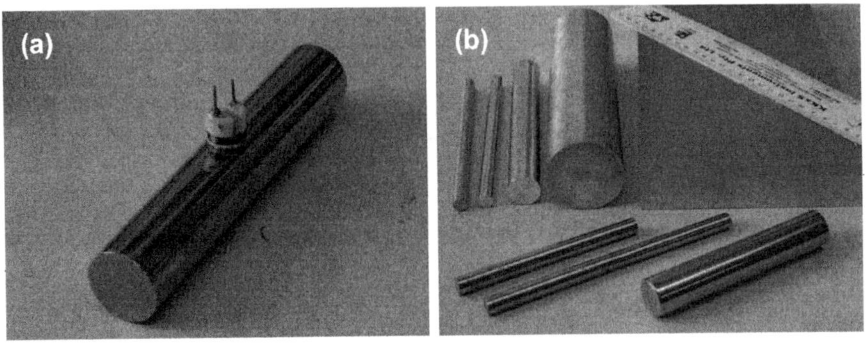

Fig. 2. Photographs of (a) the coil situated on NAB cylinder N3 and (b) specimens used in the experimental measurements of ΔZ.

measurements of ΔR. The experimental results were corrected for the effect of stray capacitance using the procedure described by Harrison, Jones and Burke [10].

The resistivity ρ of the test specimens was determined either by 4-point DC measurements or via a conductivity meter in the case of the A4 cylinder and the plate. The effective relative magnetic permeability of the NAB specimens was determined by fitting the results of independent encircling coil impedance measurements using the theory of Dodd and Deeds [8], with the permeability as the only free parameter. The coil and cylinder parameters are summarized in Table I.

Table I. Experimental coil and cylinder parameters (see Fig.1).

COIL		TEST SPECIMENS			
		Al Alloy	b (mm)	ρ (μΩ cm)	μ_r
r_1 (mm)	3.015	A3	5.0	4.50	1.0
r_2 (mm)	5.46	A4	25.22	3.65	1.0
Length h_2-h_1 (mm)	2.94	A5	11.01	5.60	1.0
Wire turns N	900	A6	5.31	5.60	1.0
Lift-off h_1 (mm)	1.32	Plate	–	6.08	1.0
		NAB			
Inductance L_0(mH)	6.03	N1	4.98	16.1	1.18
Resistance R_0(Ω)	104	N2	5.10	18.3	1.30
Resonance f_R(kHz)	587	N3	11.11	18.3	1.30

3.3 Results

The experimental ΔZ data are compared with the results of the parameter-free theoretical calculations in the impedance plane plots shown in Fig. 3. Following the usual practice in eddy-current nondestructive evaluation (NDE), the results are normalized to the isolated coil reactance $X_0 = \omega L_0$ where L_0 is the coil inductance in air. The corresponding experimental and theoretical normalized impedance data for the same coil above a thick Al alloy plate are also plotted in Figure 3a.

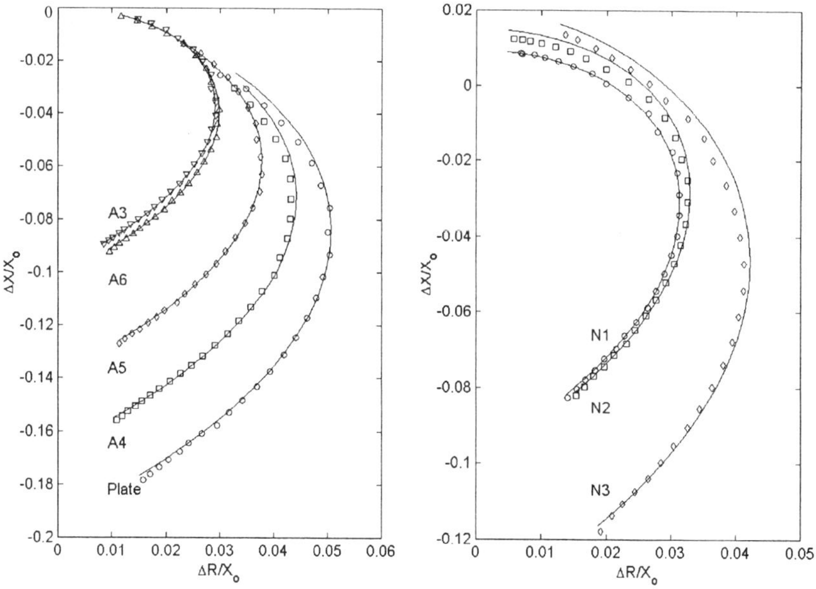

(a) **(b)**

Fig. 3. Normalized impedance plane showing the frequency variation locus of ΔZ for a horizontal coil outside a conducting cylinder. Solid curves are the theoretical results and the symbols denote the experimental data. (a) Al alloy cylinders and Al alloy plate, (b) NAB cylinders

The theoretical calculations of ΔZ due to a horizontal coil outside a conducting cylinder are in excellent agreement with the experimental results and provide confidence in the validity of the numerical approach for calculating C_s and ΔZ in this geometry. The observed level of agreement between theory and experiment for the conducting cylinder geometry is equal to that obtained in the benchmark case where the coil is situated above a conducting half-space and the theory is well established. Excellent agreement between theory and experiment is also obtained when the normalized ΔZ data for the horizontal coil and cylinder geometry are plotted as a function of frequency rather than being displayed in the normalized impedance plane presentation.

The effect of the surface curvature can be seen in Fig.3a. The magnitude of ΔZ is smaller than that for the equivalent half-space and the coupling decreases with decreasing cylinder radius for constant coil lift-off. The opposite is true in the related geometry where the coil is in a borehole and the magnitude of ΔZ is significantly larger from the same coil with the same nominal lift-off above the equivalent half-space [6].

4. Conclusions

Using the SOVP formalism, a general closed-form expression for ΔZ due to eddy-current induction in a conducting, magnetic cylinder has been derived for an arbitrary air-cored coil. A novel approach to the computation of ΔZ has been presented in which the coil properties appear via a 2D Fourier transform of the source magnetic flux density. The theory was applied to the specific case of a cylindrical air-cored coil with the coil axis perpendicular to the cylinder axis. Excellent agreement was obtained between the results of the theoretical calculations and a series of experimental measurements of ΔZ for Al alloy and NAB cylinders for this case. The theoretical predictions did not contain any free parameters.

The general expression for ΔZ given by Eq. (16) is expected to be valuable in eddy-current NDE of solid cylinders and as an approximation in the study of convex structures containing a simple radius of curvature. As the coil characteristics enter only via a 2D Fourier transform of $\mathbf{B}^{(s)}$, the present approach is highly flexible and can deal with arbitrary coil shapes provided the free-space fields can be calculated efficiently. In the case of the cylindrical air-cored coil examined here, the extension to more general coil positions and orientations can be made by applying the appropriate coordinate transformations to transform the known magnetic flux density. While the analysis presented here is limited to the calculation of ΔZ for a single coil outside a cylinder, it can be readily extended to eddy-current driver-pickup coils.

References

1. Hannakam L. "Wirbelstrome in einem massiven Zylinder bei beliebig geformter erregender Leiterschleife" *Archiv fur Elektrotechnik* 55 (1973) pp 207-215
2. Hill D.A. and Wait J.R., "Analysis of alternating current excitation of a wire rope by a toroidal coil" *J. Appl. Phys.* 48 (1977) pp 4893-4897
3. Grimberg R., Radu E., Mihalache O. and Savin A. "Calculation of the induced electromagnetic field created by an arbitrary current distribution located outside a conductive cylinder" *J. Phys. D: Appl. Phys.* 30 (1997) pp 2285-2291
4. Sakaji N.M. "Force and eddy currents in a solid conducting cylinder due to an eccentric circular current loop" *J. Phys. D: Appl. Phys.* 33 (2000) pp 2239-2248
5. Micolau G., Pichenot G., Premel D., Lesselier D. and Lambert M. "Three-dimensional electromagnetic field in a conductive cylinder at eddy-current frequencies" Electromagnetic Nondestructive Evaluation 2002, Saarbrucken, Germany, 2002
6. Burke S.K. and Theodoulidis T.P. "Impedance of a horizontal coil in a borehole: a model for eddy-current bolthole probes" Submitted to *J. Phys.D: Appl. Phys.*
7. Theodoulidis T.P. "Analytical modelling of wobble in eddy current tube testing with bobbin coils" *Res. Nondestr. Eval.* 14 (2002) pp 111-126
8. Dodd C.V. and Deeds W.E., "Analytical solutions to eddy-current probe-coil problems" *J. Appl. Phys.* 39 (1968) pp 2829-2838
9. Urankar L.K. "Vector potential and magnetic field of current carrying finite arc segment in analytical form, Part III: Exact computation for rectangular cross-section" *IEEE Trans. Magn.*18 (1982) pp 1860-1867
10. Harrison D.J., Jones L.D. and Burke S.K., "Benchmark problems for defect sizing and shape determination in eddy-current nondestructive evaluation" *J. Nondestruct. Eval.* 15 (1996) pp 21-34

Electromagnetic Nondestructive Evaluation (VIII)
T. Sollier et al. (Eds.)
IOS Press, 2004

Uncertainties Treatment in ECT using Parallel Computing Environment

Marco CIOFFI, Alessandro FORMISANO and Raffaele MARTONE
Dipartimento di Ingegneria dell'Informazione, Seconda Università di Napoli
via Roma 29, 81031 CE Aversa, Italy

Abstract. A new concept for the search of reliable solutions in ECT analysis is presented. The approach is based on the analysis of the uncertainties effects on the Objective Function to be evaluated in the inverse problem resolution. The method is tested against a benchmark problem. The need for higher computational power raising from statistical analysis is tackled using parallel computation on a cluster multiprocessors machine.

1. Introduction

Goal of an Eddy Current Testing (ECT) process is to provide accurate and reliable reconstructions of defects in conductive structures [1]. The unavoidable uncertainties, of both systematic (*e.g.* inaccuracies in the knowledge of measurement system parameters) and random (*e.g.* measurement noise) nature have an impact on the reliability of the ECT reconstruction, which can be quantified in terms of the *Probability Of Detection* (POD) and the *Probability of False Indication* (PFI) [2].

As a matter of fact, apart from the measurement noise, the signal from ECT probes is usually calibrated before the flaw detection by acquiring a reference field map of undefected parts and by removing possible offsets; the quality of the reconstruction highly depends on the accuracy of such calibration. The reliability of the ECT reconstruction also depends on the precision of the geometrical data, on the accuracy of the model and on the precision of both direct and inverse procedures used in the reconstruction process. Therefore an estimation of the reliability for the ECT results is required. In this paper, this issue is formalised in rather general terms and discussed also with the aid of some numerical exemplifications. In addition, a new stochastic approach to reduce the PFI of ECT analysis is presented.

2. Solution Reliability and Robustness

To be appealing in real applications, an ECT procedure has to be reliable. The reliability concept in Non Destructive Testing (NDT) has a quite wide meaning with several different definitions in the scientific literature [2, 3]. In any case, the reliability concept includes the accuracy and the false indication prediction capability, together with the solution robustness with respect to the procedure uncertainty, including measurement noise, model errors, computing inaccuracies and partial system failures.

An accurate analysis of an ECT procedure reliability is very useful for different reasons. First, hopefully, because it can provide estimations of the accuracy ranges of each parameter to identify and, therefore, of their "observability". In addition, it can suggest criteria for the optimal use of computing models included in the procedure. Finally, it can provide guidelines for the improvement of the excitation and detection system.

To illustrate these elements, it should be reminded that ECT procedures base the flaw detection in the sample on the analysis of a *reaction* to an electromagnetic excitation. Such reaction can be one of the components of the magnetic field in one or more measurement points or the impedance variation of the excitation coil itself when scanning the part. For the reconstruction success, it is required that the reaction should be depending by the defect or, in other words, the defect must be observable using the chosen measurement within the actual system accuracy.

Together with the reaction measuring system, any ECT technique incorporates a computational module (a *direct problem solver),* able to compute the reaction in presence of the "trial defects": the solving algorithm can be analytic, for the simplest cases, numerical otherwise. The ECT procedure provides, as an estimation of the real defect, the "trial defect" whose computed reaction results most "similar" to the measured one.

The concept of "similitude" between measured and computed reaction is mathematically expressed by a suitable norm; therefore, the reconstruction reduces to the minimisation of such a norm. For example, it can be chosen a quadratic norm such as:

$$d\left(\mathbf{r}_m,\mathbf{r}_c\right)=\left\|\mathbf{K}\ \left(\mathbf{r}_m-\mathbf{r}_c\right)\right\|_{l^2} \tag{1}$$

where $\mathbf{r}_m$ and $\mathbf{r}_c$ are arrays of dimension N_m and represent, respectively, the measured and the computed reaction in the N_m measurement points, and $\mathbf{K}$ is a preconditioning matrix.

Ideally it can be supposed that $\mathbf{r}_m$ could be noiseless and that the direct algorithm could be able to represent exactly all possible defects and to give exact values of $\mathbf{r}_c$. In such an ideal case, the distance (1) would be zero for the real defect. By assuming that (1) would have a single minimum and, in addition, that the direct algorithm could analyse all possible actual defects, then the search will be always successful and the solution unique and error-free.

Regrettably, in the real applications, any ECT technique is affected by several relevant restrictions:

a) $\mathbf{r}_m$ is affected by noise;

b) $\mathbf{r}_c$ is affected by inaccuracies due, by the way, to:

- errors on the system parameters, errors on the probes position, errors on the sample shape and properties, electrical and geometrical characteristics of exciter, and so on;
- computational errors: round-offs, discretisation, impact of the mathematical model simplifying hypotheses, and so on.

As a consequence, the error function (1) will not vanish also in the presence of the exact solution and, therefore, the search becomes a minimisation process and not a zero search. In addition, the solution will be affected by errors and the uncertainty range depends on the measurement noise and on the computing and modelling inaccuracy.

2.1 Mathematical model and nominal solution

To consider the measuring errors in (1), $\mathbf{r}_m$ has to be written as:

$$\mathbf{r}_m=\mathbf{r}_{mo}+\Delta\mathbf{r}_m \tag{2}$$

where $\mathbf{r}_{mo}$ is the true values array (the *nominal measurements*) and $\Delta\mathbf{r}_m$ is the measuring error, in general composed by a systematic and a stochastic component. It should be observed that, depending by the error model assumed, $\Delta\mathbf{r}_m$ could be either fixed with respect to the measurement value (absolute error) or proportional to the measured value (relative error).

To formalise the impact of computing inaccuracies, it is useful in advance to formally separate, among the independent variables of $\mathbf{r}_c$, the N_p parameters $\mathbf{p}$ characterising the possible defects geometry and the N_q parameters $\mathbf{q}$ describing the system and the computing algorithm: therefore $\mathbf{r}_c = \mathbf{r}_c\ (\mathbf{p}, \mathbf{q})$. Then, to consider the inaccuracy on $\mathbf{q}$, it can be stated that:

$$\mathbf{q}=\mathbf{q}_o+\Delta\mathbf{q} \tag{3}$$

where $\Delta\mathbf{q}$, error on the system parameters, should be assumed as random, with a suitable probability density function (pdf) for each component. Expression (1) becomes therefore:

$$d=\left\|\mathbf{K}\left[\mathbf{r}_{m0}+\Delta\mathbf{r}_m-\mathbf{r}_c\left(\mathbf{p},\mathbf{q}_o+\Delta\mathbf{q}\right)\right]\right\| \tag{4}$$

and the defect search turns into the minimisation of d with respect to $\mathbf{p}$. In the case of $\Delta\mathbf{r}_m=\mathbf{0}$ and $\Delta\mathbf{q}=\mathbf{0}$, (4) becomes:

$$d_o = \left\| \mathbf{K}\left[\mathbf{r}_{mo} - \mathbf{r}_c(\mathbf{p}, \mathbf{q}_o)\right]\right\| \tag{5}$$

which can be called the *nominal distance*. The *nominal solution* $\mathbf{p}_o$ is consequently the value of $\mathbf{p}$ which minimises d_o. It should be observed that, if the real defect is included in the selected class represented by $\mathbf{p}$, the minimum of d_o is exactly zero, and $\mathbf{p}_o$ is also the "exact" solution. In the general case, the minimum of d gives an "approximate solution":

$$\mathbf{p}_a = \mathbf{p}_o + \Delta\mathbf{p} \tag{6}$$

The error $\Delta\mathbf{p}$, depending on the random variables $\Delta\mathbf{r}_m$ and $\Delta\mathbf{q}$, is random itself. The estimate of the dependence of $\Delta\mathbf{p}$ by $\Delta\mathbf{r}_m$ and $\Delta\mathbf{q}$ has an important operative effectiveness. In the following, for the sake of simplicity, the impacts of $\Delta\mathbf{r}_m$ and $\Delta\mathbf{q}$ will be evaluated independently.

2.2 Impact of the measuring errors

By assuming $\Delta\mathbf{q}=\mathbf{0}$, (4) becomes:

$$d_{qo} = \left\| \mathbf{K}\left[\mathbf{r}_{m0} + \Delta\mathbf{r}_m - \mathbf{r}_c(\mathbf{p}, \mathbf{q}_o)\right]\right\| \tag{7}$$

The *nominal noisy solution* will consequently be defined as the $\mathbf{p}$ value which minimises (7): it is an approximate solution of (4) for $\mathbf{q}=\mathbf{q}_0$ and, therefore, it will be noted $\mathbf{p}_{a,q_0}$. The analysis of the dependence of $\mathbf{p}_{a,q_0}$ on $\Delta\mathbf{r}_m$ has to be performed for the generic k-th component of $\mathbf{p}_{a,q_0}$ (by keeping $p_l=p_{lo}$ for $l\neq k$) and for the generic h-th component of $\Delta\mathbf{r}_m$ (by keeping $\Delta r_{ml}=0$ for $l\neq h$). In Fig. 1 a typical dependence of d_{qo} on the k-th parameter p_k is reported, together with the uncertainty range $2E_h$ on Δr_{mh}. It is easy to observe that the uncertainty interval on p_k is the one indicated by $I_{k,h}$. Note that $I_{k,h}$ is determined by the intersection of the lower bound with the minimum level line of the upper bound.

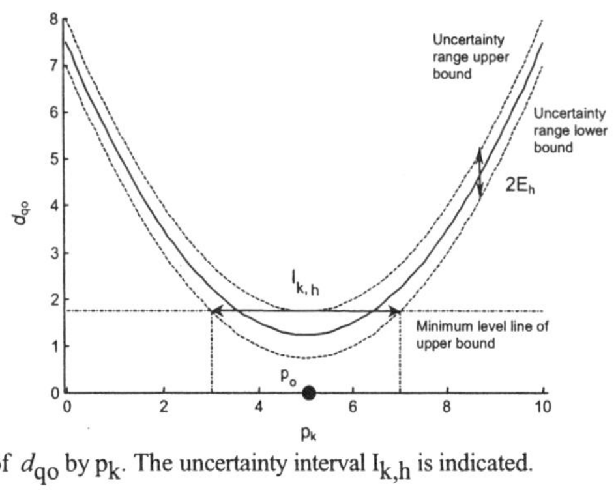

Figure 1. Dependence of d_{qo} by p_k. The uncertainty interval $I_{k,h}$ is indicated.

Fig. 2 shows that $I_{k,h}$ increases with the error amplitude E_h, while Fig. 3 shows that $I_{k,h}$ depends on the geometric properties of the function.

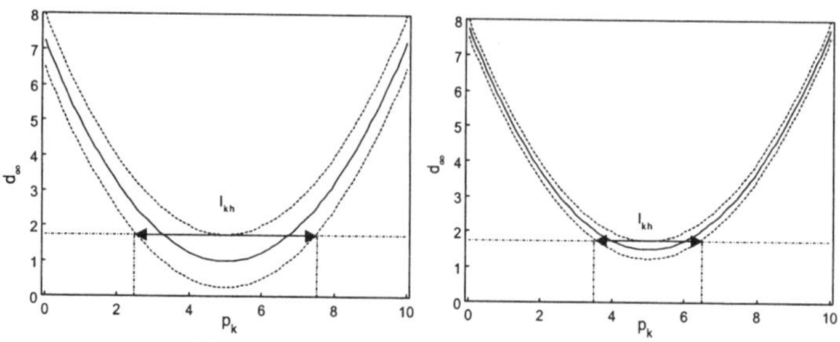

(a) (b)

Figure 2 a-b. Dependence of $I_{k,h}$ by error amplitude: increased (a) and decreased (b) with respect to Fig. 1

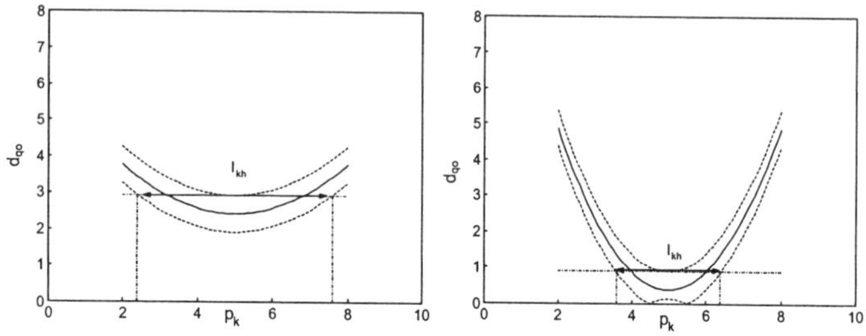

(a) (b)

Figure 3 a-b. Dependence of $I_{k,h}$ by q_o shape: decrease (a) and increase (b) of curve convexity.

The pdf of p_k in the interval $I_{k,h}$ depends on the pdf of Δr_{mh} in $2E_h$ and on the system itself. In particular, if the latter is uniform, the former is not necessarily uniform too, and, as a general rule, it must determined by means of specialised methods (*e.g.* Monte Carlo approach). The set of the intervals $I_{k,h}$ ($k=1,\ldots,N_p$) defines also the *uncertainty domain* of the solution $\mathbf{p}_{a,q_0}$ due to the h-th measure. The intervals $I_{k,h}$, normalised on p_k and divided by the relative error E_h/r_h provide a robustness index of the p_k reconstruction through the r_h measure. The product of the normalised intervals (*i.e.* the volume of the uncertainty normalised domain) provides therefore an index of the global effectiveness of the measure h. Similar domains can be calculated for the other components of the measuring system: the intersection of the N_m uncertainty domains gives the overall measuring systems uncertainty domain.

2.3 Impact of the computing errors

On the other hand, when assuming $\Delta\mathbf{r}_m = \mathbf{0}$, (4) becomes:

$$d_{mo} = \left\| \mathbf{K} \left[\mathbf{r}_{m0} - \mathbf{r}_c \left(\mathbf{p}, \mathbf{q}_o + \Delta\mathbf{q} \right) \right] \right\| \tag{8}$$

The $\mathbf{p}$ value, which minimises (8), will be consequently called *approximate noiseless solution*: this is the solution of (4) for $\mathbf{r}_m = \mathbf{r}_{mo}$, and it will be noted $\mathbf{p}_{a,m_0}$. By expressing $\mathbf{r}_c$ in the form:

$$\mathbf{r}_c\,(\mathbf{p}, \mathbf{q}_o + \Delta\,\mathbf{q}) = \mathbf{r}_{co}\,(\mathbf{p}, \mathbf{q}_o) + \Delta\mathbf{r}_c \tag{9}$$

expression (8) becomes

$$d_{mo} = \left\| \mathbf{K} \left[\mathbf{r}_{m0} - \mathbf{r}_{co} \left(\mathbf{p}, \mathbf{q}_o \right) - \Delta\mathbf{r}_c \right] \right\| \tag{10}$$

which is very similar to (7) and therefore it is possible to evaluate also the uncertainty intervals of each component of $\mathbf{p}_{a,m_0}$ due to each component of $\mathbf{q}$.

In analogy, it is possible to define the robustness of the reconstruction of p_k with respect to the generic computational model parameter and, in addition, the uncertainty domains determined by the uncertainties on such parameters. This allows to perform an assessment of the various system parameters influence on the result quality, concentrating consequently the model resolution power (which amounts to saying computational power) on the most relevant ones. As an example, the most effective mesh can be determined for a fixed number of finite elements. It should be observed that $\Delta\mathbf{r}_c$ must be computed by specialised statistical procedures taking into account the parameters pdf's, as the usual sensitivity analysis may provide too conservative indications. The combined analysis of the uncertainty domains due to both the measuring noise and to the $\mathbf{q}$ errors gives the overall reconstruction reliability.

3. Numerical exemplification

To illustrate the exposed considerations, a very simple example (JSAEM Problem #2) is here presented by using a computing procedure based on FEM. In the JSAEM Problem #2, the test specimen is a square plate, with dimensions 140x140x1.25 mm (Fig. 4), made of nickel-based non-magnetic superalloy INCONEL 600 MA with a relative permeability $\mu_r = 1$ and a conductivity $\eta = 10^6$ S/m. The plate presents in the centre a rectangular crack (length = 10 mm; width = 0.2 mm; depth varying from 40 % to 100 % of plate thickness); both inner and outer defects are considered. The eddy current probe is a cylindrical coil, with inner and outer diameters equal to 1.2 mm and 3.2 mm respectively and 0.8 mm height. The lift-off from plate is assumed equal to 0.5 mm. The coil is excited with AC current with RMS value equal to 1/140 A at 300 kHz working frequency, which corresponds to a skin depth of 0.92 mm.

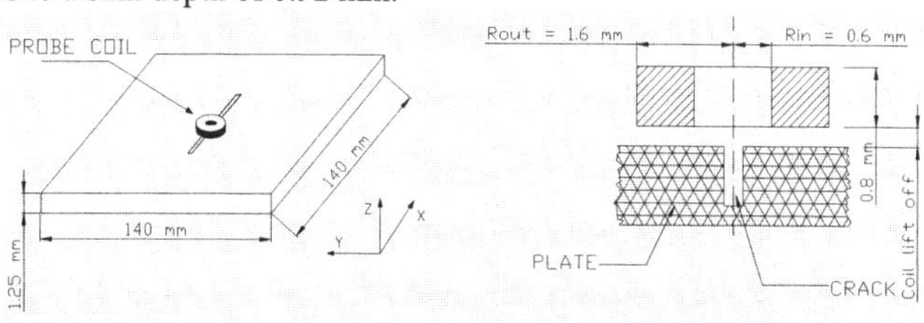

(a) (b)

Figure 4. Schematics for JSAEM benchmark problem #2: Schematics of the sample (a) and dimensions of the crack and the probing coil (b).

The solver for the direct problem is based on a finite edge-element integral code developed for the analysis of the eddy current induced in a non-magnetic and linear conductor using a 3D integral formulation based on the electric vector potential [4]. The integral formulation adopted by the solver allows to discretise only the region where an eddy current density is induced. The tested specimen is modelled by a brick-elements mesh.

By assuming the crack aligned to one direction (x-axis), the non-uniform mesh topology already introduced in [5] has been here adopted: this mesh achieves almost the same spatial resolution along the two x-y directions. A generic crack is therefore described by five integer parameters: the row ny position along y-axis, the initial (nx_{in}) and the final (nx_{fin}) element along x-axis, the initial (nz_{in}) and the final (nz_{fin}) element along the vertical z-axis on the mesh.

3.1. Reconstruction uncertainty

The influence of the "quality" of the adopted mesh on the problem solution has to be checked in order to evaluate its sensibility: the mesh topology and its size have to be properly selected to suitably define the class of flaws expected to be found. A preliminary sensitivity analysis has been performed by using three different meshes: a coarse one (*mesh A*) with 648 elements, a medium one (*mesh B*) with 1296 elements and a refined mesh (*mesh C*) with 3840 elements (Table I). Mesh A resulted too coarse, while mesh C showed well known numerical instabilities, filtered by mesh B. Consequently, the calculations have then been performed on the mesh B.

Table I. Adopted meshes for JSAEM #2 test problem (Lc is the crack length).

Mesh	Elements	Elements along the thickness z	Elements length along x in the fine region	Fine elements along x and y
A	648	3	Lc/2	6
B	1296	6	Lc/2	6
C	3840	6	Lc/4	12

As an example of the measurement noise effect on the uncertainty of the reconstructed parameters, a noise level of 5% has been assumed on the field measurements. The effect on the Objective Function (OF) of the reconstruction procedure, defined as the l_2 distance between measured and simulated magnetic field in the measurement points, is reported in Fig. 5 as a function of the number of FEM elements along the crack depth (acting as a discrete parameter p_k). Note that Fig. 5 has the same meaning of Fig. 1 in the case of JSAEM problem with the OF defined as in (1).

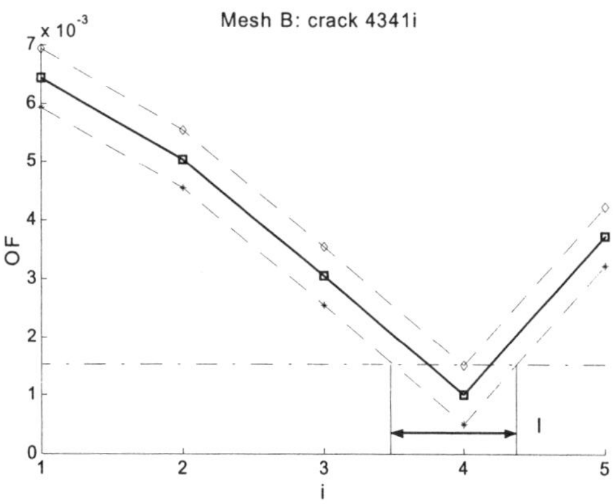

Figure 5. Evaluation of the dependence of an absolute measuring OF error of ± 5% on the crack depth uncertainty for a *4341i*-type crack (*i* is number of elements along the crack depth).

4. Uncertainties Treatment using Parallel Computing Environment

The previous concepts, in the framework of an inverse problem resolution environment based on the parallel global model genetic algorithm [5], have been exploited to design a robust ECT data treatment system.

The effects of the detection system uncertainty can be effectively considered in the inverse problem resolution if it is included in the objective function a suitable term related to the parametric sensitivity of performances. The OF definition can be modified in such a way to take into account some information about the OF behaviour in a neighbourhood of the candidate solution, defined by the expected uncertainty on the system parameters. Similar concepts of robust optimisation have been introduced for the electromagnetic optimal design [6-8]

When considering the problem parameters uncertainty, the performances of a feasible solution have to be redefined taking into account the uncertainty impact. A possible OF modification can be to take its statistical average <OF>; for the case where just system **q** parameters are affected by uncertainty, OF can be redefined as [9]:

$$\langle OF(\mathbf{p},\mathbf{q}_o)\rangle = \int_{\Omega(\mathbf{q})} P(\mathbf{q}')\, OF(\mathbf{p},\mathbf{q}')\, d\mathbf{q}' \tag{11}$$

where $P(\mathbf{q})$ is the pdf of the configuration described by the parameters vector $\mathbf{q}$ and $\Omega(\mathbf{q})$ is the uncertainty range around the nominal configuration $\mathbf{q}_O$.

The integration of (11) for a multi-variable general pdf requires suitable integration techniques to be adopted, because the computational cost of multidimensional integration increases exponentially with the number of parameters considered. A simple possible, but computationally heavy, approach for multiple parameters is to consider statistical independence among uncertainties, assuming the joint probability $P(\mathbf{q})$ as the product of the single marginal probabilities $P_i(q_i)$ of each parameter. In the general case, there could be some dependencies among system parameters or their uncertainties: the number of independent parameters therefore decreases, leading to a reduction of the computational cost for evaluation of (11). For example, the knowledge of the parameters statistical properties can be exploited by a Bayesian-type approach.

In the following, just a one-parameter model will be considered, by assuming that the probe lift-off is affected by an uncertainty modelled with a Gaussian pdf. A confidence interval of 99% is adopted (*i.e.* the percentage of times in repeated sampling that the interval contains the true value of the unknown parameter). The procedure to compute the random variable standard deviation is reported in [9].

To be applied in the framework of a detection system, (11) has to be properly discretised. Different approaches can be used for the approximate evaluation of (11). Here, for the one-variable case, the integral is evaluated by the rectangular rule with three equally spaced points (Fig. 6). The discretised probabilistic densities (usually referred as Probability Mass Functions) are evaluated in such a way to conserve the total area [9].

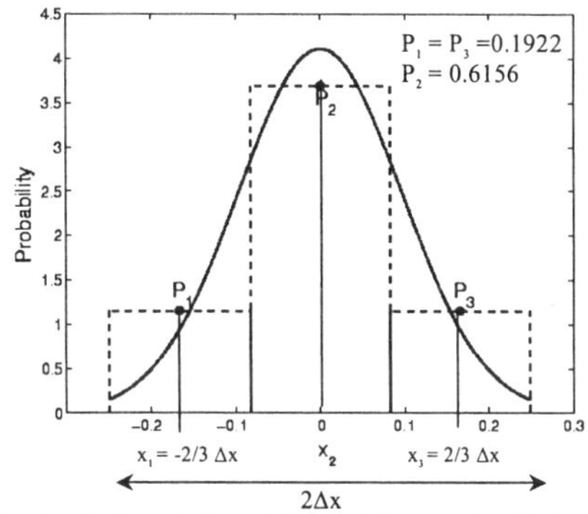

Figure 6. Continuous (continuous line) and discrete (dashed line) probabilistic density functions.

Finally we can express the <OF> as a weighted sum in the three integration points x_i:

$$\langle OF(\mathbf{x})\rangle = P_1\, OF(\mathbf{x}_1) + P_2\, OF(\mathbf{x}_2) + P_3\, OF(\mathbf{x}_3) \tag{12}$$

where the weights P_i have been normalised with respect to their sum. Results are reported in Fig. 7. Note that when the actual lift off is 0.3 mm, while the computation is performed assuming the nominal value, two local minima appear in the unmodified OF. This situation is recovered when assuming for the OF the definition (11).

5. Conclusions

The analysis of the uncertainties ranges is an useful tool for the design of an ECT reconstruction system. Some design criteria can be stated as follow:

1) The choice of the measurement system parameters can be guided by the dimension of the uncertainty ranges. The measurements with a poor efficiency should be avoided in favour of others with higher effectiveness. "Specialised" measurements for the detection of particular parameters can be included, if necessary.

2) The uncertainty ranges due to the system parameters **q** help to identify the critical aspects of the detection system and of the computing model, to which particular attention should be paid, whereas other aspects, with less influence on system performance, can be treated more coarsely.

3) The introduction of the modified OF, defined as a statistical average, can improve the identification ability in presence of noise or uncertainties but heavily increases the computational burden of the solving procedure: therefore the adoption of a high performance parallel computing environment becomes mandatory.

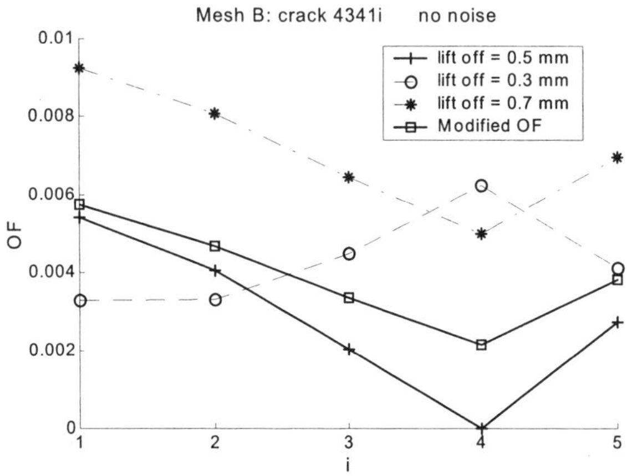

Figure 7. OF dependence by the crack depth for different coil lift-off and modified OF for a *4341i*-type crack.

Acknowledgements

This work has been partly sponsored by Italian Education and Research Ministry (MIUR) in the framework of the Relevant National Interest Research Project "MADEND".
The authors want to thank R. Albanese and G. Rubinacci for the use of the Cariddi code for the present work.

References

[1] Skordev, A., "Uncertainties and errors in Non-Destructive Characterisation of Discontinuities", *Proc. of 15th WCNDT*, Roma, Italy, October 15-21, 2000, http://www.ndt.net/article/wcndt00/toc/rel.htm.

[2] Pairazaman, C., Keller, S., and Rummel, W.D., "Reliability Assurance for Automated Eddy Current Inspection Systems", *Proc. of 15th WCNDT*, Roma, Italy, October 15-21, 2000, http://www.ndt.net/article/wcndt00/toc/rel.htm.

[3] Ericsson, L., and Stepinski, T., "Robust Signal Processing for Material Noise Suppression in Ultrasonic NDT", *Rev. of Progress in QNDE*, vol. 12A, 1993, pp. 695-702.

[4] Albanese, R., and Rubinacci, G., "Finite Element Methods for the Solution of 3D Eddy Current Problems" in *Advances in Imaging and Electron Physics*, Academic Press, vol. 102, 1998.

[5] Cioffi, M., Formisano, A., and Martone, R., "Analysis of Concurrent Multi-Frequency Eddy Current Testing Data", *Proc. of ENDE 02*, Saarbrücken, Germany, June 13-14, 2002.

[6] Jung, D.H., and Lee, B.C., "Development of a simple and efficient method for robust optimisation", *Int. J. Numerical Methods in Engineering*, vol. 53, issue 9, March 2002, pp. 2201-2215.

[7] Yoon, S.B., Jung, I.S., and Hyun, D.S., "Robust Shape Optimization of Electromechanical Devices", *IEEE Trans. on Magnetics*, vol. 35, n. 3, May 1999, pp. 1710-1713.

[8] Alotto, P., Magele, C., Renhart, W., Steiner, G., and Weber, A., "Robust Target Function in Electromagnetic Design", *Compel*, vol. 22, n. 3, 2003, pp. 549-560.

[9] Cioffi, M., Formisano, A., and Martone, R., "Increasing Design Robustness in Evolutionary Optimisation", *Proc. of PMAPS 2002*, Napoli, Italy, September 2002, to be published on *Compel*.

Electromagnetic Nondestructive Evaluation (VIII)
T. Sollier et al. (Eds.)
IOS Press, 2004

Numerical Modelling of Eddy Current Non-Destructive Evaluation by Integral Formulation and Parallel Implementation*

Ermanno CARDELLI, Antonio FABA
Department of Industrial Engineering, University of Perugia, Perugia, Italy.
Raffaele FRESA, Salvatore VENTRE
DIFA, University of Basilicata, Potenza, Italy. DAEIMI University of Cassino, Cassino, Italy.

Abstract. In this paper we present some results drawn from a research project of national interest named MADEND (Methods and Applications of Non-destructive Electromagnetic Diagnostic). In particular we show a comparison between two numerical models: eddy current integral formulation on parallel computer systems and differential formulation of finite element method. The numerical results and accuracy are confirmed by experimental tests.

1. Introduction

In Italy, since January 2002, some University research groups have been working on a so called "Project of Remarkable National Interest" of the National Italian Ministry of Education about methods and applications of electromagnetic diagnostic based on eddy current phenomena. The name of the project is MADEND and its purpose is to present and to develop the research possibilities in this field. Some results dealing with the MADEND programme are reported in this paper. In particular we present here a numerical model based on integral formulation and parallel implementation that allows a great saving of calculation time with respect to traditional numerical models. The benchmarks defined for this comparison are the MADEND benchmarks # 1, 2 and 3. The benchmark #1 is an aluminium plate with a thickness of 4 mm. On this plate there is an artificial defect with cylindrical shape, with a diameter of 1 mm and a height of 4 mm, (reduced respectively to 3mm and 2 mm for the benchmarks #2 and #3). The used probe is a coil placed on the metallic surface under test. For the benchmark # 2 and 3 the probe coil is placed on the opposite side with respect to the cylindrical defect. This coil is electrically linked to a suitable balanced bridge circuit and the voltage variation on the bridge allows the crack identification.

2. Integral formulation and parallel implementation

Integral formulation

The design and the analysis phases of modern devices for engineering applications, including among others the electromagnetic non-destructive evaluation, require accurate numerical solutions of Maxwell equations in the Quasi Stationary Magnetic limit. In the last decades many scientists focused their research activity on this topic. In the usual classification the

* This work has been supported by the Italian Ministry of Education (MIUR, Project MADEND) and EURATOM/ENEA/CREATE.

proposed approaches are divided in the two categories of integral and differential formulations.

The integral approaches present several advantages with respect to the differential ones because they require only the discretization of the space domain where the sources and the materials are located; moreover the regularity conditions at infinity for unbounded domains are implicitly included in the formulation, so that the overall discretization cost is very limited. Moving parts can be easily taken into account, while the presence of non linear magnetic materials require an additional moderate effort.

For linear materials, the numerical model is given by a system of linear algebraic equations with a dense coefficient matrix. In this paragraph we recall the salient points of the formulation adopted for the development of Cariddi, a volume integral code intensively used in the sector of the nuclear fusion for the solution of 3D eddy current problems in the time or in the frequency domain [1].

From the mathematical point of view the code solves the Maxwell equations in the QSM limit.

With reference to the time domain formulation and adopting the usual formalism, let's assume that an impressed current density is present in the domain V_s and the fixed conducting domain V_c is characterized by the $\mathbf{E} = \eta\mathbf{J}$ constitutive relationship with the normal component of current density equal to zero on the boundary ∂V_c.

Posing $\mathbf{B} = \nabla \times \mathbf{A}$ and $\mathbf{E} = -\frac{\partial \mathbf{A}}{\partial t} - \nabla \varphi$, the solenoidality of the flux density **B** and the Faraday law are automatically imposed. Moreover, the vector potential **A** can be expressed in terms of both impressed and unknown current density sources as:

$$\mathbf{A}(\mathbf{r},t) = \frac{\mu_0}{4\pi}\int_{V_c} \frac{\mathbf{J}(\mathbf{r}',t)}{|\mathbf{r}-\mathbf{r}'|} dV' + \frac{\mu_0}{4\pi}\int_{V_s} \frac{\mathbf{J}_s(\mathbf{r}',t)}{|\mathbf{r}-\mathbf{r}'|} dV' = \frac{\mu_0}{4\pi}\int_{V_c} \frac{\mathbf{J}(\mathbf{r}',t)}{|\mathbf{r}-\mathbf{r}'|} dV' + \mathbf{A}_s(\mathbf{r},t) \quad (1)$$

The Ohm law can be satisfied in weak form finding $\mathbf{J} \in Q$ such that $\int_{V_c} (\eta\mathbf{J} - \mathbf{E}) \cdot \mathbf{w}\, dV = 0$ with $\mathbf{w} \in Q$, where $Q = \{\mathbf{q} \in L^2_{div}(V_c) | \nabla \cdot \mathbf{q} = 0, \hat{\mathbf{n}} \cdot \mathbf{q} = 0 \ on\ \partial V_c\}$ is the functional space of admissible solutions **q** and $\mathbf{L}^2_{div}(V_c) = \{\mathbf{a} \in \mathbf{L}^2(V_c) \,|\, \nabla \cdot q \in L^2(V_c)\}$.

The position $\nabla\times\mathbf{T}=\mathbf{J}$, where **T** is the electric vector potential, assures the solenoidality of **J** and the continuity of its normal component. Finally, we obtain:

$$\frac{\mu_0}{4\pi}\int_{V_c}\int_{V_c} \frac{\partial(\nabla\times\mathbf{T}(\mathbf{r}',t))/\partial t \cdot \mathbf{w}(\mathbf{r})}{|\mathbf{r}-\mathbf{r}'|} dV'dV + \int_{V_c} \eta(\nabla\times\mathbf{T}(\mathbf{r}',t)) \cdot \mathbf{w}\, dV + \int_{V_c} \nabla\varphi \cdot \mathbf{w}\, dV = -\int_{V_c} \frac{\partial \mathbf{A}_s}{\partial t} \cdot \mathbf{w}\, dV \quad (2)$$

Since $\mathbf{J}, \mathbf{w} \in Q$ in V_c, results $\int_{V_c} \nabla\varphi \cdot \mathbf{w}\, dV = -\int_{\partial V_c} \varphi\, \mathbf{w} \cdot \hat{\mathbf{n}}\, dV = 0$

Furthermore, the electric vector potential can be expanded in terms of the basis functions $\mathbf{N}_k$ associated to the edges of the domain mesh : $\mathbf{T}(\mathbf{r},t) = \sum_{k=1}^{n} I_k(t)\mathbf{N}_k(\mathbf{r})$.

The boundary condition $\mathbf{J} \cdot \hat{\mathbf{n}} = 0$ on ∂V_c can be directly imposed on the weight functions $\mathbf{w}_k$, while their solenoidality can be automatically obtained by choosing $\mathbf{w}_k = \nabla \times \mathbf{N}_k$. In order to assure the uniqueness of the solution, the decomposition tree-cotree of the edges of the mesh is applied and only the basis functions associated with the edges of the cotree are retained as unknowns. In conclusion, the eq. (2) becomes:

$$\underline{\underline{L}}\frac{d\underline{I}}{dt}+\underline{\underline{R}}\,\underline{I}=\underline{V} \quad (3)$$

where $\underline{\underline{R}}$ is a sparse positive semidefinite symmetric matrix, while $\underline{\underline{L}}$ is a dense N×N symmetric positive definite matrix:

$$R_{i,j}=\int_{V_c}\eta\mathbf{w}_i\cdot\mathbf{w}_j\,dV\ ;\ L_{i,j}=\frac{\mu_0}{4\pi}\int_{V_c}\int_{V_c}\frac{\mathbf{w}_i(\mathbf{r}')\cdot\mathbf{w}_j(\mathbf{r})}{|\mathbf{r}-\mathbf{r}'|}dV'\,dV\ ;\ V_i=-\int_{V_s}\frac{\partial\mathbf{A}_s}{\partial t}\cdot\mathbf{w}_i\,dV \quad (4)$$

Adopting the theta time-stepping method for the time integration, the differential system (3) can be reduced to a linear algebraic system:

$$\underline{\underline{A}}\,\underline{I}(t_{k+1})=\underline{\underline{B}}\,\underline{I}(t_k)+\theta\underline{V}(t_{k+1})+(1-\theta)\underline{V}(t_k) \quad (5)$$

where $\underline{\underline{A}}=\frac{1}{\Delta t}\underline{\underline{L}}+\theta\underline{\underline{R}}$; $\underline{\underline{B}}=\frac{1}{\Delta t}\underline{\underline{L}}-(1-\theta)\underline{\underline{R}}$.

The formulation can also be developed in the frequency domain, leading to the complex linear algebraic system [1]:

$$\underline{\underline{Z}}\,\underline{I}=\underline{V} \quad (6) \qquad \text{where } \underline{\underline{Z}}=j\omega\underline{\underline{L}}+\underline{\underline{R}}$$

Parallel implementation

The implementation of Cariddi requires several tasks to be carried out :

1) Preprocessing (for instance, initialization and computation of the incidence matrices)
2) Computation of matrices $\underline{\underline{L}}$ and $\underline{\underline{R}}$ and vector $\underline{V}$, by means of volume integration
3) LU decomposition of the $\underline{\underline{A}}$ and Z matrices respectively for the solutions in time (eq. 5) and frequency (eq. 6) domains.
4) Computation of the solutions in time or frequency domain by back-substitution
5) Postprocessing (for instance, computation of integral quantities of interest like power, energy, etc)

The number of unknowns, N, used to discretize conductors and sources, affects the accuracy of the numerical solution. If N is increased to improve the accuracy, the memory occupation scales as $O(N^2)$, while the computational cost scales in a different way according to the nature of the task. The fast scaling of memory and CPU-time requirements set a severe limit on the maximum value of N that can be considered on standard Pcs. In alternative to the utilization of supercomputers, the parallel implementation of the code represents a convenient way to overcome the above mentioned limitations and to obtain accurate solutions by meshes with finer discretization.

Among the various steps required from the algorithm, for large values of N, the most critical ones in terms of CPU time are related to the computation of the $\underline{\underline{L}}$ matrix and to the LU decomposition operations: a simple analysis shows that they scale with the unknown number N respectively like $O(N^2)$ and $O(N^3)$. As ScaLAPACK libraries allow to perform the latter task in a very efficient way, we'll dedicate the remaining part of this section to illustrate shortly the algorithms proposed in [2] for the numerical computation of the matrix elements L_{ij} on a parallel computation system.

We recall that :

$$L_{ij} = \sum_{e_i \in E_i} \sum_{e_j \in E_j} \sum_{G_i=1,N_G} \sum_{G_j=1,N_G} \mathbf{w}_i(\mathbf{r}_{G_i}) \cdot \mathbf{w}_j(\mathbf{r}_{G_j}) J_{e_i}(\mathbf{r}_{G_i}) J_{e_j}(\mathbf{r}_{G_j}) / |\mathbf{r}_{G_i} - \mathbf{r}_{G_j}| \qquad (7)$$

where E_i is the set of elements that share the edge i, J_{ei} is the Jacobian of the element e_i computed at the Gauss point r_{Gi} and N_G is the number of Gauss points.

The direct use of this definition leads to a lack of efficiency because the inner loops over the elements have to be repeated several times, as each element contributes up to twelve unknowns. The redundancy can be avoided by subdividing the computation by elements on the different processors. On the other side, the above strategy imposes a cost in terms of communication overhead, because the contribution to a single matrix element is due to volume elements allocated on different processors.

These considerations lead to the following algorithm (Local Computation And Communication, LCAC).

1) **Initialization**

To obtain the best performances, the available processors, in a number equal to the square of an integer number, are allocated in a square grid. The matrix coefficients are partitioned between the processors in a two-dimensional block cyclic way [2].

On the other side the volume elements of the mesh are distributed by rows among different processors in order to have balanced loads and taking into account the symmetry of interaction among the elements. Pointers from rows of the local part of the L matrix to the same rows of the global L matrix and vice-versa are computed. Let's note that the same element L_{ij} can be present in different processors, since the edges i and/or j can belong to elements stored in different processors. In addition, for this reason, the memory required for storing the computed data could be, for some processors, much larger than the memory necessary for storing the local part of L.

2) **Computation of local matrices**

```
For each element of current_process Iel1 do
        for each element Iel2 with Iel2 ≥ Iel1 do
                Compute the edge-edge interactions associated to the couple of
                elements (iel1-iel2) taking into account the symmetry of the matrix,
                using indirect addressing.
        end for
end for
```

3) **Redistribution of local matrices**

The local contributions to the processors are summed according to the allocation of the unknowns required by the solver.

The last step requires an all-to-all communication among the processors that can become a bottle neck if the speed of the network is not very high. To obviate this situation we proposed a second algorithm (Local Algorithm, LA), that minimizes the data communication cost, by limiting it to the exchange of the data between processors at symmetrical positions in the processor grid; the price to be paid is a redundancy in the computation phase. On the other side, differently from what happens for the LCAC algorithm, the memory required for each processor is now exclusively related to the size of the local part of L.

In [2] it 's also showed that a suitable renumbering of the unknowns increases in general the computational efficiency.

The proposed algorithms have been implemented on a “beowulf system” (BWS) (consisting of a cluster of sixteen Pcs with a Pentium III 450 MHz CPU, 512Mb Ram, 8 GB EIDE hard disk, connected via a 16 port fast ethernet switch) and a “shared memory system”, (i.e.a Sun Fire 880, equipped with 4 Ultra Sparc-3 750Mhz CPU, 12 Gb Ram, 100GB SCSI hard disk). The executed benchmarks showed a speed-up of 10 or 8 (respectively for complex or real systems) on the16 processors Beowulf system and of 3.3 or 3.6 (respectively for complex or real systems) on the 4 processors SUN workstation.

The LA algorithm performs better than LCAC on BWS because of the low communication speed. Instead, on the shared memory system SUN, the time devoted to communications is much shorter and the two algorithms show almost the same speed-up.

3. Differential formulation of finite element method

The finite element analysis is widely used in resistance and reactance calculation [3][4]. In this work we have used a differential formulation and two kind of unknowns: magnetic vector potential on the total region, and the electric scalar potential only inside the metallic plate (eddy current region). The basic equations are:

$$\nabla\times\left[\frac{1}{\mu}\right]\nabla\times\mathbf{A}+[\sigma]\frac{\partial\mathbf{A}}{\partial t}+[\sigma]\nabla\cdot \mathrm{V}=0 \tag{1}$$

$$\nabla\cdot\left(-[\sigma]\frac{\partial\mathbf{A}}{\partial t}-[\sigma]\nabla\cdot \mathrm{V}\right)=0 \tag{2}$$

$$\nabla\times\left[\frac{1}{\mu}\right]\nabla\times\mathbf{A}=\mathbf{J} \tag{3}$$

where **A** is the magnetic vector potential, V is the electric scalar potential and **J** the applied source current density vector. The basic algebraic equation shape is:

$$[\mathrm{B}]\{\dot{\mathbf{u}}\}+[\mathrm{C}]\{\mathbf{u}\}=\{\mathbf{J}\} \tag{4}$$

where $[\mathrm{B}]$ and $[\mathrm{C}]$ are a suitable coefficient matrix that takes into account the element shape functions, {u} is the vector of the unknowns (magnetic vector potential **A** and electric scalar potential V) and {**J**} is the vector of applied loads [5]. We have modeled the probe coil, the metallic plate as well as the empty space around them using a mesh of hexahedrons nodal elements. We have obtained meshes using an extruding technique that produces automatically a three-dimensional mesh from the two-dimensional map of the system constituted by the coil and the plate with the defect (see figure 1). In order to show clearly the used mesh type the figure 3 is limited to the region around the defect and the probe. The equivalent resistance and equivalent reactance of the system constituted by the probe coil and the plate with the defect are obtained by computing the current density and the magnetic vector potential. The current density has been evaluated in the center of the hexahedrons, while the magnetic vector potential in the center of the element has been derived by interpolating the values in the nodes. We have obtained satisfactory results using a cylindrical symmetry mesh centered in the probe coil (see figure 1). In order to avoid an excessively time-consuming re-meshing, it is necessary that different locations of the probe coil with respect to the crack can be modeled with the same grid or some parts of it, suitably connected. In this case we have solved the problem simply by moving the crack on different positions with respect to the plate mesh, while the position of the probe coil was fixed. The position of the crack is determined by the material assignment (air or metal) for each volume cell, along the search path. The boundary conditions used are $\mathbf{A} = 0$ in a region with cylindrical shape and with a thickness up to ten times larger than the probe coil diameter. We have stopped the increasing of the diameter of the bounding region when we have attained no changes in the numerical solution.

4. Experimental set-up

The detection technique based on the impedance variation of a probe coil is well known, and widely studied and used for many applications. In particular it is possible to detect a thin crack in a metallic body by indirect measurement of the impedance variation of a probe coil, using the electrical circuit represented in figure 2. V is an a.c. voltage source, R is a bridge resistance, Z1 is the impedance of a probe coil placed on a metallic plate, Z2 is the impedance of a reference coil. The probe coil has 400 turns, 10 mm of height, 12 mm of inner diameter, 18 mm of outer diameter and 0.5 mm of average lift-off. The reference coil is identical to the probe coil described above, and is placed in a region of the metallic plate without defects. The value of Z2 is the same of Z1 when the probe coil is far from the defect position. This way, by moving the probe coil on the metallic plate surface and by measuring the voltage V_{BC} between the nodes B and C, it is possible to determine the defect locations in the metallic plate (see figure 2). This particular topology of electric circuit gives an output signal V_{BC} immune with respect to the change of the voltage source value and to the thermal phenomena of the probe coil.

Critical points for the experimental procedure are the positioning system for the probe coil with respect to the cracks on the metallic plate and the parameter acquisition system. We have faced these questions with the experimental set-up including a two-dimension automatic positioning system with a resolution of 0.1 mm, a signal voltage analyzer with a resolution of 1 μV for the VBC measurement, an amplified signal generator for the electrical circuit supply and dedicated software + digital control. This experimental set-up is represented in figure 3:

1 Automatic positioning system.
2 Signal voltage analyzer.
3 Signal generator.
4 Dedicate software + personal computer control for 1 and 2.

In particular the probe coil path and the acquisition time of each point of measure can be computer-controlled.

In our work we have analyzed three kinds of artificial defects realized on an aluminum plate. Shapes and dimension of these defects are represented in the figure 4. In the case b) and c) the probe coil is placed on the opposite side with respect to the cylindrical defect.

5. Numerical and experimental results

When the probe coil is placed on a given point around the defect, the voltage variation on the balanced bridge is calculated by means of both the integral formulation above and a classical differential formulation of finite element method. The results obtained and the experimental data [6] are compared in the figures 5, 6 and 7. Some details about the cited numerical models, calculation time and computers used are reported in the table 1. It can be noted that the integral formulation on parallel systems gives more accurate results with reduced calculation time in relation to the classical finite element approach.

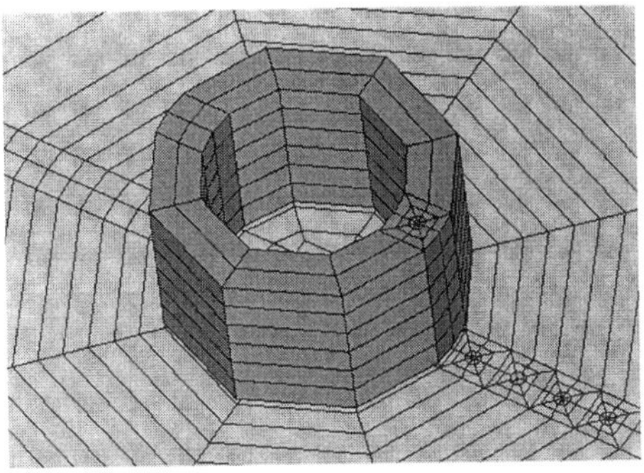

Fig. 1. Detail of probe coil + plate mesh.

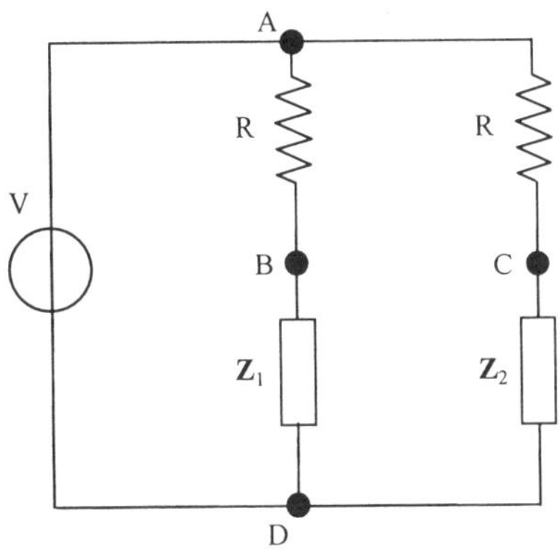

Fig. 2. Electrical circuit linked to probe coil.

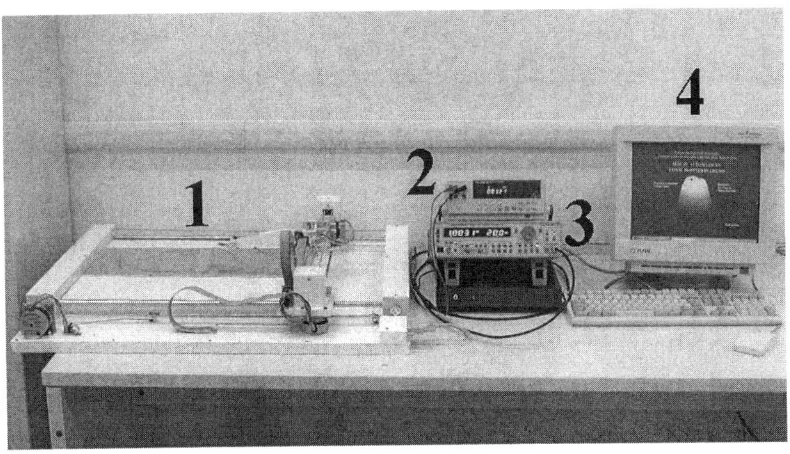

Fig. 3. Picture of experimental set-up.

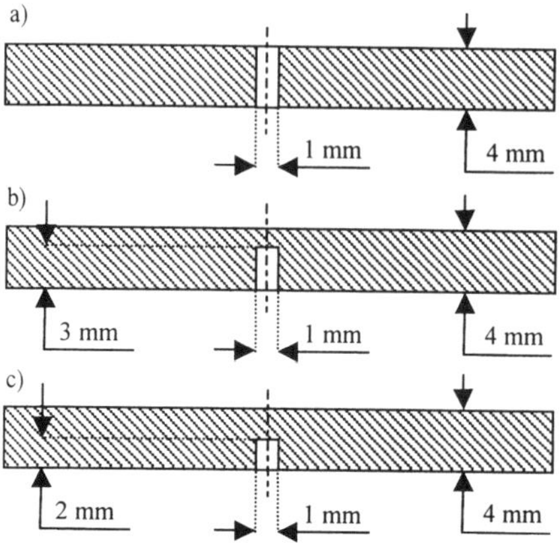

Fig. 4. Shape and dimension of defects.

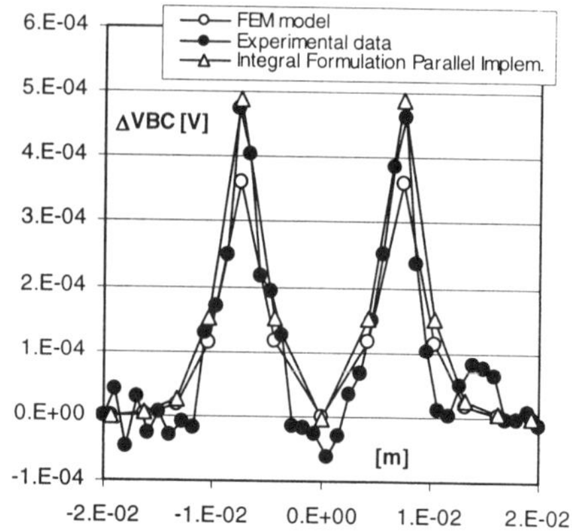

Fig. 5. Comparison between numerical and experimental data for defect a) described in figure 4.

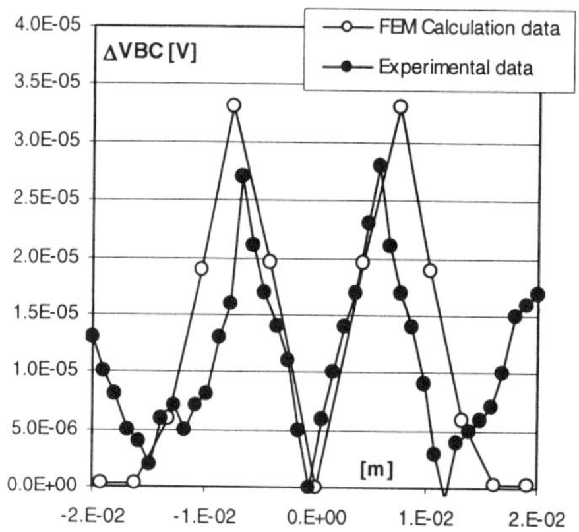

Fig. 6. Comparison between numerical and experimental data for defect b) described in figure 4.

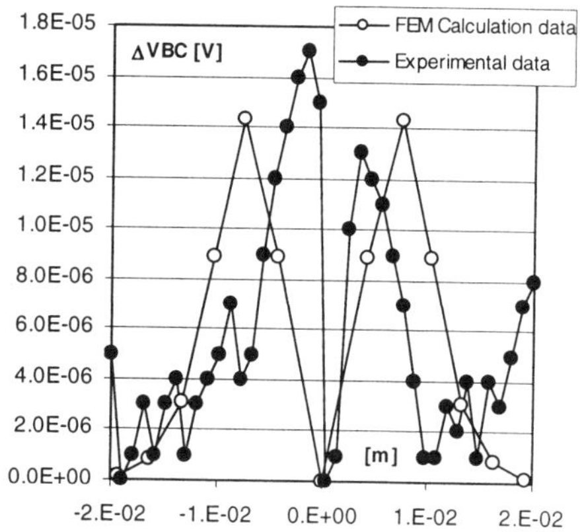

Fig. 7. Comparison between numerical and experimental data for defect c) described in figure 4.

Table 1: Numerical model performances and hardware details.

Numerical method	Number of elements	Calculation time results of fig. 1 [min]	Processor [MHz]	RAM Memory [Mb]	Hard Disk Memory [Gb]
Eddy current integral formulation on 16 nodes parallel systems	4224	164	Intel Pentium III© 450 (for each node)	512 (for each node)	8 (for each node)
Differential formulation of finite element method	6194	630	Intel Pentium III© 450	128	4

References

[1] R. Albanese, G. Rubinacci, "Finite element methods for the solution of 3D eddy current problems," *Advances in Imaging and Electron Physics*, vol. 102, pp. 1-86, 1998

[2] R. Fresa, G. Rubinacci and S. Ventre, "An eddy current integral formulation on parallel computer system", submitted for publication on the International Journal for Numerical Methods in Engineering.

[3] E. Cardelli, A. Faba, A. Massinelli, "FEM Applied to the Detection of Extremely Thin Cracks via Eddy Current Non Destructive Test", The 8th Joint MMM – Intermag Conference 2001, San Antonio, Texas, USA January 7 – 11, 2001.

[4] P. Burrascano, E. Cardelli, A. Faba, S. Fiori, A. Massinelli, "Numerical Analysis of Eddy Current Non-Destructive Testing (JSAEM Benchmark, Problem #6 – Cracks with different shapes)", Proc. of the E'NDE Conference, pp. 333 – 340, Budapest, Hungary, June 28 – 30, 2000.

[5] O. Biro and K. Preis, " On The Use of The Magnetic Vector Potential in the Finite Element Analysis of Three-Dimensional Eddy Currents", IEEE transactions on Magnetics, Vol. 25, No. 4, pp 3145-3159 (July 1989).

[6] E. Cardelli and A. Faba "FEM Analysis of Thin Cracks in Metallic Plates", Accepted for presentation to ISEM Conference, Versailles, Paris 2003.

[7] Rubinacci G., Tamburrino A., Villone F. Circuits/Fields Coupling and Multiply Connected Domains in Integral Formulations. IEEE Transactions on Magnetics 2002; 38: 581 –584.

Electromagnetic Nondestructive Evaluation (VIII)
T. Sollier et al. (Eds.)
IOS Press, 2004

Magnetic Sensor Array for Eddy Current Field Measurement with a Racetrack Coil Excitation

J. R. Bowler and Vipul Katyal
Iowa State University, Center Nondestructive Evaluation,
Applied Sciences Complex II, 1915 Scholl Road, Ames Iowa 50011, USA.

Abstract. Calculations have been performed to determine the response of an eddy current probe consisting of a racetrack coil and a linear array of solid state sensors for detecting perturbations in the magnetic field due to subsurface corrosion in layered aircraft structures. The sensor array allows field measurements to be made at a number of closely spaced locations without moving the probe and thereby accelerates the inspection process. A equivalent magnetic source is used to represent the probe coil and to determine the electric field in the unflawed structure. Multi-sensor flaw responses have been calculated, using a volume element calculation, to simulate the inspection capabilities of the probe.

1 Introduction

In eddy current inspection, an induction coil is often used both to induce current in a conducting component and to detect magnetic field perturbations due to flaws. For subsurface defects, a low frequency of excitation ensures an adequate depth of field penetration. However, at lower frequencies the effectiveness of the coil as both inducer and sensor is diminished since electromagnetic induction depends on the rate of change of magnetic flux. To overcome the limitations of the induction coil as a low frequency field sensor, a solid state device, such as a giant magneto-resistor or Hall sensor, can be used instead. A coil used only as driver can be larger than otherwise without compromising the spatial resolution of the measurements. The large coil can produce a greater field while good spatial resolution of the field measurements is guaranteed by using small magnetic field sensors.

This article gives the analysis of an eddy current probe for the detection of subsurface flaws in multilayered structures such as aircraft skins. The probe contains a racetrack coil with semi-circular bends and linear straights, Figure 1. The magnetic field between the straights is measured using a linear array of magnetic field sensors. The sensor array samples the magnetic field at multiple sites without moving the probe and hence reduces the inspection time. The overall objective of this work is to evaluate the capabilities of array probes and assess their performance for the detection of cracks, material loss and surface roughness due to corrosion. Here we focus on the details of the coil field calculation.

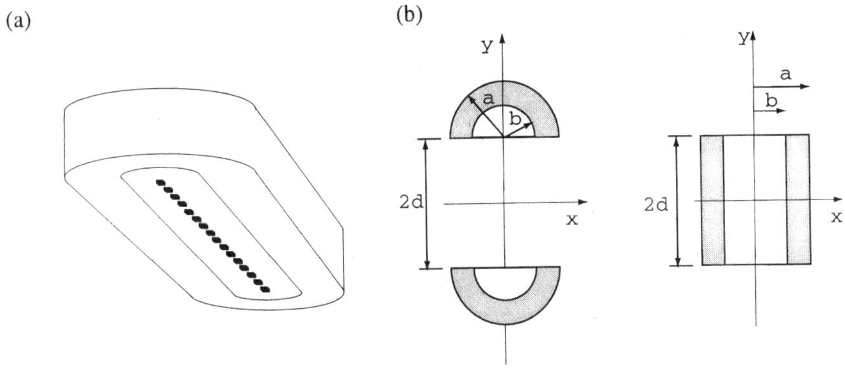

Figure 1: (a) Racetrack coil and magnetic field sensor array. (b) The field of the coil is computed by subdividing it into bends and straight sections.

2 Coil Field Evaluation

The theory for computing the electromagnetic field of a racetrack coil, Figure 1, has been developed by determining separately the electric field due to the two bends and the combined field of the straights, Figure 1(b), and summing the three components to obtain the total field. The current path of the straight sections is closed by joining the ends of the straights by current filaments but ultimately the effect of these filaments is cancelled by similar filaments added to the current path of the bends. Results for a field in layered conductors have been obtained and readily reduce to the well-known formulae of Dodd and Deeds [1] in the case where the length of the straight section vanishes.

The magnetic dipole formulation is used to derive the unperturbed field in the layered structure, the z-direction being normal to the layers. Thus, the transverse source current $\mathbf{J}(\mathbf{r})$, having zero divergence, is written in transverse scalar form as

$$\mathbf{J}(\mathbf{r}) = \frac{1}{\mu_0}\nabla \times [\hat{z}M(\mathbf{r})], \tag{1}$$

where the function $M(\mathbf{r})$ represents the current source in terms of a magnetic dipole density, the orientation of the polarization being in the $z-$direction. This is an adaptation of the magnetic shell model which represents a filamentary current loop in terms of magnetic shell bounded by the loop. The electric field is similarly represented in terms of a scalar function as

$$\mathbf{E}(\mathbf{r}) = -\jmath\omega\mu\nabla \times [\hat{z}\psi(\mathbf{r})] \tag{2}$$

where ψ is the transverse electric (TE) scalar potential.

The translational invariance of the workpiece means that the one can write the TE potential as a convolutional integral operator acting on $M(\mathbf{r})$ with a suitable Green's kernel. Thus

$$\psi(\mathbf{r}) = \frac{1}{\mu_0}\int_{\Omega_0} G(\mathbf{r},\mathbf{r}')\, M(\mathbf{r}')\, d\mathbf{r}'. \tag{3}$$

A further consequence of the translational invariance is that the field can be conveniently expressed in terms of a two dimensional Fourier transform with respect to x and y, written as

$$\tilde{f}(u,v) = \int_{-\infty}^{\infty}\int_{-\infty}^{\infty} f(x,y)e^{-iux-ivy}\,dx\,dy. \tag{4}$$

The two dimensional Fourier transform of (3) is then written as

$$\tilde{\psi}(u,v,z) = \frac{1}{\mu_0}\int_{h-c}^{h+c} \tilde{g}(\kappa,z,z')\,\tilde{m}(u,v,z')\,dz', \tag{5}$$

where $\tilde{g}$ and $\tilde{m}$ are the Fourier transforms of G and M respectively, $\kappa = \sqrt{u^2+v^2}$ and the integration is between the lower and upper limits of the source coil. Here, it is assumed that the coil is uniform between $z = h-c$ and $z = h+c$. Also, the Green's kernel is viewed as representing the TE potential for a point source and therefore satisfies the same interface conditions as ψ. In order to guarantee the continuity of the tangential electric and magnetic field at the interface, we require continuity of μG and $\partial G/\partial z$, respectively. Thus the Green's function for calculating the TE potential in a plate of thickness D for an external source above the plate is derived by taking into account the continuity conditions at both upper and lower surfaces of the plate to give

$$\tilde{g}(\kappa,z,z') = \frac{1}{\gamma+\mu_r\kappa}e^{\gamma z-\kappa z'}\frac{1+\Gamma e^{-2\gamma(D+z)}}{1-\Gamma^2 e^{-2\gamma D}} \qquad \text{for } 0 \geq z \geq -D \tag{6}$$

where $\gamma = \sqrt{\kappa^2 + j\omega\mu\sigma}$, taking the root with a positive real part and

$$\Gamma = \frac{\gamma-\mu_r\kappa}{\gamma+\mu_r\kappa}.$$

Note that the Fourier transform of the half-space kernel can be recovered from (6) by letting $D\rightarrow\infty$. Performing the integration in (5) gives

$$\tilde{\psi}(u,v,z) = 2nI\tilde{f}(u,v)\tilde{g}(\kappa,z,h)\sinh(\kappa c)/\kappa, \qquad z<0, \tag{7}$$

where I is the coil current, n the turns density for unit cross-sectional area and we have introduced a function $\tilde{f}$ defined such that $\tilde{m}(u,v) = nI\tilde{f}(u,v)/\mu_0$. Hence, from (2), the Fourier transform of the electric field is given by

$$\tilde{\mathbf{e}}(u,v,z) = -2\omega\mu_0 nI\frac{v\hat{x}-u\hat{y}}{\kappa}\tilde{f}(u,v)\tilde{g}(\kappa,z,h)\sinh(\kappa c), \tag{8}$$

which can be inverted numerically using a fast Fourier transform for a given coil function, $f(x,y)$.

3 Straight Coil Sections

The y-component of the current density in the straight elements of the source coil, Figure 1, is written as

$$J_y(\mathbf{r}) = \begin{cases} nI\,\mathrm{sign}(x), & h-c\leq z\leq h+c, \quad b\leq|x|\leq a, \quad |y|\leq d, \\ 0, & \text{otherwise} \end{cases} \tag{9}$$

where the current is deemed to flow in a counter-clockwise direction viewed from above. It can be deduced from (1), by writing

$$M(x,y,z) = \begin{cases} \mu_0 n I f(x,y), & h-c \le z \le h+c \\ 0, & \text{otherwise} \end{cases} \tag{10}$$

that

$$f(x,y) = \begin{cases} a-b, & 0 \le |x| \le b, \quad |y| \le d, \\ a-|x|, & b \le |x| \le a, \quad |y| \le d, \\ 0, & \text{otherwise.} \end{cases} \tag{11}$$

Because $f(x,y)$ is even in x and y, the Fourier transform may be written in the form of the double cosine integral as

$$\begin{aligned} \tilde{f}(u,v) &= 4\int_0^\infty \int_0^\infty f(x,y)\cos(ux)\cos(vy)\,dx\,dy \\ &= -\frac{4}{u^2 v}\left[\cos(ua) - \cos(ub)\right]\sin(vd). \end{aligned} \tag{12}$$

which is used together with (8) to calculate the electric field due to the straight sections.

4 D-Coil Field

Two D-coils are used to represent the bends of the racetrack coil, Figure 1. Using essentially the same formulation that was used for the linear coil, ψ for the D-coil is written as in (3). It is convenient to express the Green's function in cylindrical polar coordinates, as

$$G(\mathbf{r},\mathbf{r}') = \frac{1}{2\pi}\sum_{m=0}^{\infty} \epsilon_m \cos[m(\phi-\phi')] \int_0^\infty J_m(\kappa\rho)J_m(\kappa\rho')\tilde{g}(\kappa,z,z')\kappa\,d\kappa, \tag{13}$$

which can be derived using an approach given by Morse and Feshbach [2]. In (13), ϵ_m is the Neumann factor: $\epsilon_0 = 1$ and $\epsilon_m = 2$ $(m = 1,2,3,\ldots)$. For the interior of a conducting plate, $\tilde{g}$ is given by (6). In order to evaluate (3), the explicit form of $M(\mathbf{r})$ appropriate for the D-coil must be found. This form is developed as follows.

The azimuthal counter-clockwise current in a D-coil is written as

$$J_\phi(\mathbf{r}) = \begin{cases} nI, & h-c \le z \le h+c, \quad 0 \le \phi \le \pi, \quad b \le \rho \le a \\ 0, & \text{otherwise.} \end{cases} \tag{14}$$

where I is the current and n the number of turns per unit area. It can be deduced from (1) with (14), by writing

$$M(\rho,\phi,z) = \begin{cases} \mu_0 n I f_D(\rho), & h-c \le z \le h+c, \quad 0 \le \phi \le \pi, \\ 0, & \text{otherwise} \end{cases} \tag{15}$$

that

$$f_D(\rho) = \begin{cases} a-b, & 0 \le \rho \le b, \\ a-\rho, & b \le \rho \le a, \\ 0, & \text{otherwise} \end{cases} \tag{16}$$

It is now possible to obtain ψ by substituting equations (13),(15) and (16) into (3). Integration with respect to ρ', ϕ' and z' gives a summation of integrals with respect to κ:

$$\begin{aligned}\psi(\mathbf{r}) &= nI\left[\frac{1}{\pi}\sum_{\lambda=0}^{\infty}\frac{4}{2\lambda+1}\sin[(2\lambda+1)\phi]\right.\\ &\quad\times\int_0^\infty J_{2\lambda+1}(\kappa\rho)\mathcal{F}_{2\lambda+1}(a,b,\kappa)\tilde{g}(\kappa,z,h)\sinh(\kappa c)\,d\kappa\\ &\quad\left.+\int_0^\infty J_0(\kappa\rho)\mathcal{F}_0(a,b,\kappa)\tilde{g}(\kappa,z,h)\sinh(\kappa c)\,d\kappa\right]\end{aligned} \tag{17}$$

where

$$\begin{aligned}\mathcal{F}_\nu(a,b,\kappa) &= \int_0^a f_D(\rho)J_\nu(\kappa\rho)\rho d\rho\\ &= \frac{1}{\kappa^2}\left[a\mathcal{J}_\nu^{(1)}(\kappa a)-b\mathcal{J}_\nu^{(1)}(\kappa b)\right]-\frac{1}{\kappa^3}\left[\mathcal{J}_\nu^{(2)}(\kappa a)-\mathcal{J}_\nu^{(2)}(\kappa b)\right]\end{aligned} \tag{18}$$

with f_D given by (16) and

$$\mathcal{J}_\nu^n(z) = \int_0^z x^n J_\nu(x)\,dx. \tag{19}$$

These functions are evaluated for $\nu > 3$ with the aid of a recursion relationship

$$(\nu-n)\mathcal{J}_{\nu+1}^n(z) = -2\nu z^n J_\nu(z) + (\nu+n)\mathcal{J}_{\nu-1}^n(z) \tag{20}$$

derived using Eq 11.3.6 of reference [3]. The integrals with respect to κ must be computed numerically and the summation in (17) truncated at a suitable order depending on the required accuracy of the result. A truncated series of five terms is sufficiently accurate in most cases. This number of terms was also found to be adequate for a double-D differential filamentary loop [5].

From equation (2), the electric field in cylindrical coordinates is

$$\mathbf{E}(\mathbf{r}) = -\jmath\omega\mu_0\left(\hat{\rho}\frac{1}{\rho}\frac{\partial}{\partial\phi}-\hat{\phi}\frac{\partial}{\partial\rho}\right)\psi'(\mathbf{r}). \tag{21}$$

The components of the electric field are therefore,

$$\begin{aligned}E_\rho(\mathbf{r}) &= -\frac{4\jmath\omega\mu_0 nI}{\pi\rho}\sum_{\lambda=0}^{\infty}\cos[(2\lambda+1)\phi]\\ &\quad\times\int_0^\infty J_{2\lambda+1}(\kappa\rho)\mathcal{F}_{2\lambda+1}(a,b,\kappa)\tilde{g}(\kappa,z,h)\sinh(\kappa c)\,d\kappa\end{aligned} \tag{22}$$

and

$$\begin{aligned}E_\phi(\mathbf{r}) &= \jmath\omega\mu_0 nI\left\{\frac{1}{\pi}\sum_{\lambda=0}^{\infty}\frac{4}{2\lambda+1}\sin[(2\lambda+1)\phi]\right.\\ &\int_0^\infty\left[\kappa J_{2\lambda}(\kappa\rho)-\frac{2\lambda+1}{\rho}J_{2\lambda+1}(\kappa\rho)\right]\mathcal{F}_{2\lambda+1}(a,b,\kappa)\tilde{g}(\kappa,z,h)\,\sinh(\kappa c)\,d\kappa\\ &\left.-\int_0^\infty \kappa J_1(\kappa\rho)\mathcal{F}_0(a,b,\kappa)\tilde{g}(\kappa,z,h)\sinh(\kappa c)\,d\kappa\right\}\end{aligned} \tag{23}$$

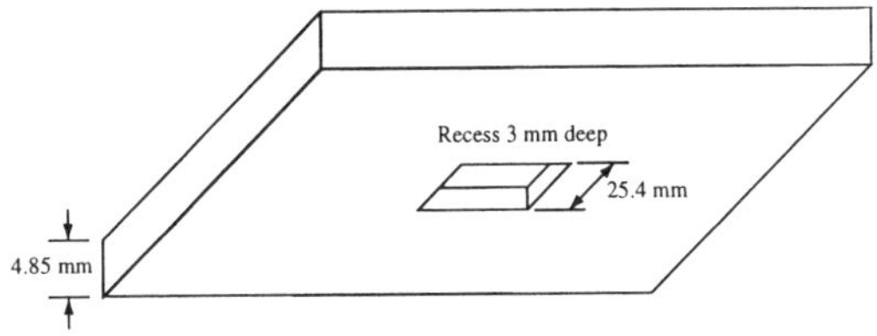

Figure 2: Conducting plate with a square recess.

TABLE 1. Parameters for magnetic sensor measurements.

	Coil		
Outer radius	3.89 mm	Inner radius	1.55 mm
Axial length	5.99 mm	Straight length	27.0 mm
Nominal lift-off	2.5825 mm	Number of turns	517 ± 1
Number of sensors	33	Height of sensors	0.869 mm
Distance between sensors	2.0 mm	Frequency	2000 Hz
Current	1 Amp		
	Plate		
Conductivity	1.82×10^7 S/m	Thickness	4.85 mm

respectively. Equations (22) and (23) give the electric field due to one D-coil. By a displacement and 180 degrees rotation, the field of the second D-coil is found. The sum of the two D-coil fields added to the field due to the straight sections gives the complete racetrack coil electric field.

5 Results

The sensor signals due to a square recess in the bottom surface of a plate of thickness 4.85 mm, Figure 2, with the racetrack coil above the plate providing the excitation field has been calculated using a volume element code [4]. The dimensions of the recess are 25.4 mm × 25.4 mm × 3 mm. These and other parameters are given in Table 1.

The sensors measure the magnetic field component normal to the surface of the conductor. Figure 3, compares the variation with probe position of the normal magnetic field at a central sensor. Field values, normalized to a coil current of 1 Amp, due to the back surface recess is plotted for four different racetrack excitation coils. The coils are similar except that the straight sections are respectively 64 mm, 24 mm, 4 mm and zero. Results from the zero straight section coil calculated using computer code for the racetrack analysis agree with results found for a dedicated circular coil calculation [1]. The absolute value of the z-component of the field found by simulating the response of a 33 element sensor array to the square recess is shown in Figure 4.

6 Conclusion

The theory for a racetrack coil in the presence of a planar conductor has been given in two parts based on a formulation using a magnetic dipole representation of the effect of the coil.

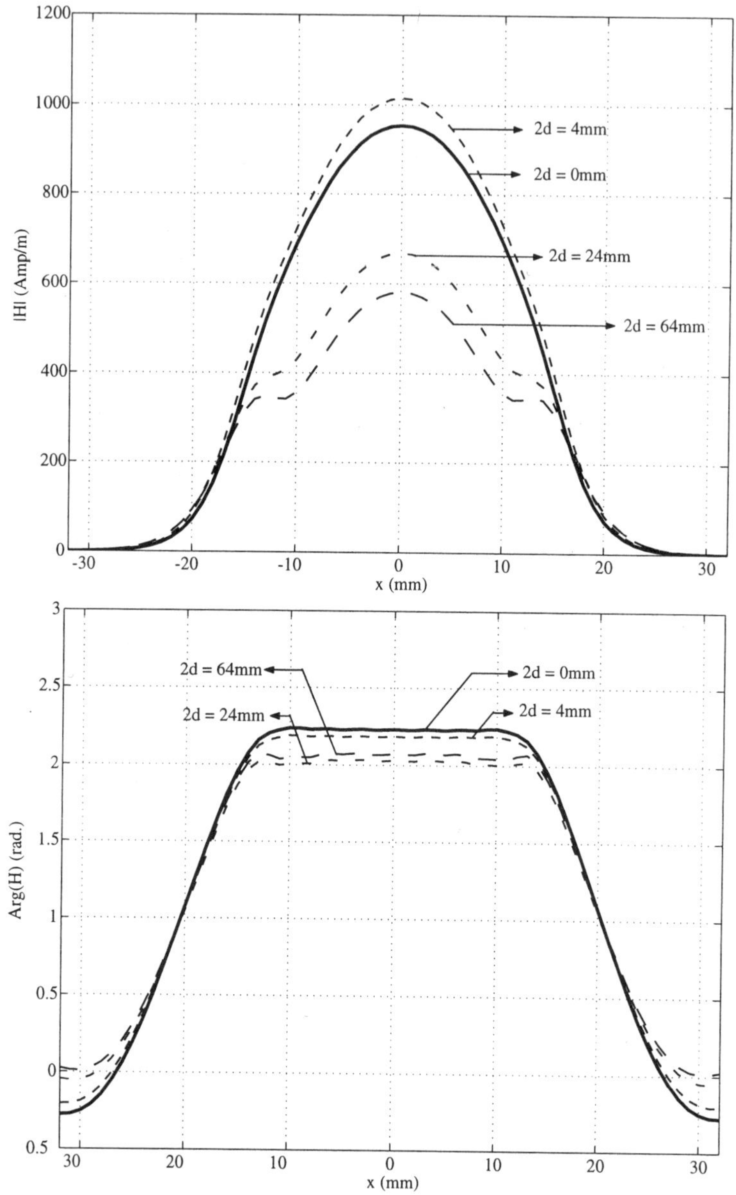

Figure 3: Variation with probe position of the magnitude and phase of the magnetic field at the site of a central sensor 0.89 mm from the surface of the material.

In the first part, the field due to the straight sections of the track are found using a two dimensional Fourier transform and in the second part, the field due to the semicircular bends

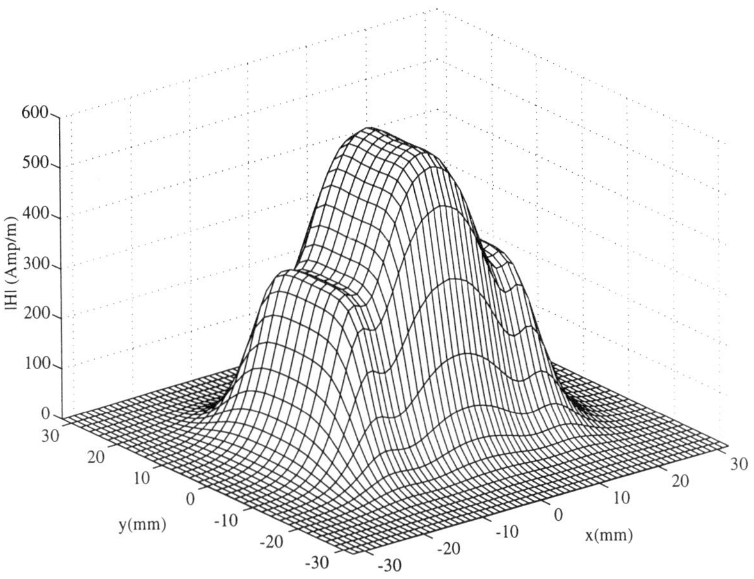

Figure 4: Magnitude of the magnetic field at 33 sensor sites, 2 mm apart in the y direction, due to racetrack coil excitation of a metal plate containing a back surface recess, Figure 2. The excitation frequency is 2000 Hz, the straight sections of the racetrack are 64 mm long and the other probe dimensions are as given in Table 1.

of the track are determined using integrals containing Bessel functions. By using a magnetic sensor array together with a racetrack coil, the aim is to produce probes that are compact, easy to manipulate and can reduce inspection times.

Acknowledgement This work is supported by the AFRL Materials and Manufacturing Directorate at Wright-Patterson Air Force Base, contract F33615-97-D-5271-0031 sponsored by Dr. Thomas J. Moran.

References

[1] C.V Dodd and W. E. Deeds, J. Appl. Phys., 39, p. 2829, (1968).

[2] Morse, P. and Feshbach, H., *Methods of Theoretical Physics*, McGraw-Hill, New York, 1953, p. 1263.

[3] Abramowitz, M. and Stegun, I. A., *Handbook of Mathematical Functions with Formulas, Graphs and Mathematical Tables*, 9th printing, Dover, New York, 1970.

[4] Bowler, J. R., Jenkins, S. A., Sabbagh, S. A. and Sabbagh, S. A., J. Appl. Phys., **70**, p. 1107, (1991).

[5] D. McA. McKirdy, S. Cochran, G. B. Donaldson and A. McNab, Theoretical Consideration of Fatigue Crack Detection and Characterisation Using SQUID Sensors. In: R Collins W D Dover J R Bowler and K Miya (eds.), Nondestructive Testing of Materials. IOS Press, Amsterdam, p. 185, (1995).

Electromagnetic Nondestructive Evaluation (VIII)
T. Sollier et al. (Eds.)
IOS Press, 2004

A Fast Model of Eddy Current Ferrite-Cored Probes for NDE

F. Buvat †, G. Pichenot †, D. Lesselier ‡, M. Lambert ‡, H. Voillaume +
† : *SISC/LCME CEA Saclay 91191 Gif-sur-Yvette France.*
‡ : *DRÉ-L2S (CNRS-SUPÉLEC-UPS), Plateau de Moulon, 91192 Gif-sur-Yvette France.*
+ : *EADS CRC DCR/SE/IND, 92152 Suresnes France.*

Abstract The modelling of the response of a time-harmonic, eddy-current ferrite-cored probe to a 3D flaw in a conductive multilayered, non-magnetic workpiece is aimed at. The first step, which is considered herein, is the calculation of the impedance of an axisymmetric probe and of the corresponding primary field distributions in the magnetic core and the tested workpiece in the absence of flaw.

1 Introduction

Eddy-current nondestructive evaluation of conductive materials is of importance in many domains of industry and transportation. A simple yet typical testing configuration may consist of an axisymmetric magnetic-cored probe which is placed above a planarly layered workpiece (conductivity σ varying typically from 0 between the metal parts to 60 MS/m inside them), and which is operated in the time-harmonic regime at some low frequency chosen in accord with the problem at hand (typically between a few tens Hz and a few MHz).

The interest of the core, in this case, is to better focus the magnetic fields onto certain regions of the workpiece. However the air-cored probe model put forth by Dodd and Deeds [1] cannot be used anymore and other models are in need. Those include a semi-empirical approach [2], which uses experimental (rather than theoretical) values of the Hankel transform of the magnetic field produced by the ferrite-cored probe to calculate the impedance of the probe, as well as an analytical approach for which the boundary value problem involves eigenfunctions expansions of the magnetic vector potential in the various homogeneous zones of the probe and tested layering [3].

In this paper, a solution method based on a rigorous volume integral formulation which is involving Green dyadic functions is investigated, and put together within the framework of the so-called CIVA platform [4], in order to calculate in both reliable and quick fashion the driving-point impedance of the probe and the electromagnetic field distribution in the (linear) magnetic core and in the workpiece.

2 Theoretical approach

A typical configuration is shown in Fig.1. An axially symmetric probe is placed above and parallel to a metal block (or a layered structure). It is made of a ferrite core, typically circular cylindrical or E-shaped, excited by a coaxial coil comprising n_c turns and carrying electrical currents of density $\mathbf{J_e}$ and total intensity I, their time-harmonic dependence $Ie^{j\omega t}$ being henceforth implied.

The above probe in effect consists of two sources of current, an electrical one, which is the one imposed by the user in the windings, and a magnetic one, which is induced in the core in the presence of the workpiece. The fields due to the current-carrying windings and the corresponding probe impedance are easily modeled [1, 4]. On the other hand, the fields due to the magnetic core and their contribution to the impedance are more difficult to compute, even though, as the excitation of the core itself remains weak, the problem may be treated as a linear one (hysteresis is neglected).

To carry out this calculation properly, closely following in that matter the pioneering investigation in [5], one introduces a core magnetization $\mathbf{M}$, the magnetic field $\mathbf{B}$ within the core then satisfying the state (or coupling) equation

$$\mathbf{B}(\mathbf{r}) = \mathbf{B_{inc}}(\mathbf{r}) + 2\pi \int_{core} \left(1 - \frac{\mu_0}{\mu}\right) \overline{\overline{\mathbf{G}}}_{11}^{(mm)}(\mathbf{r},\mathbf{r}')\cdot\mathbf{B}(\mathbf{r}')d\mathbf{r}', \quad \mathbf{M}(\mathbf{r}) = \left(\frac{1}{\mu_0} - \frac{1}{\mu}\right)\mathbf{B}(\mathbf{r}), \quad (1)$$

where μ_0 and μ respectively are the permeabilities of the vacuum and of the core, where $\mathbf{B_{inc}}$ is the primary field induced by currents $\mathbf{J_e}$, and where $\overline{\overline{\mathbf{G}}}_{11}^{(mm)}(\mathbf{r},\mathbf{r}')$ is an appropriate magnetic-magnetic Green dyad (field point $\mathbf{r}$ and source point $\mathbf{r}'$ are both located in air above the workpiece).

Let us notice that electrical conduction is neglected in the core but that μ may be taken as complex-valued to account for magnetic losses. As for the planar stratification of the workpiece (if any, since it can also reduce to a half space), it is taken into account by introducing generalized reflection and transmission coefficients at each interface in the Green dyads [7] (in the above equation and those next).

A discrete solution of the coupling equation is obtained by application of a Galerkin variant of the Method of Moments, which involves a suitable discretization of the core into elementary N_{cel} cells in which the field quantities are assumed constant. Then, once $\mathbf{M}$ is computed within each cell, the vector potential $\mathbf{A_i}(\mathbf{r})$ in a given i^{th} layer of the workpiece easily follows by integration over the regions occupied by the sources (coil and core) from a discrete counterpart of

$$\mathbf{A_i}(\mathbf{r}) = \mu_0 \int_{coil} \overline{\overline{\mathbf{G}}}_{i1}^{(ee)}(\mathbf{r},\mathbf{r}')\cdot\mathbf{J_e}(\mathbf{r}')d\mathbf{r}' + \mu_0 \int_{core} \overline{\overline{\mathbf{G}}}_{i1}^{(ee)}(\mathbf{r},\mathbf{r}')\cdot\nabla\times\mathbf{M}(\mathbf{r}')d\mathbf{r}' \quad (2)$$

where $\overline{\overline{\mathbf{G}}}_{i1}^{(ee)}(\mathbf{r},\mathbf{r}')$ is an appropriate electric-electric Green dyad (for which the field point $\mathbf{r}$ lies in air above the workpiece and the observation point $\mathbf{r}'$ lies in the ith layer).

The probe impedance in the presence of the workpiece follows from

$$Z_{probe} = \frac{j\omega 4\pi^2 n_c \mu_0}{I} \int_{coil} \left\{ \int_{coil} \overline{\overline{\mathbf{G}}}_{11}^{(ee)}(\mathbf{r},\mathbf{r}')\cdot\mathbf{J_e}(\mathbf{r}')d\mathbf{r}' + \int_{core} \overline{\overline{\mathbf{G}}}_{11}^{(ee)}(\mathbf{r},\mathbf{r}')\cdot\nabla\times\mathbf{M}(\mathbf{r}')d\mathbf{r}' \right\} d\mathbf{r} \quad (3)$$

As for the electric field $\mathbf{E_i}(\mathbf{r})$ within the ith layer, it satisfies

$$\mathbf{E_i}(\mathbf{r}) = -\frac{\partial \mathbf{A_i}(\mathbf{r})}{\partial t} = -j\omega\mu_0 \int_{coil} \overline{\overline{\mathbf{G}}}_{i1}^{(ee)}(\mathbf{r},\mathbf{r}')\cdot\mathbf{J_e}(\mathbf{r}')d\mathbf{r}' - j\omega\mu_0 \int_{core} \overline{\overline{\mathbf{G}}}_{i1}^{(ee)}(\mathbf{r},\mathbf{r}')\cdot\nabla\times\mathbf{M}(\mathbf{r}')d\mathbf{r}' \quad (4)$$

The axial symmetry of the sensor and workpiece is such that the vector potential $\mathbf{A_i}(\mathbf{r})$, from which the probe impedance and the fields in the workpiece are derived, is azimuthal (ϕ component only). So the magnetic filed $\mathbf{B}$ has two components (radial and longitudinal) only. As for the Green dyads, they are reduced to scalar functions which are expressed in the form of semi-infinite integrals involving the product of the superposition of the amplitudes of upgoing and downgoing waves in the ith layer and of oscillating functions (first-order Bessel functions).

3 Illustrative results

To illustrate the behavior of the approach, three different configurations are studied below, which are involving a cylindrical or a E-shaped ferrite-cored probe above a half space or a multilayered workpiece.

3.1 Cylindrically cored probe above a conductive half space

The first configuration under study is the one of a probe made of a coil (15.3 turns) partly encircling a circular cylindrical ferrite core (with relative permeability $\mu_r = 1000$) as is shown in Fig. 1, this probe being placed slightly above an aluminum half space ($\sigma = 35\ MS/m$). The core is discretized into 2 cells along the radial direction and 64 cells along the vertical direction. The amplitude of the magnetic field in the core, defined as $\mid \mathbf{B} \mid = \sqrt{\mid \mathbf{B}_r \mid^2 + \mid \mathbf{B}_z \mid^2}$, and the one of the induced electric field $\mid \mathbf{E} \mid$ inside the workpiece ($I = 1\ mA$) for a 2 kHz frequency are depicted in Fig.1.

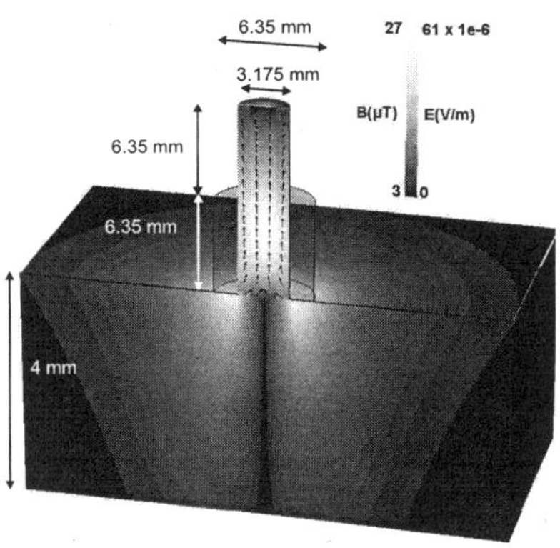

Figure 1: Fields in the core and the workpiece.

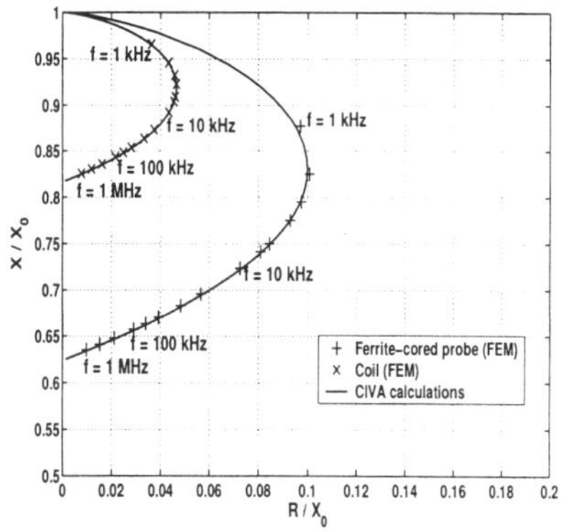

Figure 2: Normalized impedance diagram of the cylindrical ferrite-cored probe and of the coil alone.

The probe performance can be assessed to some extent from the variation of its impedance above the workpiece, $Z = R + jX$, with frequency. The latter is usually considered via a normalized impedance diagram, illustrating how $R_n = R/X_0$ and $X_n = X/X_0$ vary with frequency, X_0 being the free-space reactance, R and X the resistance and the reactance of the probe above the workpiece. In Fig.2, the said diagrams obtained for the coil alone and the ferrite-cored probe are compared to results provided by a commercial finite-element solver from Vector-Fields, frequencies of operation here varying from 1 kHz to 1 MHz. [1] In all

[1]However, for high frequencies, losses due to coupling in coil windings or due to the hysteresis and/or conduction in the core should be taken into account. Experimental data in that case are often corrected by evaluating the coil and the core losses determined experimentally.

cases, the agreement of the numerical results obtained with our model (CIVA) and the finite-element solver (FEM) are better than 2%, both for the real and imaginary parts. A significant result is also the increase of the probe sensitivity when a ferrite core is introduced inside it. Indeed, the field lines become much more confined, and the coupling between the probe and the workpiece improves, resulting in an increase of R_n and X_n.

In addition, in Table 1, one is providing the driving point impedance of the ferrite-cored probe at four different frequencies (1, 10, 100, and 1000 kHz) obtained either by CIVA (left column) or by FEM (right column). Again agreement of real and of imaginary parts is better than 2 %.

Table 1: Driving-point impedance.

Probe in air		
Frequency in Hz	CIVA in Ohms	FEM in Ohms
100 kHz	1.87j	1.88j
Probe on workpiece		
Frequency in Hz	CIVA in Ohms	FEM in Ohms
1 kHz	$1.79 \times 10^{-3} + 1.63 \times 10^{-2}$j	$1.82 \times 10^{-3} + 1.65 \times 10^{-2}$j
10 kHz	$1.34 \times 10^{-2} + 1.34 \times 10^{-1}$j	$1.36 \times 10^{-2} + 1.36 \times 10^{-1}$j
100 kHz	$5.32 \times 10^{-2} + 1.22 \times 10^{+0}$j	$5.45 \times 10^{-2} + 1.23 \times 10^{+0}$j
1000 kHz	$1.81 \times 10^{-1} + 1.18 \times 10^{+1}$j	$1.85 \times 10^{-1} + 1.19 \times 10^{+1}$j

Finally, the electric field induced in the workpiece by the probe with the cylindrical core has been computed at a 2 kHz frequency for two different depths (at the interface between air and metal and inside the workpiece at 2.5 mm depth). The amplitude of the $\mathbf{E}$ field (in V/m) as a function of the radius (in m) varying from 0 to 16 mm are displayed in Fig. 3 and in Fig. 4. Here agreement is better than 1 %.

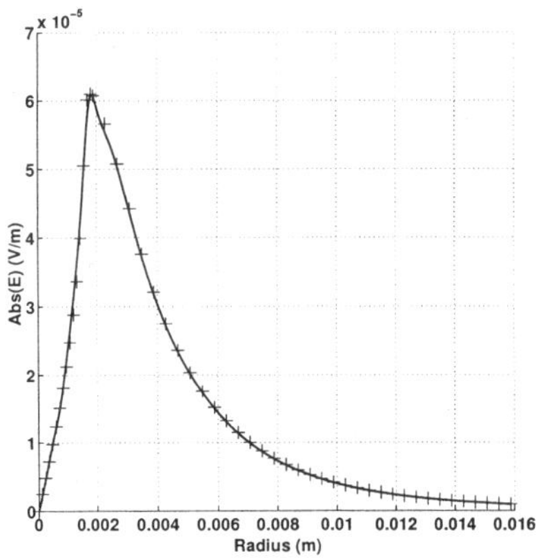

Figure 3: E field at $z = 0\ mm$ with a cylindrical core probe: CIVA (solid line) and FEM (+) results

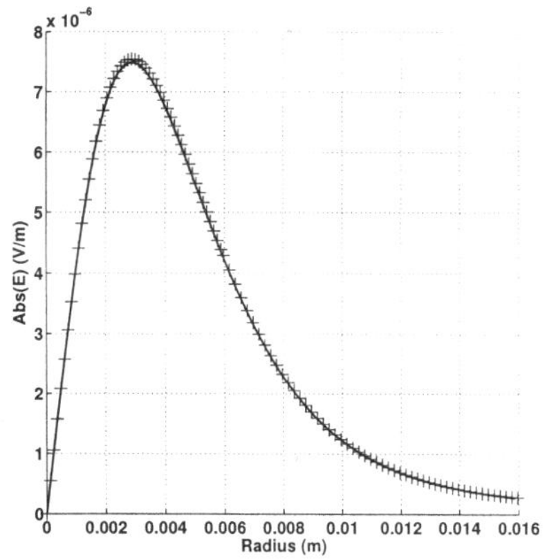

Figure 4: E field at $z = 2.5\ mm$ with a cylindrical core probe: CIVA (solid line) and FEM (+) results

3.2 E-shaped cored probe above a conductive half space

In the second configuration, the cylindrical ferrite core has been replaced by a E-shaped core (keeping same relative permeability $\mu_r = 1000$). The lift-off is also assumed to be close to zero and the workpiece is still a half space made of aluminum ($\sigma = 35\ MS/m$). The core is now discretized into 24 cells in the radial direction and 32 cells in the longitudinal direction. The amplitude of the magnetic field $\mathbf{B}$ in the core and the electric field inside the workpiece ($I = 1\ mA$) for a 2 kHz frequency are depicted in Fig. 5.

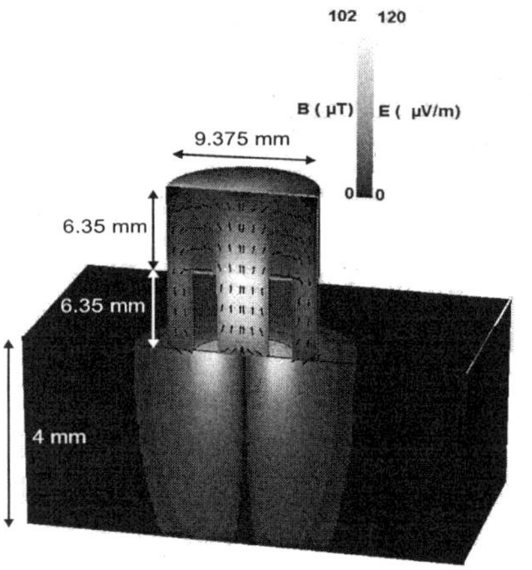

Figure 5: Fields in the core and the workpiece with a cup-core probe.

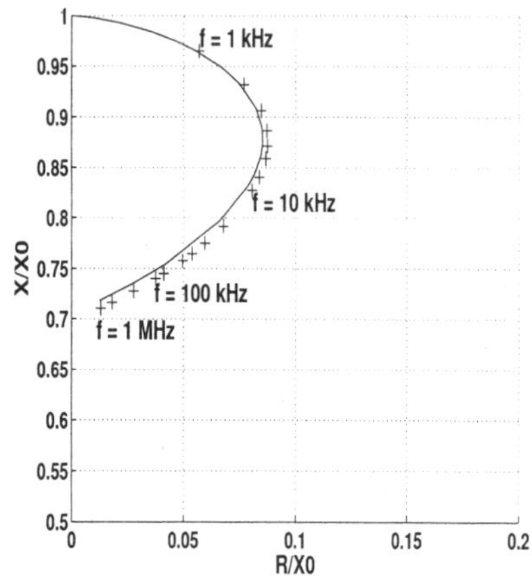

Figure 6: Normalized impedance diagram of the E-shaped core probe, obtained with CIVA (solid line) and FEM (+).

At this frequency, the $\mathbf{E}$ field amplitude in the workpiece is almost two times larger (124 $\mu V/m$ instead of 61 $\mu V/m$) when the cylindrical core is replaced by the E-shaped one, and the $\mathbf{B}$ field amplitude is multiplied by a factor of 4 inside the core. Also, the $\mathbf{E}$ field produced by the cup-core inside the workpiece is much more confined near the probe than the one which is produced by the cylindrical core. Indeed there is considerably more "leakage flux" produced by the cylindrical core (since it has no return path) than by the E-shaped core.

The normalized impedance diagram of the E-shaped core probe is given in Fig.6. Discrepancies between the CIVA and FEM impedance calculations are now lower than 8 % for frequencies varying from 1 kHz to 1 MHz (maximum discrepancies are about 10% for the real part and 7 % for the imaginary part).

Two slices of the $\mathbf{E}$ field have also been computed for two different depths: at the interface air-metal and at $z = 2.5\ mm$; they are displayed in Fig.7 and in Fig.8. The results exhibit discrepancies lower than 7 %, both for the real and imaginary parts when comparing the peak values of the fields. For example, at the interface air-metal, the maximum value of the electric field amplitude for both methods is obtained at $r = 1.8\ mm$ (124.5 $\mu V/m$ with CIVA and 133 $\mu V/m$ with FEM). At 2.5 mm depth, these values are 11.2 $\mu V/m$ with CIVA and 11.9 $\mu V/m$ with FEM, both obtained at $r = 2.3\ mm$.

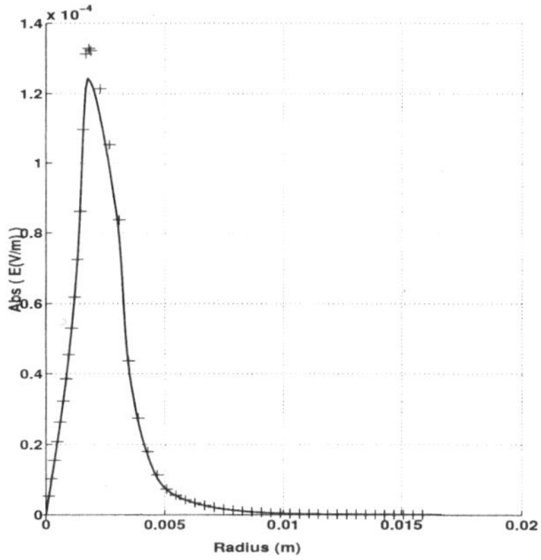

Figure 7: Electric field produced by the E-shaped cored at the interface air-metal: CIVA (solid line) and FEM (+) results.

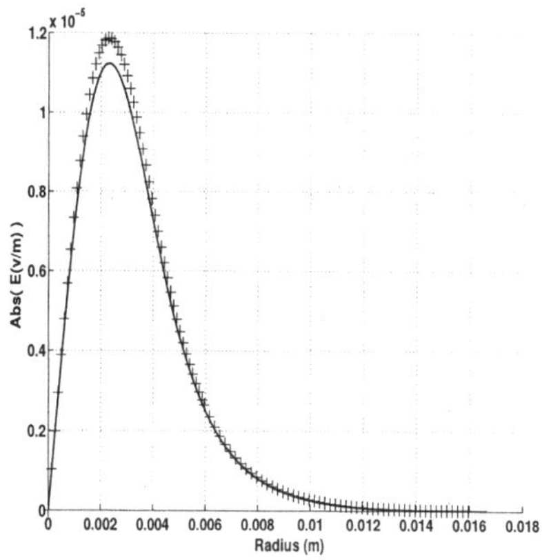

Figure 8: Electric field produced by the E-shaped cored at $z = 2.5\ mm$: CIVA (solid line) and FEM (+) results.

3.3 Cylindrically cored probe above a conductive layering

The third configuration involves a coil (328 turns) encircling a cylindrical core ($\mu_r = 1000$), discretized into 8 cells along both radial and longitudinal directions [2], and placed above a four-layer medium as is depicted in Fig.9. Layers 1 and 3 are metal slabs with respective conductivities $\sigma = 1\ MS/m$ and $\sigma = 10\ MS/m$, and respective thicknesses of 0.5 mm and 0.2 mm. Layer 2 is air and its thickness is 0.2 mm (the last layer, a half space, is also air). In this configuration, the lift-off is equated to 0.1 mm.

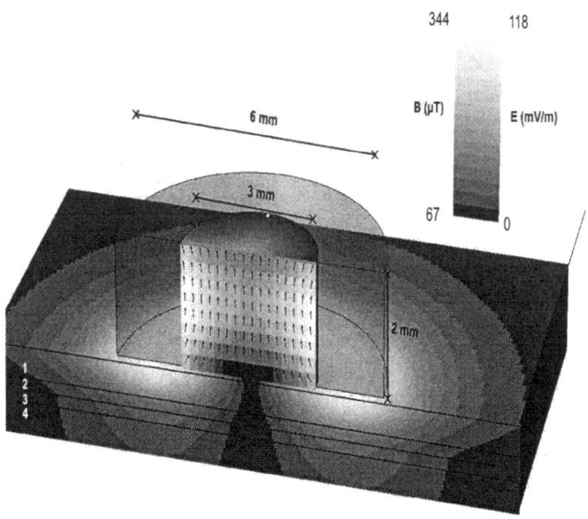

Figure 9: Fields in the core and the workpiece with a cylindrical core above a multilayered medium.

Two slices of the electric field (with $I = 1\ mA$) have been computed for a frequency of

[2]For the three configurations, the discretization is chosen so as numerically convergent results be reached.

300 kHz and compared to the FEM results in the middle of each layer (i.e. $z = 0.25\ mm$, and $z = 0.8\ mm$) are depicted in Fig.10 and in Fig.11.

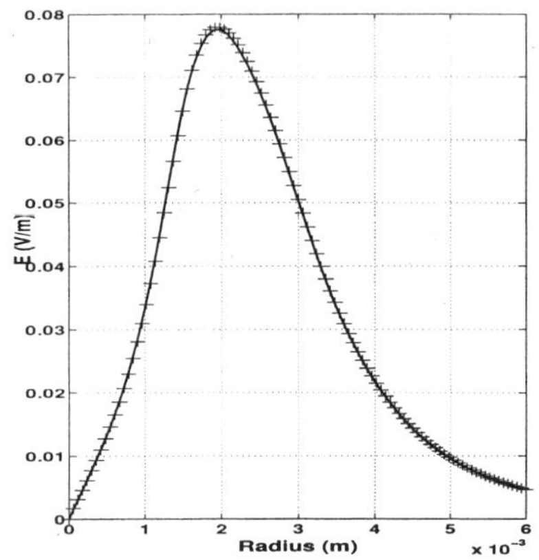

Figure 10: Electric field produced by the cylindrical core above a multilayered medium at $z = -0.25\ mm$: CIVA (solid line) and FEM (+) results.

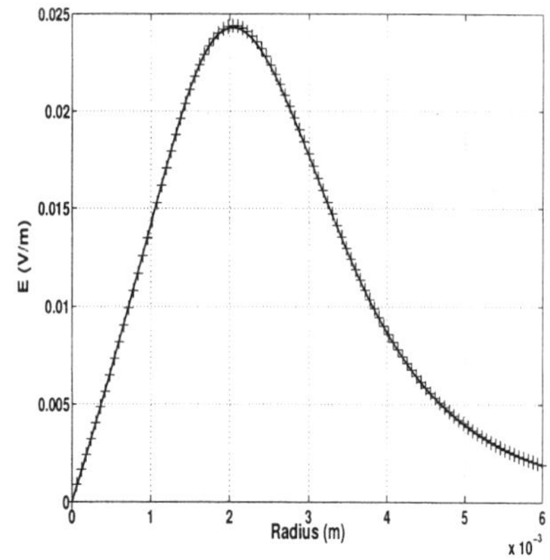

Figure 11: Electric field produced by the cylindrical core above the multilayered medium at $z = -0.8\ mm$: CIVA (solid line) and FEM (+) results.

The relative error between FEM and the CIVA model calculations appears to be better than 1 % for the two different depths. The stratification of the workpiece appears well accounted for via the transmission and reflection coefficients within the Green dyads.

4 Concluding remarks

The model developed herein is based on a volume integral method which has the advantage that one only discretizes the ferrite core in order to solve the problem and not the entire space. The results appear reliable and are obtained rather quickly also (the code is fully vectorized). For example, on a modern PC (2 GHz , 528 Mo), 3 minutes are sufficient to calculate all quantities of interest for the last configuration.

In a near future, the model will be coupled to another one, also developed at CEA and based on a similar integral formalism, which takes into account an arbitrary 3D defect affecting a flat workpiece, in order to compute the ferrite-cored probe response (assuming weak coupling, i.e., the primary field due to the core magnetization is not affected by the presence of the defect). Results will also be validated by comparison with experimental data for several axisymmetric ferrite-cored probes (including E-shaped and cylindrical cores of any size).

As an extension, the non axially symmetric ferrite-cored probe will be investigated in order to handle more general geometries and to develop a model of a sensor working in emission-reception as well. To do so, one will follow a very similar path (solving for the magnetic field $\mathbf{B}$ by applying the Method of Moments with a 3D discretization, and then calculating the $\mathbf{E}$ field and the driving-point of impedance of the probe), being said that the complexity of the calculations should be much higher due to the fact that 3-component vector magnetization and fields and full 9-component Green dyads will have to be tackled properly.

References

[1] C. V. Dodd and W. E. Deeds, Analytical solutions to eddy-current probe coil problems, J. Appl. Phys, 39 (1968) 2829-2838.

[2] S. K. Burke, A semi-empirical model for eddy-current NDE using ferrite-cored probes, Nondestr. Test. Eval, 6 (1992) 267-277.

[3] T. P. Theodoulidis, Model of ferrite-cored probes for eddy-current nondestructive evaluation, J. Appl. Phys, 93 (2003), 3071-3078.

[4] G. Pichenot and T. Sollier, Eddy current modelling for nondestructive testing, Proc. 8th European Conf. on Nondestructive Testing, Barcelona, June 2002.

[5] H. A. Sabbagh, A model of eddy-current probe with ferrite cores, IEEE Trans. Magn., 23 (1987) 1888–1904.

[6] R. F. Harrington, The Method of Moments in electromagnetics, J. Electromagn. Waves. Applicat., 1 (1987) 181-200.

[7] W. C. Chew, Waves and Fields in Inhomogeneous Media, IEEE Press, Pictaway (1995).

Electromagnetic Nondestructive Evaluation (VIII)
T. Sollier et al. (Eds.)
IOS Press, 2004

Tools for the Design of FLUXSET Probes Addressed to the Detection of Deep Defects

Raffaele ALBANESE, Raffaella BARRESI, Marco CARBONE
Assoc. EURATOM/ENEA/CREATE, DIMET, Univ. Mediterranea di Reggio Calabria
Via Graziella, Loc. Feo di Vito, I-89128 Reggio Calabria, Italy

Antal GASPARICS
Research Institute for Technical Physics and Materials Science
H-1525 Budapest, PO. Box 49, Hungary

Abstract. We present a set of tools for the FLUXSET probe optimization in this paper. This probe was designed for eddy current testing of deep defects. The objective function to be described is envisaged to be related to the signal-to-noise ratio. We present a method for the evaluation of the response due to a spherical defect embedded in a metallic plate. We identify in the probe tilting the major source of the noise, and provide an accurate approximation of the tilting effect. These tools allow the prediction and the comparison of the different probe capabilities.

1. Introduction

This work is addressed to the development of nondestructive techniques based on eddy currents analysis, with the aim of investigating the possibility to detect spherical defects with small size (about 10 mm^2 cross-section) in thick plates (10 to 50 mm thick), dealing in particular to the problem of the on-line retrieval of welding defects (e.g., gas bubbles) occurring during the welding procedure of superconducting coil casings.

The activity is focused primarily on the development of simulation tools able to predict the magnetic field perturbations due to these defects and the effect of the main sources of noise (i.e., probe and sensor alignment errors), in view of the definition of a procedure for the optimization of the probes.

The numerical predictions are then compared to the experimental data obtained with a Prototype FLUXSET probe [1] for the analysis of deep defects.

2. The FLUXSET Probe

The investigated FLUXSET probe [1] consists of an exciting coil and a high sensitivity magnetic field sensor (the FLUXSET sensor [2]) placed at the center of the exciting coil, in parallel way to the plane of the specimen to be tested (Fig. 1). In principle, in the absence of defects and far from the specimen edges, the output signal should be zero due to the symmetry. This feature should improve the capability to detect "deep" defects, i.e. beyond the penetration depth of the conventional ECT inductive coils.

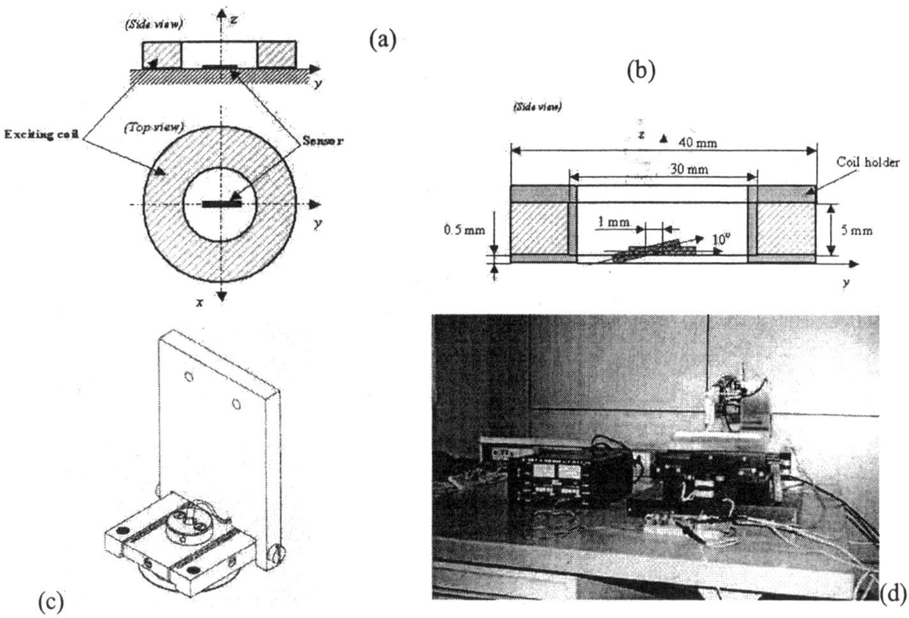

Fig. 1. A FLUXSET probe. The exciting coil has 126 turns, 32 and 37 mm inner and outer diameters, respectively, 4.7 mm height and a minimum lift-off of 0.65 mm; frequency up to 50 kHz, current up to 300 mA. The magnetic field sensor is a FLUXSET #7B, 5 mm long, with a lift-off of 0.65 mm and a resolution of 30 pT: a) layout; b) details; c) holder; d) scanning system.

3. Signal due to a small spherical defect embedded in a metallic plate

The unperturbed eddy current distribution induced in an infinite plate with finite thickness can be computed either numerically or analytically [3].

Once the eddy current density is known at the defect location, the effect of the flaw can be simulated by superimposing a small current flowing in the opposite direction. The eddy current and magnetic field distributions can then be obtained via superposition using the dyadic Green functions. At a distance from the defect, the field perturbation generally approaches the dyadic Green function multiplied by a suitable factor k.

In the case of a small spherical defect not too close to the surface (e.g., due to a gas bubble trapped in the metal during a welding process), at low frequencies, this factor is:

$$k=3J_0V_d \tag{1}$$

where J_0 is the unperturbed current density at the defect location, and $V_d=4\pi a^3/3$ is the volume of the spherical defect (Fig. 2).

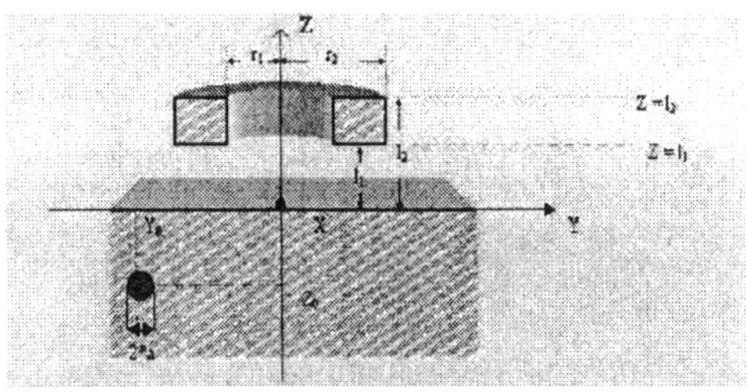

Fig. 2. Coil above a metallic plate with a small spherical defect.

4. Verification with numerical and experimental results

The tangential component of the signal provided by the FLUXSET sensor during a linear horizontal scan can be analytically estimated for a given depth of the flaw. Fig. 3 depicts a comparison between 3D numerical calculations using the CARIDDI eddy current code [4] and the analytic solution illustrated in Section 3.

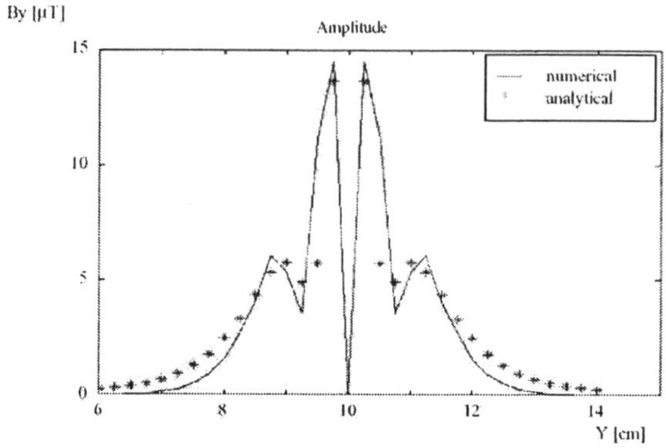

Fig. 3. Perturbation of the horizontal magnetic field on the axis of the inducing coil at z = l_1 due to a small spherical defect (see Fig. 2): numerical vs analytical simulations.

There is also a satisfying agreement with noisy experimental data referring to a cylindrical defect having the same volume as the sphere (see Fig. 4). The signal, which is a complex quantity should be antisymmetric in principle; its amplitude should then be symmetric with respect to y. The nonsymmetric behavior of the experimental data shows the presence of offset in the signal, probably due to coil and/or sensor tilting.

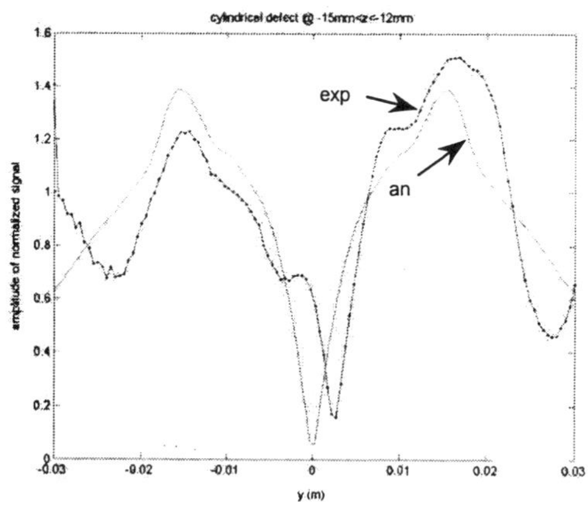

Fig. 4. Perturbation of the horizontal magnetic field on the axis of the inducing coil at z = l_1 due to a small defect (see Fig. 2): experimental data vs analytical results.

5. Effects of defect shape, location and extent

As for as the shape is concerned, experimental data as well as 3D numerical analysis show that a cylindrical defect having radius r and height h is roughly equivalent to a sphere having the same volume, provided that $r \leq h \leq 2r$.

As indicated in (1), for small spheres the signal scales with the volume, rather than its cross section.

The influence of the vertical location of the defect on the signal has been studied at various frequencies. The results shown in Fig. 5 refer to the FLUXSET probe shown in Fig. 1 placed above a 50 mm thick stainless steel (SS) plate with a spherical defect having a radius of 1.5 mm located at 14 mm from the coil axis.

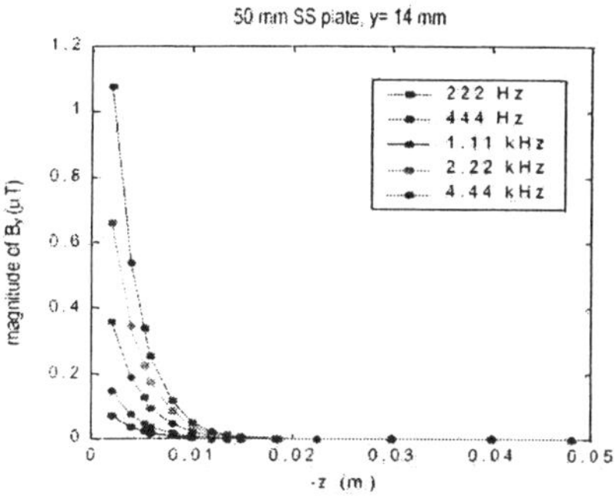

f	|By| (nT) z=-13.5 mm	|By| (nT) z=-18.5 mm	|By| (nT) z=-30 mm	|By| (nT) z=-48 mm
222 Hz	1.66	1.40	0.38	0.0521
444 Hz	3.11	2.06	0.47	0.0501
1.11 kHz	7.51	2.73	0.37	0.0213
2.22 kHz	12.16	2.75	0.17	0.0044
4.44 kHz	13.28	1.98	0.043	0.0003

Fig. 5. Influence of frequency and vertical location of the defect. Perturbation of the horizontal magnetic field on the axis of the inducing coil at z = 0.65 mm due to a spherical defect having a radius of 1.5 mm located in a 50 mm thick SS plate at 14 mm from the coil axis (FLUXSET probe of Fig. 1 with a 117 mA current).

6. Effects of tilting and positioning error of sensor and probe

6.1 Influence of tilting and positioning error of the probe on a plate without defects

The major source of noise is assumed to be related to the tilting of the probe. This effect can be estimated either analytically or numerically. The analytical estimations are trivial at zero and infinite frequency (Fig. 6). Analytical estimations are also possible at finite frequencies using the dyadic Green functions, but the computation is rather awkward, as the integration has to be carried out over the region covered by the coil turns.

The numerical calculation is cumbersome as well, as the tilting makes the problem three-dimensional. Fig. 6 also shows the numerical results obtained with CARIDDI. The order of magnitude is in good agreement with the analytical estimate at high frequency; the magnitude is lower in CARIDDI due to a higher equivalent lift-off because of mesh size, which is limited by memory constraints. The results shown in Fig. 6 at 2.22 kHz and the experimental data collected when scanning plates with and without defects, allows to estimate that the variation of the tilting angle during a scan is in the order of 0.01 deg.

(a)

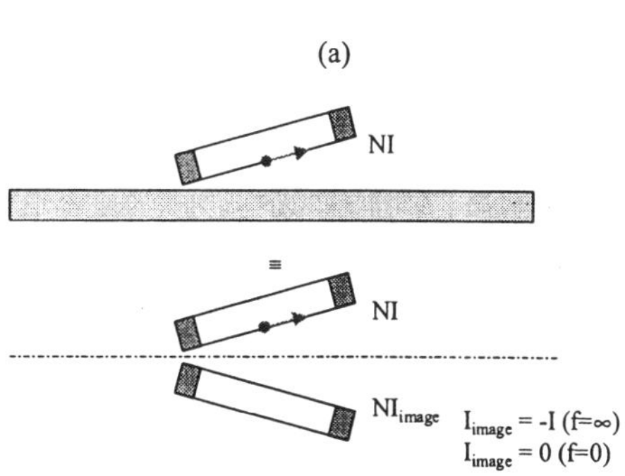

(b)

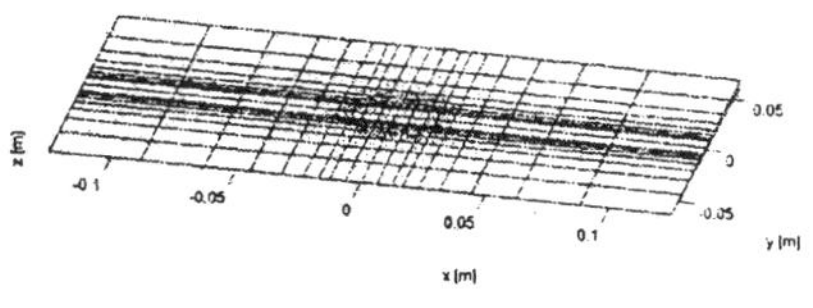

f	thickness	By (μT)
0	50 mm	0
2.22 kHz	50 mm	-2.7-4.4j
∞	50 mm	-14.4
∞	50 μm	-18.0

Fig. 6. FLUXSET probe of Fig. 1 (I_{exc}=117 mA): a) analytical estimation of the tilting effect at high frequency (18 μT/deg); b) numerical results given by CARIDDI (the high frequency case is better modeled with a thin layer on the surface).

6.2 Influence of tilting and positioning error of the sensor on a plate without defects

The sensor signal variations for given displacements and tilting angles can readily be computed for the case of a plate without defects (Table I).

Table I. Influence of tilting and positioning error of the sensor on a plate without defects on the horizontal field (FLUXSET probe of Fig. 1 with a 117 mA current).

f	Δy=0.2mm	Tx=1 deg	Δy=0.2mm Tx=1 deg
0	-1.27	9.21	8.27
222 Hz	-1.32-0.29i	9.01-0.69i	8.02-0.97i
444Hz	-1.42-0.52i	8.74-1.16i	7.65-1.67i
1.11 kHz	-1.8-1.01i	7.9-2.08i	6.45-3.06i
2.22 kHz	-2.44-1.42i	6.69-2.81i	4.62-4.18i
4.44 kHz	-3.43-1.52i	4.99-3.18i	1.97-4.66i
∞	-3.14	0.36	-2.58-0i

6.3 Influence of tilting and positioning error on the signal due to a defect

Horizontal displacements simply introduce offsets. Vertical displacements increase lift-off and slightly reduce the signal. For instance, there is a 5% signal reduction for a 0.2 mm (30%) increase in the lift-off of the sensor for a 13.5 mm deep defect in a 50 mm SS plate at 2.22 kHz.

Coil tilting T_x modifies the induced eddy currents $\Delta J_0/J_0 \approx O[\sin(T_x)]$, whereas sensor tilting T_x modifies the signal. In any case these effects are much lower than those depicted in section 6.1.

7. Optimization of the FLUXSET probe

The methods and the results described in the previous section show that it is possible to estimate the signal-to-noise ratio (S/N) at a given frequency as a function of the vertical position of the defect. It is then possible to optimize the parameters of a given coil and propose alternative probe designs, like the compensation coil probe, for instance, where the inducing field is provided by two concentric solenoids yielding a low horizontal and vertical field in the region around the axis where the sensor has to be located.

The expected magnetic field distribution and signal can be determined analytically, numerically and/or experimentally for the investigated specimens/defects with and without the presence of the displacement/tilting error s(Δx, Δy, Δz, Tx, Ty, Tz). The error suppression objective function S/N can for instance be defined as the reciprocal of the first order signal energy derivative, e.g. for a Δx displacement:

$$S/N_{\Delta x} = \frac{1}{\dfrac{\sqrt{E[s(\Delta x,0,0,0,0,0)] - E[s(0,0,0,0,0,0)]}}{\sqrt{E[s(0,0,0,0,0,0)]}\cdot \Delta x}}, \quad E[s] = \int_{\Gamma} s^2 \, d\Gamma \tag{2}$$

8. Conclusions

The agile tools, presented in this paper, allow us to give a quick but reliable estimate of the signal-to-noise ratio as a function of frequency, expected tilting error, defect extent and depth.

The limits of present design in terms of S/N ratio for given defect and depth have been established and found to be due to small signal below 20 mm and large noise due to probe tilting.

A further optimization is possible. Prediction of both signal and noise are possible for new probe concepts using the analytic simulation tools. In this way it is possible to examine the performance of alternative probe designs (e.g., compensation solenoid, double "D" probe, etc.). In particular, to improve the ability of detecting deep defect, the vertical field at sensor position should be reduced to limit the noise, and the induced current density at deeper layers should be increased to enhance the probe signal.

Acknowledgments

This work has been carried out in the framework of the EFDA Task TW1-TMS/MMTFRD. The authors would like to thank various colleagues from CREATE-Italy (S. Calcagno, E. Coccorese, R. Fresa, G. Leone, G. Paleologo, G. Rubinacci, A. Tamburrino, M. Valentino, S. Ventre, F. Villone), HAS-Hungary (J. Pavo, G. Vertesy) and EFDA-Germany (C. Sborchia, A. Portone, E. Salpietro) for their useful support.

References

[1] A. Gasparics, Cs.S. Daroczi, G. Vertesy, J. Pavo, "Improvement of ECT probes based on Fluxset type magnetic field sensor", in "Electromagnetic Nondestructive Evaluation (II.)" pp. 146-151, R. Albanese et al. (Eds.), IOS Press, 1998.

[2] A. Gasparics et al., "FLUXSET Type Sensor for Eddy Current Testing and DC to Medium Frequency Field Measurement", *Proc. of ISEM '97*, May 12-14 (1997), Braunschweig Germany, TPA2-20.

[3] C.V. Dodd., W. E. Deeds, "Analytical Solutions to Eddy-Current Probe-Coil Problems", *J. Appl. Phys.*, Vol. 39, No. 6, pp. 2829-2838 (1968)

[4] R. Albanese, G. Rubinacci, "Integral Formulation for 3D Eddy Current Computation Using Edge Elements", *IEE Proceedings*, Vol. 135, Pt. A, No. 7, Sept. 1988, pp. 457-462.

Inverse Problem

Electromagnetic Nondestructive Evaluation (VIII)
T. Sollier et al. (Eds.)
IOS Press, 2004

An Electrical Resistive Tomography Based on Direct Information

Fabrizio FERRAIOLI, Alessandro FORMISANO and Raffaele MARTONE
Dipartimento di Ingegneria dell'Informazione, Seconda Università di Napoli
via Roma 29, 81031 CE Aversa, Italy

Abstract. An approach to non destructive testing suitable for solving a class of particular inverse problems whose peculiarity is to have the defect electrically accessible is proposed. The impact of this feature on the mathematical formulation of the problem and the advantages in terms of improvement of the sensitivity of the measurement system are discussed through the evaluation of the shape and the dimension of the uncertainty band. An application oriented to the diagnostic for underground concrete structures is then presented.

1. Introduction

In the last years, the Electrical Resistive Tomography (ERT), has been applied in several fields from geotechnical surveys [1] to clinical diagnostics on human body [2], and has been considered lively in the international scientific community. The classical ERT is based on the measurement of resistances among electrodes arranged externally to the specimen under test. The resistances obtained in this way depend on the shape and the position of the electrodes as well as on the distribution of the electrical conductivity inside the domain. Therefore, by making use of the mathematical techniques for the inverse problems resolution, it is possible to look for the electrical conductivity profile, and reconstruct from this property other relevant and related information [3], e.g. the eventual presence of "inclusions" with different conductivity.

The accuracy of the method and its resolution depend on the sensitivity of the measured resistances with respect to the parameters to be identified. Unfortunately, such sensitivity is generally low; therefore, especially when the aim is to localize "inclusions" whose dimensions are small compared with the dimensions of the measurement system, the performances are typically limited.

The overall performance of the method would be greatly improved if it was possible to "directly" measure the inclusions properties. As an example, in some cases of remarkable applicative interest, the inclusion is characterized by a high conductivity compared with the surrounding domain and, in addition, is "electrically accessible". In such cases, the defect can be included among the electrodes of the measurement system. This work is aimed to explore the possibilities offered by this extension of the ERT, that can be defined *Direct Electrical Resistive Tomography* (DERT), highlighting its peculiarities and showing, by means of a simple example, how it can improve and simplify the estimation process.

The example considered is the detection of damages due to micro-landslides during the concrete casting for a reinforced concrete pile. As a matter of fact, under unfavourable circumstances, especially in soft grounds, micro slides during the casting can affect the shape of the pile, producing defects (inclusions of ground) that seriously impact the pile reliability and eventually interrupt its continuity. In this latter case, which is the most relevant from the viewpoint of security, the reinforcement in the concrete comes to direct contact with the ground. For grounds in which the landslides are likely to happen, the electrical conductivity is rather higher than the concrete one, and the direct contact of the reinforcement with the ground creates a low resistivity channel that can be easily detected. Note that also in the case of partial ground penetration, the overall resistivity is lowered.

Of course, a number of techniques do exist to prevent such phenomena, but nevertheless the possibility of performing a diagnosis of the pile status could be quite useful for the responsible of the building.

The described application has been chosen for its ease of treatment from the numerical point of view, and for the possibility to easily carry out a set of experimental tests aimed to obtain real world data to be processed, although the cases really requiring the application of the method could be in small amount.

On the other hand, the DERT approach could be applied to a wide set of non destructive diagnostic methods, namely in all the cases when some electrical access to the specimen under testing is available; in addition the on-line diagnostic of some structures (e.g. composite materials) could be performed by merely adding conducting wires inside of them.

2. ERT problem and DERT problem formulation

2.1 Measurement systems description

In both ERT and DERT for the application at hand, the measurement system consists in the same components. The ERT basically performs self and mutual resistance measurements by connecting an electrical source between the reference electrode and each of the others in turn; the self resistance is defined as the ratio between the current and the voltage of the connected electrode, whereas the mutual resistances are defined as the ratio between the voltage of the remaining electrodes and current in the connected electrode. The obtained data are processed in order to find the electrical conductivity distribution which generates a set of resistances best fitting the measured ones. The DERT performs resistance measurements by connecting the electrical source between one of the electrodes and the reinforcement itself instead of another electrode.

The systems to be analysed are sketched in Fig.1. The pile is modelled as a cylinder with a cut (the defect D) and the metal reinforcement as a single perfectly conducting wire (A). A number M of measurement bars are considered to form the measurement system , together with an electrical source; each bar contains N_M electrodes dipped into the ground, for a total of N independently accessible electrodes. The ground is modelled as a homogeneous indefinite conductive half space and its electrical conductivity σ_T is assumed to be much higher than the concrete one σ_C.

The shape of real defects is quite complex; it is assumed here, for the sake of simplicity, that the defect D is described by its position alongside the pile Z_D, by its thickness ΔZ_D and by its "angular aperture", defined in terms of angle centre θ_D and angle width $\Delta\theta_D$ (Fig. 2).

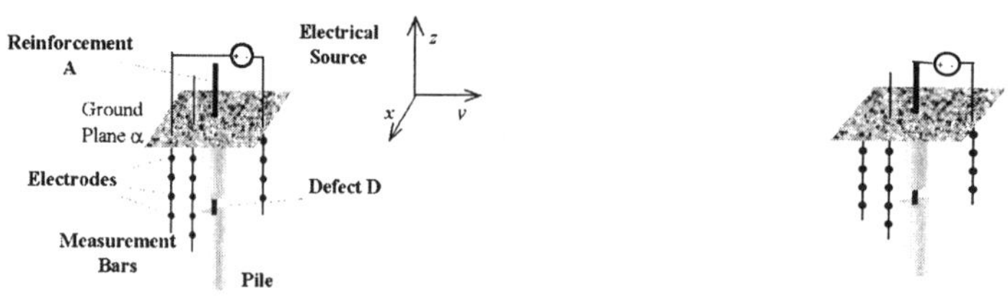

Fig. 1 - Schematic view of the ERT (left) and of the DERT (right) technique applied to reinforced concrete piles diagnostic

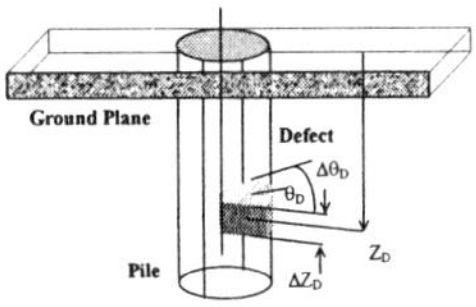

Fig. 2 - Geometric parameters to be identified

Note that, in the following, to help appreciating the improvements offered by DERT, attention will be focused on the ability of the two techniques in detecting the presence of defects and in estimating just one of the parameters, namely Z_D. The defect will consequently be assumed as a fixed thickness cylinder, ($\Delta\theta_D=2\pi$), to be localized along the pile. In both the schemes the self resistances are expected to be the terms with the highest sensitivity with respect to the defect parameters, and will consequently be analysed with some detail.

2.2 Mathematical Formulation

The description of any non destructive testing technique is typically divided into two different steps, a direct problem formulation, describing in a suitable mathematical language the problem of determining the measurements once given the sources and the system parameters, and an inverse problem formulation, describing the strategy adopted to find a set of sources and system parameters providing, trough the direct analysis, a set of simulated measurements, best matching the real ones.

In the case of DERT the direct problem can be stated as a combination of a forward operator F transforming an electrical conductivity distribution function $\sigma(P)$ defined in an half space S delimited by a plane α (the ground surface) into an electrical potential distribution function φ, defined in the measurement points (if assuming a current source) and a functional G, transforming the potentials into a set of integral parameters such as the resistances.

Note that in the example considered the electrical conductivity of the metal reinforcement is much higher than those of the concrete and of the ground, and the reinforcement is consequently modelled as an equipotential wire A; similarly, the electrodes in each measurement bar are modelled as lumped current sources. The operator F can then be defined by means of the following equation:

$$F:\sigma \rightarrow \{\varphi_k\}: \begin{cases} \nabla\left(\sigma\nabla\varphi_k\right) = I\delta(x-x_k, y-y_k, z-z_k) - I\delta(x-x_A, y-y_A)p_A(z) & in\, S \\ \sigma\dfrac{\partial\varphi_k}{\partial n} = 0 \ \ on \ \ \partial S \end{cases} \tag{1}$$

where I is the source current, k=1, 2…N, (x_k, y_k, z_k) is the position of the k-th electrode, (x_A, y_A) is the position of A on the ground plane, and $p_A(z)$ is equal to $1/\Delta Z_D$ only along the reinforcement section corresponding to the cut, and zero otherwise. A constant electric potential value is imposed along the reinforcement. In the case of ERT the formulation only differs in that $-I\delta(x-x_A,y-y_A)p_A(z)$ must be substituted with $-I\delta(x-x_j,y-y_j, z-z_j)$, where $j\neq k$. In each of the problems above defined a homogeneous Neumann condition is applied on α and $\partial\varphi_k/\partial n$ is considered vanishing at infinity. The adoption of "point" and "line" current sources has been suggested by the need of formulating the problem in a way that can be easily translated into a numerical formulation.

The functional G simply reduces in this case to assembly the matrix of potential differences taken at the electrodes, and normalizing each row to the related source current value.

A significant simplification in the overall problem is attainable if some special sets of σ distributions are considered. Particularly interesting are the spaces of σ's associated to a single integral cut or a partial cut of constant thickness and constant conductivity (as those considered above), because each defect corresponds to an array $\boldsymbol{D}$ of real numbers ($\boldsymbol{D}$=[Z_D, ΔZ_D] for a total cut and $\boldsymbol{D}$=[Z_D, ΔZ_D, θ_D, $\Delta\theta_D$] for a partial cut; see Fig. 2 for a visual description of the associated cut). By restricting the research space to one of those above described, G^*F can be considered as a function defined in a subset of $\boldsymbol{R}^2$ or $\boldsymbol{R}^4$ instead of a functional.

The direct problem is solved with a finite element approach, therefore it is reasonable to model an electrode as a node and the reinforcement as a set of element edges neglecting all the voltage drops inside of it. Being the described problem an open boundary one, a layer of boundary elements has been employed to take into account this peculiarity.

In both the approaches the inverse problem can be formulated as

$$\min_{\mathbf{D}} \| \boldsymbol{R}^m - \boldsymbol{R}(\boldsymbol{D}) \| \qquad (2)$$

where $\boldsymbol{R}^m$ is the matrix of the measured resistances and $\boldsymbol{R}(\boldsymbol{D})$ is the correspondent trial configuration matrix, numerically evaluated by the direct solver. Thanks to the linearity and reciprocity properties, the number of independent elements of $\boldsymbol{R}$ is only N(N+1)/2 in the DERT case.

3. Comparison of performance

Aim of this discussion is just to assess the improvement of performance of the "direct" approach called *DERT* with respect to the indirect schemes usually adopted in *ERT*. As a consequence, a quite simple problem will be considered as test bed for this analysis, not considering for the moment the full direct problem and limiting to the direct problem analysis in order to highlight the relevant properties of $\boldsymbol{R}(\boldsymbol{D})$. On the other hand, once assessed the differences and the similarities of DERT and ERT on a simple problem, the additional features introduced by more complex defect geometries are rather similar in the two approaches both for the direct and for the inverse problems. The simplest problem to be examined consists in identifying the position of a fixed thickness cut of the pile, i.e. assuming D=[Z_D]; this is an effective choice to highlight some relevant remarks concerning both the minimum number of electrodes necessary to the estimation process and the sensitivity of the two approaches. For the conciseness sake the sensitivity of the resistance matrix with respect to just one parameter (Z_D) has been considered, but similar properties for the remaining parameters as have been numerically verified. All the other geometric parameters of the pile and of the reinforcement are known, as well the electrical conductivities of all the materials. No uncertainty is considered for the moment in these information, and no measurement noise is considered. The geometrical and electrical data are reported in Table 1.

TABLE 1 – SYSTEM CHARACTERISTICS

Radius of the pile r_p [m]	0.2
Length of the pile h_p [m]	6
Distance between the electrodes and the reinforcement [m]	0.40
Thickness of the cut [m]	0.2
Distance between the electrodes [m]	0.50
Electrical conductivity of the ground σ_T [$\Omega^{-1}m^{-1}$]	8.0e-3
Electrical conductivity of the concrete σ_C [$\Omega^{-1}m^{-1}$]	1.0e-7
Position of measurement bar in ground plane for DERT	x = 0.4, y = 0.3
Position of second bar for ERT	x = -0.4, y = -0.3

The attention in this section will be focused on just self resistances, as the behaviour of other elements of the matrix $\boldsymbol{R}$ is quite similar. A scalar function $R_d(Z_D)$ can then be defined for an electrode placed in the k-th source point (x_k, y_k, z_k) in the case of DERT. R_d provides the ideal values of the (self) resistance between the electrode and the reinforcement when varying the position of the cut Z_D. Similarly, a function $R_e(Z_D)$ can be defined in the ERT scheme providing the resistance measured between two electrodes belonging to different measurement bars, but having the same depth. Three quotes of the electrode have been considered; results for both DERT and ERT are reported in Fig. 3a and 3b respectively (a grey triangle indicates on all the plots the quote of the electrodes for each function).

The first feature to be noticed is the percentage variation of the mean values with respect to the calculated value of the resistance in absence of cut, which is reported in the plots for ERT, while is in the order of 50kΩ for DERT. As a consequence of the values assigned in Table 1, $\Delta R_e\%$ is less than 1% for ERT, whereas in the DERT case $\Delta R_d\%$ is greater than 98% for all the electrodes considered; it is therefore evident the better attitude of the DERT scheme to detect the presence of the defect. It should also be noted that the codomain of the functions is much wider for DERT than in ERT; consequently the sensitivity of the DERT with respect to the position of the cut Z_D is much higher than the ERT one.

Note that eq. (2) reduces to $\left|R_d(Z_D)-R_d^m\right|$ and $\left|R_e(Z_D)-R_e^m\right|$ for DERT and ERT respectively for the scalar case considered, but in the following the algebraic differences $\left(R_d(Z_D)-R_d^m\right)$ and $\left(R_e(Z_D)-R_e^m\right)$ will be considered, in order to help detecting some features of the method. A first relevant feature of the error functions associated to the considered cases consists in having, for both techniques, a local minimum when $Z_D=z_k$, but in being biunique only locally for certain values of z_k. Consequently, for some values of the measured resistance, there are many possible positions of the cut among which the actual one has to be found. In Fig. 4a an example referring to a DERT case with two possible positions is reported. This simple argument demonstrates the impossibility of calculating the position of the most elementary defect of the pile by considering just a single electrode in the DERT or just a couple of electrodes in the ERT, unless the cut has the same quote of the electrode or Z_D belongs to the interval where R_d or R_e is biunique (see fig. 4(b)).

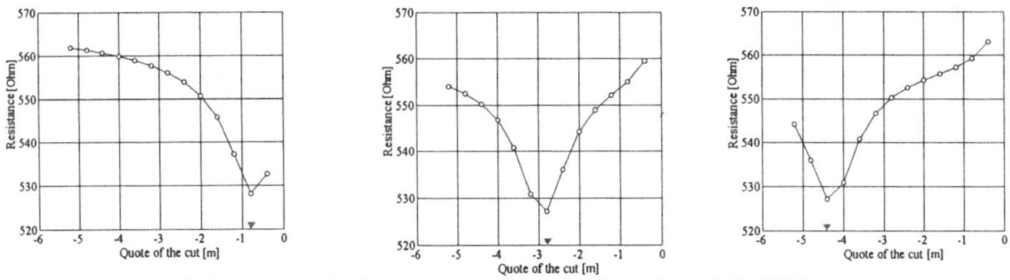

Fig. 3a - Variation of self resistance R_d for various positions of the k-th electrode in DERT

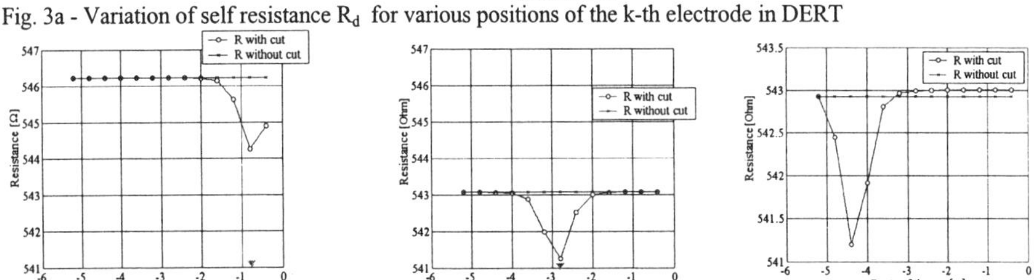

Fig. 3b - Variation of self resistance R_e for various positions of the electrodes in ERT

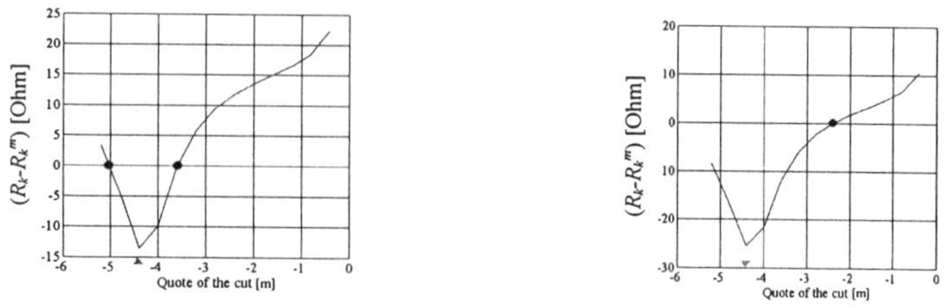

(a) (b)

Fig. 4 – Occurrence of multiple solutions for some values of the measured resistance

The false solutions disappear in the DERT case if a couple of electrodes (e.g. k=1 and k=2) at different quotes z_1 and z_2, are inserted into the ground. As a matter of fact, the knowledge of corresponding functions R_{d1} and R_{d2} is enough to calculate the position of the cut. If noiseless measured resistances R_{d1}^{m} and R_{d2}^{m} are considered, this property can be easily demonstrated being the position of the cut Z_D the common solution of the two non-linear equations $\left(R_{d1}(Z_D) - R_{d1}^{m}\right) = 0$ and $\left(R_{d2}(Z_D) - R_{d2}^{m}\right) = 0$ (see fig. 5 (a)). Similarly, in the ERT measurement scheme two couple of electrodes are necessary for uniquely evaluating the position of the cut (fig.6 (c)).

4. Effect of Uncertainties

Up to now, perfect knowledge of system parameters and noiseless measurements have been considered. Reality is quite different, and the performance of the two approaches must be compared with reference to actual devices, characterized by an uncertainty band on the parameters, and by a noise level on the measurements (see [4] for a wider discussion about this relevant issue).

In the case of DERT and ERT, an analysis of the approaches robustness against uncertainties can be performed by modelling the effect of the uncertainties as an uncertainty band U_{ZD}, from which the estimate of Z_D must be extracted.

In the following, an absolute error model will be considered; consequently, an uncertainty band with constant amplitude will affect the functions R_d and R_e. The amplitude of the uncertainties on the resistances can typically be easily computed by considering the limit values corresponding to the limits of the uncertainty bands on the system parameters. This assumptions drive to a "sensitivity analysis" approach based on the evaluation of the derivatives of the error function with respect to the parameters, neglecting the effect of the probability distribution of the parameters inside the uncertainty band. Other, more sophisticated approaches, such as Montecarlo analysis, should be recommended to obtain more reliable estimates [5].

On the other hand, the interest here is just on the effect of such uncertainties, and a simpler approach will be pursued, by considering as uncertainty level on the measurement E the 1% of the mean value. This absolute error takes into account all the contributions, including those due to measurements and parameters uncertainties, as well the numerical ones (truncation and rounding). The effect of uncertainties is to generate a range of values for Z_D, corresponding to the range of values for the trial function R_d or R_e. The uncertainty band affecting Z_D can be obtained by considering as possible solutions all the values corresponding to a resistance function value $R_d(Z_D)$ differing from R_d^{m} less than E (to a resistance function value $R_e(Z_D)$ differing from R_e^{m} less than E in the case of ERT). Results obtained considering two electrodes (k=1, 2) are reported in Fig. 5b for DERT and Fig. 5d for ERT respectively, for electrodes at (x_1=0.4, y_1=0 z_1=-4.4) and (x_2=0.4, y_2=0 z_2=-2.8).

By examining fig. 5 (d), it is evident in the ERT approaches there are no chances of identifying the position of the cut, being the uncertainty band affecting Z_D comprehensive of all the interval where R_e is defined. A second characteristic is that, due to the uncertainties on Z_D, different solutions corresponding to the same measured resistance, may collapse. This is not a desirable circumstance, being the width of the interval and the corresponding uncertainty on Z_D generally considerably larger than the uncertainty associated to the single solutions. (see fig. 6).

Finally, there is a further property of the function R_d (and R_e) to be noticed, consisting in the relationship between their first derivatives and the uncertainty level on Z_D. As Figs. 7 (a)-(b) show, small values of $\frac{dR_d}{dZ_D}$ correspond to wide uncertainty on Z_D itself and consequently the estimation of the parameter is critical.

The exposed remarks can be extended to a more general inverse problem consisting in estimating both the position and the thickness of the cut, but the geometric approach became obviously less immediate to apply, and the ease of interpretation has been the first concern in this discussion.

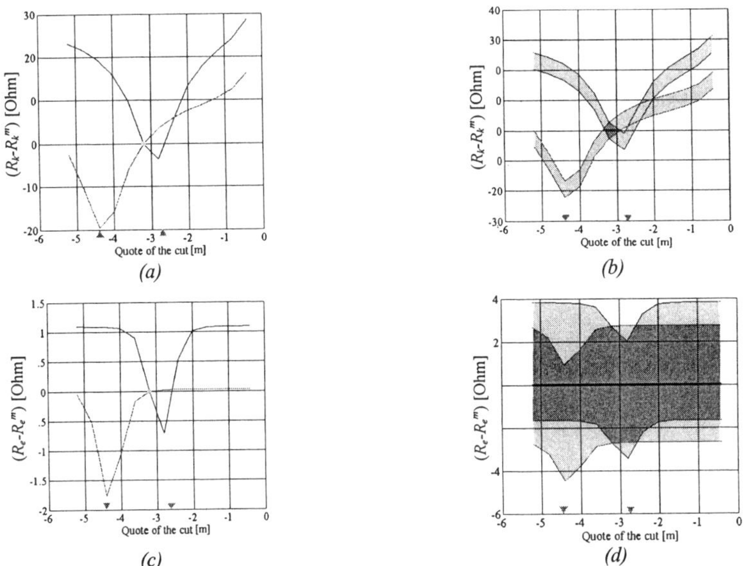

Fig. 5 – Evaluation of the position of the cut when considering data unaffected ((a)-(c)) and affected((b)-(d) by uncertainties for DERT case (a,b) and for ERT case (c,d).

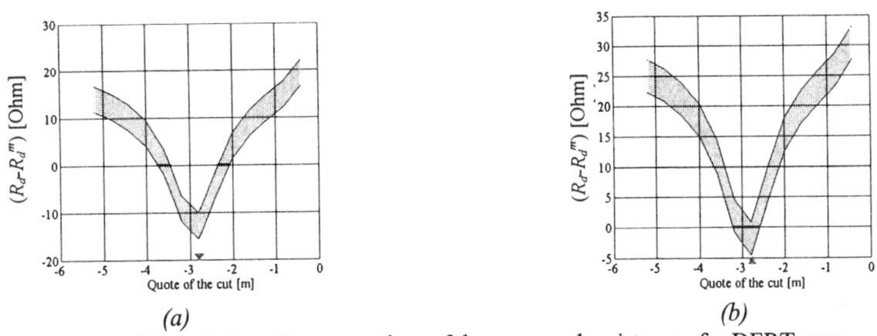

Fig. 6 – Degeneracy of the solutions for some values of the measured resistances for DERT case.

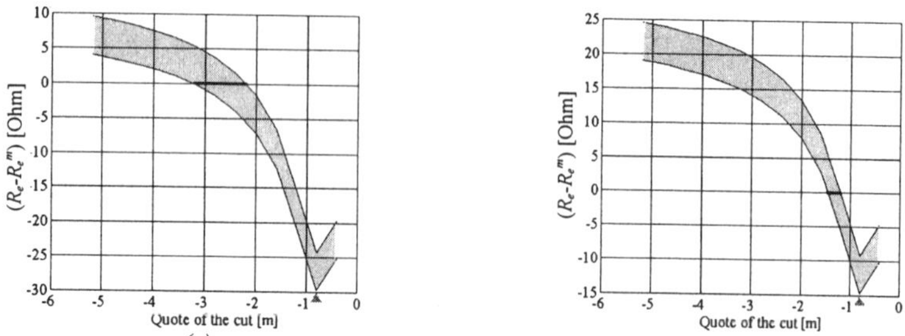

(a) (b)

Fig. 7 – Narrowing of the uncertainty level on Z_D as the first derivative of R_d increase for DERT case.

5. Conclusions

The detection of conductivity variations in some systems by means of resistance tomography can be improved if electrical access to the interior of the system is available. The concept of Direct Electrical Resistance Tomography, taking advantage of this circumstance, has been discussed in terms of sensitivity and robustness. The application to a very simple example has been used to highlight some of the properties, that can be nevertheless easily extended to more complicated and realistic situations.

Given the extremely low sensitivity of the ERT even in the identification of a simple but of consistent dimension defect, the advantages of the DERT have been demonstrated and quantified.

The future steps of the inquiry will consist first of all in performing the estimation of more complex defects by solving a minimization problem formulated as in (2) and subsequently in testing the method on some sets of experimental data.

The extension of the technique to different kind of structures is probably the most interesting aspect of the enquiry and will eventually contribute to make popular the method.

Acknowledgements

This work has been partly sponsored by Italian Education and Research Ministry (MIUR) in the framework of the Relevant National Interest Research Project "MADEND".

References

[1] J.G. Webster: Electrical Impedance Tomography, Adam Hilger, Bristol, 1990.

[2] B.H. Brown, D. C. Barber (eds.) (1992): Electrical Impedance Tomography, 207 pp. The Institute of Physical Sciences in Medicine, York. (Clinical Physics and Physiological Measurement, Vol. 13, Suppl. A).

[3] F. Ferraioli, A. Formisano, R. Martone: Assessment of Performance for Direct Electrical Resistance Tomography, accepted for presentation at "EMF 2003", Aachen (Germany), October 2003.

[4] M. Cioffi, A. Formisano, R. Martone, "Uncertainties Treatment in ECT using Parallel Computing Environment", in *Proceedings of E'NDE 2003*, Saclay (France), May 2003.

[5] M. Cioffi, A. Formisano, R. Martone, "Increasing Design Robustness in Evolutionary Optimisation", in *Proceedings of PMAPS 2002 Conference*, Naples (Italy), September 2002.

Electromagnetic Nondestructive Evaluation (VIII)
T. Sollier et al. (Eds.)
IOS Press, 2004

Design Method for Optimal Database Used for Inverse Problems in ECT

József PÁVÓ, Szabolcs GYIMÓTHY
Department of Electromagnetic Theory, Budapest University of Technology and Economics, H-1521 Budapest, Hungary, e-mail: pavo@evtsz1.evt.bme.hu

Abstract. A novel method is proposed for generating a database, by which the defect reconstruction problem of eddy current testing can be solved, and which is optimal in a certain sense. The method applies to a given eddy current testing experiment, and a defect type characterized by a predefined set of parameters. The database itself is an adaptive anisotropic simplex mesh generated in the abstract space that is spanned by the defect parameters. Each node of the mesh refers to a defect prototype, and contains the corresponding measured or simulated data of the testing experiment. Thereby the reconstruction of an unknown defect from its signal is reduced to a search and interpolation in the mesh database. The mesh is optimized for its interpolation properties, so its structure and density depend much on the applied interpolation method. Several mesh adaptation strategies are discussed, and a four dimensional example is presented in the paper. In addition to defect reconstruction, the database can be used to investigate the capabilities of the testing experiment, and the adequacy of the chosen defect parameters.

1. Introduction

Rigorous numerical analysis of the defect-probe interaction in eddy current testing (ECT) arrangements is a time consuming task. Consequently the computation time that is needed for reliable defect reconstruction based on optimization might be also quite large, especially if stochastic optimization methods are applied. One possible way to speed-up the inversion is the application of a pre-calculated or measured database that contains the probe responses of selected defect prototypes [1-2]. Applying this, in every step of the optimization loop one simply retrieves the appropriate signal stored in the database instead of doing the whole time consuming numerical analysis. It is obvious however, that the quality of the reconstruction depends much on the actual database used in the procedure. In this paper we present a method for creating a database, which is optimized for size, reconstruction speed and quality.

The method can be summarized briefly as follows. One has to specify in advance the testing experiment, the output data (signal) of the measurement – the latter can be either real or simulated –, as well as the type of the analysed defect together with the set of parameters that can characterize it. The n parameters of the defect span an n-dimensional abstract space, inside which the acceptable parameter combinations frame a closed domain. This domain will be subdivided into a simplex mesh. Each node of the mesh refers to a defect prototype – the one which is described by the node coordinates – and at the same time it stores the signal of the measurement for that particular defect. This mesh constitutes the database.

The signal of a given defect not contained by the database can be estimated by doing interpolation above the corresponding simplex element using its stored nodal values. The proposed method optimizes the mesh in such a way that the interpolation properties of each simplex "be the same". More precisely, the error of interpolation should not depend on the location (in the domain), at which it is taken. Of course the choice of the applied interpolation method and the measure of the error strongly influence both the structure and the density of the resulting adaptive mesh.

We understand that it is after all the quality of the experimental data that limits the accuracy of defect reconstruction. In other words, unless we consider methods that are able to utilize the statistical properties of experimental data, a reconstruction method cannot make difference between two defects, the signals of which differ only within the noise limit of the measurements. Even in the case of simulated measurements one has to consider some practical level of uncertainty. The noise of the measurement should restrict the density of the mesh database, because it has no reason to provide an interpolation error being lower than the noise itself.

In this paper we discuss first the key issues of mesh generation and describe several mesh adaptation strategies according to the applied interpolation methods. Following this, a sample database of a four-parameter crack type is presented. Finally, the usage of the database for gathering relevant information on the test experiment and the related inverse problem is discussed.

2. Database generation procedure

The mathematical problem targeted in this paper is to find a multivariate multi-component relationship between defect parameters and measured data, which relationship is easy and fast to evaluate in both directions. Similar problems arise in many areas of applied mathematics and engineering, and have a rich literature [3-4]. Point-to-point correspondence of the mentioned relationship in one direction can be given by solving the so-called direct problem, which means the calculation of the probe signal due to a known defect. A generalized framework for the solution of such problems is presented among others in [5-6], while the particular technique used by the authors for the purpose is described in [1].

The look-up table, which is one classical manifestation of the above relation used in NDT, requires some optimization both for estimating the output signal for defect parameters not being in the table (forward search), as well as for finding the defect parameters for given measured data (backward search or inverse problem). Contrary to this, if we had a *functional* type of relationship it would allow the forward search – and in case it was invertible: also the backward search – to be carried out in a direct and fast manner. The most widespread realization of such a function is the piecewise linear approximation above an n-dimensional simplex mesh. In our case the mesh should be generated on the domain spanned by the defect parameters, where to each node of the mesh belongs a set of measured or simulated data (for example the impedance change of a coil centred in some predefined scan points).

The forward search in such a mesh can be carried out in two steps. First the simplex element that contains the point specified by the defect parameters is to be found, preferably with the help of the well known and very rapid neighbour-by-neighbour search, which is based on the barycentric coordinates of the simplex [7]. Second, the queried data are linearly interpolated from the nodal data of the given simplex. The quality of interpolation can be significantly improved, if sensitivity data are also available at the nodes of the mesh.

Unfortunately, the data sets of NDT applications are usually not invertible, so the backward search cannot be carried out directly. But even in this case, the simplex mesh can help to find an optimal solution. For demonstrating this it is convenient to introduce the so called *co-domain*, which is spanned by the m components of the output (measured) data. Let us restrict our investigations to the practical case when $m > n$. The picture of the original mesh in the co-domain acts as a "surface-mesh" on some n-dimensional hyper-surface. We suppose that this mesh is regular in the sense that it contains no "tangles" and "holes". Since the point specified by the measured data (for which the defect parameters are to be estimated) is not contained in general by any of the surface-simplices, our task should be rephrased as to find the nearest point on the surface mesh. For doing this we can still use a kind of neighbour-by-neighbour search for the "nearest" simplex, and then utilize the interpolation features of that simplex to find the nearest point.

In the following we restrict our investigations mainly to the adaptive generation of an optimal mesh, and only partly to the search strategy. The mesh approach also has some known problems in higher dimensions:

a.) The sufficient number of nodes and elements grow extremely rapidly with increasing spatial dimensions, which entails high storage requirement and computational cost.

b.) The occurrence of distorted (flat, elongated, etc.) simplices is experienced to be more and more frequent in higher dimensions. The practical interpolation properties of such simplices are known to be bad, as numerical errors may arise. Several quality measures were introduced for characterizing a simplex in this respect [8].

c.) Since the sensitivity of output data may differ regarding the change of each defect parameter, we believe that a well adapted *anisotropic* mesh suits the texture of our abstract n-dimensional space better than a conventional isotropic mesh containing nearly equilateral simplices. In short: anisotropy stems from the nature of the problem. On one hand it makes more efficient meshing possible, because the anisotropic mesh usually contains much less nodes and elements than the corresponding isotropic mesh with almost the same (principal) interpolation properties. On the other hand it seems to contradict point b.) where good element quality was claimed for in order to provide acceptable practical interpolation properties. The solution might be a compromise: simplex elements with limited anisotropy [9].

d.) Most of the available mesh generators are capable only for two or three dimensions. The development of a sophisticated and specific mesh generator seems to be unavoidable, especially if anisotropic meshing is a concern.

Finally we come to the point of optimality. First, we define the error of interpolation as the difference between the "real" output signal and the interpolated one, expressed in some quadratic norm. A mesh database is deemed optimal, if the interpolation error is practically the same all over the mesh – this is called the *equidistribution* of the error –, furthermore the error in question falls in the range of the noise of the applied measurement system. This property guarantees the quality of the forward search with the smallest possible database size. But of course it influences the quality of the backward search too. We usually settle for the approximate fulfilment of error equidistribution. Depending on the interpolation method we apply, several types of mesh adaptation strategies can be worked out. In the following we present some possible solutions with their advantages and drawbacks detailed.

2.1. Zero order interpolation

In this technique the data contained by the nearest node is used without any post-processing. This method recalls a look-up table, except that the scope of one node is explicitly assigned here: in case the mesh is of Delaunay-type, the scope coincides with the dual Voronoï-cell that contains the given node. In addition to this, the knowledge about the connectivity of the mesh can speed up the search procedure comparing to a look-up table search.

We define the distance between two nodes in the m-dimensional co-domain as the Euclidean norm of the difference between the measured data contained by the nodes. A practical criterion for optimal mesh can prescribe for example that the distance r between any two neighbouring nodes – those belong to a common edge – measured in the co-domain should be approximately equal to the norm δ of some reference noise signal:

$$r_{ij} \equiv \left\| S_i - S_j \right\| \approx \delta \quad \text{for all edges,} \tag{1}$$

where i and j refer to node indices, and S represents the measured data contained by a node. Assuming that the relationship between defect parameters and output signals is "smooth" enough, this criterion promises that the error of interpolation on the mesh is resulted less or equal to the noise.

The mesh refinement method that the best suits this criterion might be Rivara's edge bisection [10]. Starting from a coarse initial mesh, edges appearing in the co-domain longer than allowed are bisected repeatedly in a predefined order, until (1) satisfies. We note however that without mesh post-processing (e.g. smoothing) perfect optimality cannot be achieved, because edge halving in itself does not provide a way to set the lengths properly.

The main drawback of this method is the rather high number of nodes and elements needed. On the other hand, this kind of "interpolation" does not require any computations, which together with edge bisection ensure that the method is not sensitive on numerical errors at all (probably in any dimensions), and thus any rate of mesh anisotropy is allowed. Furthermore this type of mesh directly reflects the resolution of the measurement system, the adequacy and role of the chosen defect parameters and other important properties.

2.2. First order interpolation

The most common way of interpolation on a simplex mesh consists in taking the appropriate linear combination of nodal data belonging to a simplex. There is already an "out-of-the-box" mesh generation method for *scalar* nodal data, by which the equidistribution of the interpolation error (using linear interpolation) can be approached [9]. This method has a particular advantage: it gives guideline for generating elongated simplex elements i.e. an anisotropic mesh. As concerning the NDT database at hand, the method can be directly applied with the *norm* of the contained nodal data as scalar:

$$s = \left\| S \right\|. \tag{2}$$

The outline of the method is the following. We introduce a local metric tensor M based on the spectral decomposition of the Hessian H of field s, as shown below for the 2D case:

$$H = \begin{pmatrix} \dfrac{\partial^2 s}{\partial x^2} & \dfrac{\partial^2 s}{\partial x \partial y} \\ \dfrac{\partial^2 s}{\partial x \partial y} & \dfrac{\partial^2 s}{\partial y^2} \end{pmatrix} = Q \begin{pmatrix} \lambda_1 & 0 \\ 0 & \lambda_2 \end{pmatrix} Q^T, \qquad M(x,y) = Q \begin{pmatrix} |\lambda_1| & 0 \\ 0 & |\lambda_2| \end{pmatrix} Q^T \tag{3}$$

Matrix Q contains the eigenvectors, while λ_1 and λ_2 denote the eigenvalues of H. The targeted simplex mesh should contain almost equilateral elements in uniform density, but in a way that edge lengths and distances are measured against the local metrics M. Mesh generators of several types can be used. The picture of such a mesh in the original domain is usually highly graded and stretched. A variant of the method was developed recently by the authors, in which the rate of anisotropy can be controlled in order to preserve good numerical interpolation properties [9].

Note that the derivatives appearing in the Hessian H may be extracted directly from the sensitivity data in case they are available. Metrics M may be rescaled in order to reach the prescribed level of interpolation error. Expectedly the number of required nodes and elements is significantly lower than with the zero order interpolation method (see above), for the same error bound. On the other hand the relationship between the mesh and the characteristics of the measurement system is not so expressive anymore.

2.3. Higher order interpolation

We realized that the sensitivity data bound to nodes, which can be expressed as partial derivatives of the measured data by the parameters, make higher order interpolation possible. For example in 2D the set of nodal data

$$s, \frac{\partial s}{\partial x}, \frac{\partial s}{\partial y} \tag{4}$$

allows an almost complete third order interpolation above a triangle.

In a first step, this relationship can be used as a correction method for the above linear interpolation. But preferably, it will be used for generating a mesh that is optimal in respect to this higher order interpolation. The extension of the adaptive mesh generation method presented in section 2.2 in this sense is underway.

3. Sample database for four-parameter cracks

We created a mesh database regarding a single crack that resides in a relatively large plate. The assumed testing experiment is described thoroughly in the JSAEM Benchmark Experiments Step 2 (Flat Plates with Cracks) [11]. The output of the experiment is the impedance change of a coil, which is measured above the plate at the points of a rectangular grid with step size 0.5 mm. 41 measurement points are taken in the direction parallel with the orientation of the crack, and 11 points in the normal direction. In our procedure the output signal of the coil was simulated by a numerical field calculation method described in [1], with the assumption of infinitesimally thin cracks.

The studied crack type is characterized by four parameters, namely the length l, depth d, and the co-ordinates of the crack-midpoint, x and y (Fig. 1). Note that the crack is oriented parallel with the y-axis. The examined domain of these parameters is framed with a dashed line on the figure. The width of the crack is negligible. Separate data sets were created for ID and OD type cracks.

Each node in the created 4D mesh database stores $m = 41 \times 11$ complex numbers as measured data for the corresponding crack prototype. The mesh was optimized for zero order

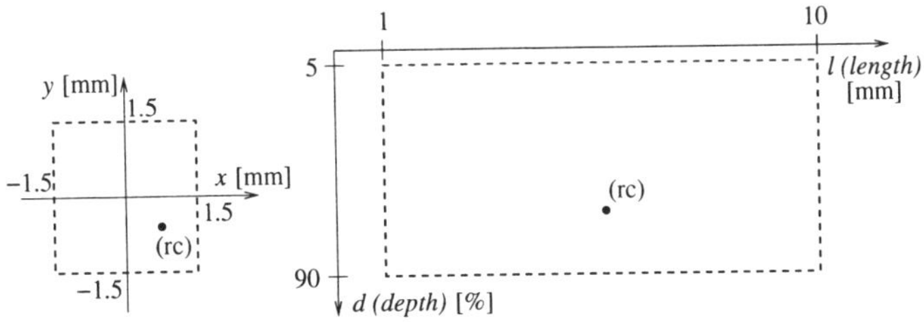

Fig. 1. Definition of crack parameters and their domains, with the location of the reference crack (rc) indicated.

Table 1. Comparison of the number of nodes for databases created with different noise levels.

Noise level	Number of nodes in ID data set	Number of nodes in OD data set
δ	7967	2321
1.4δ	2365	879
2δ	708	357

interpolation by using the method outlined in section 2.1. The noise level δ in criterion (1) was determined upon practical considerations, but higher noise levels were also prescribed for just testing the method. The number of nodes – which is the primary feature of an optimized mesh database – is presented in Table 1 for various noise levels. For comparison: if the smallest parameter subdivisions that occur in these meshes were used all over the domain, such a "completely filled" database would contain about 821250 nodes.

In order to demonstrate the applicability of the database for crack reconstruction we have chosen a reference ID crack with parameters: x = 0.75 mm, y = –0.64 mm, l = 5.5 mm and d = 80 %, as it is illustrated in Fig. 1. Figure 2 shows an l–x and an l–d cross-section in the 4D mesh of the ID database, where the remaining two parameters – not shown on the axes – were set equal to those of the reference crack. The shading of the plots represents the distance of parameter n-tuples from the reference crack, measured in the co-domain. The approximate centre of the dark hole locates the crack parameters to be found.

The generated mesh is highly graded and anisotropic. Looking at the dominant direction of element elongation (see Fig. 2, top) it is apparent that the x parameter of the crack can be retrieved from the database with higher accuracy than its length (in general, but not for very shallow cracks). More interesting is that on the bottom plot of Fig. 2 the parameters l and d seem to play similar role to some extent, just as if their sum, the *circumference* of the crack would be the really characteristic parameter.

We also investigated the search result for some defect types not covered by the database. The case of a double ID crack is presented here. Both cracks have the same length and depth (l = 5 mm, d = 80 %), and they are located at $x_1 = -1$ mm, $y_1 = 1$ mm and at $x_2 = 1$ mm, $y_2 = -1$ mm respectively. The parameters of the nearest node found in the database, which is thereby entitled the "equivalent single crack", are x = 0.19 mm, y = 0 mm, l = 4.94 mm and d = 68.5 %, which is an acceptable result. The left side plot on Fig. 3 shows the distance of nodal data from the double crack data (the way it was done on Fig. 2) above an x-y cross-section of the mesh. The position of this cross-section is given by l = 4.94 mm and d = 68.5 %. Surprisingly, cross sections at other l and d values can still contain two local minima in conformity with the double crack, as it is shown for l = 5 mm and d = 59 % on the right side plot of Fig. 3.

4. Further utilization of the mesh database

In addition to defect reconstruction, which is the primary purpose of such an NDT database, it is thought to be applicable for getting relevant information on the nature of the inverse problem.

For example it is possible to predict the reconstruction capabilities of a given testing experiment simply by measuring the noise of the measurement system. The database, which is generated using zero order interpolation (see above) with the value of δ set to the norm of the measured noise, will tell the best available resolution of the reconstruction.

Also the suitability of the parameters that describe the defect can be investigated with the proposed database. It seems to be natural that the larger the numbers of parameters the more nodes are needed to mesh the n-dimensional parameter space. But this statement may only be

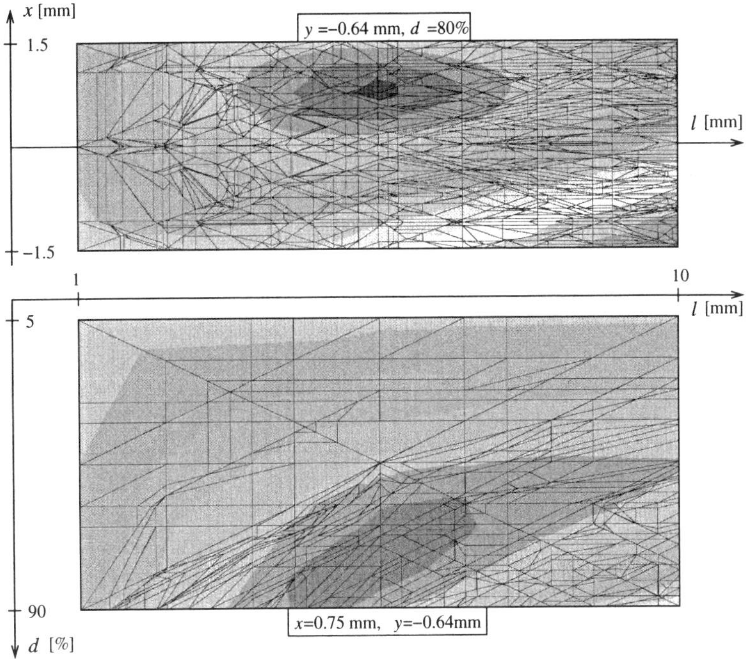

Fig. 2. Mesh cross-sections around the reference crack.

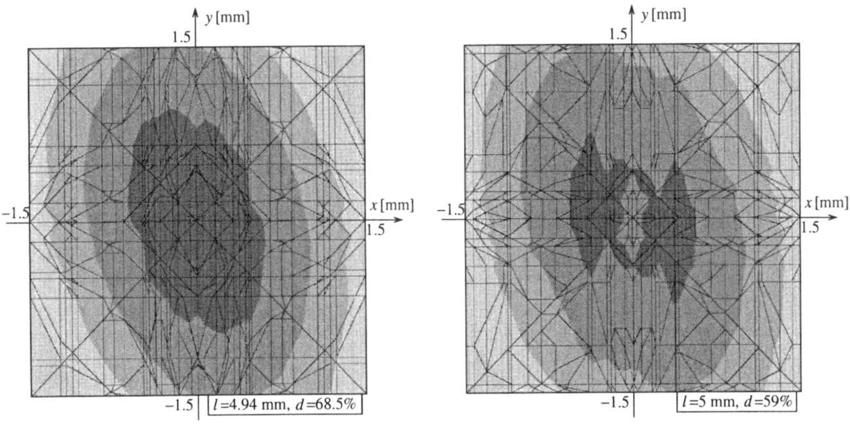

Fig. 3. Images of a double crack in a single crack database in the x-y plane.

valid for low number of parameters: while n is increased the number of nodes should reach a certain saturation number. It means that we cannot refine the reconstruction too far by introducing more and more parameters to describe the defects, because it is the given measurement that cannot distinguish between the signals of them any longer.

The structure of the database can help to reveal the hidden connections among some parameters and also to filter out the redundant ones. After all a *good* parameter set for a given measurement results defect prototypes in the database of around the mentioned saturation number, while an *optimal* parameter set can do this with the minimum number of parameters.

As a matter of fact the proposed mesh database simulates the relationship between n inputs and m outputs (and vice versa), just as an appropriate neural network could do. The difference lies in the mesh generation procedure, which acts as a systematic exploration of the parameter space, as opposed to a random training course. Nevertheless our mesh database can be used to train a neural network, which then may solve the inverse problem faster than a mesh search. We plan to prove all the mentioned capabilities of the mesh database in the near future.

5. Conclusion

A novel method has been presented for the generation of an optimal database by which the defect reconstruction problem of ECT can be solved. The database itself is an adaptive anisotropic simplex mesh generated in the abstract space that is spanned by the defect parameters. One of the most important properties of the database is that its creation is controlled by a parameter that is linked to the quality of the measured ECT signal, thereby the database can be optimized to fit the given experiment. Numerical examples for crack reconstruction and possibilities for the further utilization of the database have been also presented in the paper.

Acknowledgements

This work was supported by the Hungarian Scientific Research Fund (OTKA) under Grants T-035264 and F-030570.

References

[1] J. Pávó and K. Miya, Reconstruction of crack shape by optimization using eddy current field measurement, *IEEE Trans. on Magnetics,* **30** (1994), pp.3407-3410.

[2] Z. Badics, J. Pávó, H. Komatsu, S. Kojima, Y. Matsumoto and K. Aoki, Fast flaw reconstruction from 3D eddy current data, *IEEE Trans. on Magnetics*, **34** (1998), pp.2823-2828.

[3] R. E. Mahony, J. B. Moore and L. Dailey, Spline approximations for an arbitrary simplex grid, *International Conference on Neural Networks*, Perth, Australia, Nov 27 - Dec 1 1995, [www.syseng.anu.edu.au/papers/john/P060.pdf].

[4] R. E. Groff, P. P. Khargonekar, D. E. Koditschek, T. T. Thieret and L.K. Mestha, Modeling and control of color xerographic processes, *Proc. 38th IEEE Conf. on Decision and Control*, **2** (1999), pp. 1697-1702.

[5] Z. Badics, et al., Accurate probe-response calculation in eddy current NDE by finite element method, *J. Nondestructive Evaluation,* **14**-3 (1995).

[6] Z. Badics, et al., Rapid flaw reconstruction scheme for three-dimensional inverse problems in eddy current NDE, in *Electromagnetic Nondestructive Evaluation,* Takagi et al. (Eds), IOS Press: Amsterdam, 1997, pp.303-309.

[7] J. Frykestig, *Advancing front mesh generation techniques with application to the finite element method*, Chalmers University of Technology, Göteborg, Sweden, 1994.

[8] J. R. Shewchuk, What is a Good Linear Element? Interpolation, Conditioning, and Quality Measures, *Proc. 11th Int. Meshing Roundtable*, Sandia National Laboratories, Sept 15-18 2002, pp.115-126.

[9] Sz. Gyimóthy, A. Vágvölgyi and I. Sebestyén, Application of optimally distorted finite elements for field calculation problems of electromagnetism, *IEEE Trans. on Magnetics*, **38** (2002), pp.365-368.

[10] M. C. Rivara, Mesh refinement processes based on the generalised bisection of simplices, *SIAM J. Num. Anal,* **21** (1984), pp.604-613.

[11] T. Takagi and H. Fukutomi, Benchmark activity of eddy current testing for steam generator tubes, in *Electromagnetic Nondestructive Evaluation (IV)*, S. S. Udpa et al. (Eds.), IOS Press: Amsterdam, 2000, pp.235-252.

Electromagnetic Nondestructive Evaluation (VIII)
T. Sollier et al. (Eds.)
IOS Press, 2004

Reconstruction of 3D Volumetric Defects in Conductors of Arbitrary Shapes

Guglielmo RUBINACCI, Antonello TAMBURRINO, Salvatore VENTRE, Fabio VILLONE

Ass. EURATOM/ENEA/CREATE, DAEIMI, Università di Cassino
Via Di Biasio 43, I-03043, Cassino (FR), Italy

Abstract. We present a procedure able to reconstruct volumetric 3D defects in metallic conductors of arbitrary shapes on the basis of eddy currents inspections. Our method relies on a fast numerical technique for the solution of the direct problem based on an integral formulation of the magneto-quasi-static-problem.

1. Introduction

The problem of reconstructing 3D volumetric defects in conductors is presently receiving considerable attention [1-4]. The problem is considerably hard from a theoretical, experimental and computational point of view. From the theoretical point of view, the actual information that can be retrieved from experimental data should be assessed, for a given experimental noise level. From the numerical side, an effective inversion scheme requires on the one hand fast and efficient techniques for solving direct eddy currents problems in 3D geometries, and on the other hand suitable inversion algorithms.

In this paper we combine an efficient procedure for repetitively solving the direct problem iteratively changing the defect configuration, presented in [5], with a 0-th order inversion algorithm, which benefits from the features of the direct problem solver.

2. The direct problem

The methodology for solving the direct problem is described in more details in [5]. Here we briefly summarize the main points.

An integral formulation of the eddy currents problem is used, in terms of a two-component electric vector potential. Edge elements basis functions are employed in order to guarantee the correct continuity of the various vector quantities involved [6]. The resulting linear system in the frequency domain can be recast in the following form:

$$\underline{\underline{Z}}\,\underline{I} = \underline{V} \tag{1}$$

$$\underline{\underline{Z}} \triangleq \underline{\underline{R}} + j\omega\underline{\underline{L}} \tag{2}$$

$$R_{ij} \triangleq \int_{V_c} \mathbf{J}_i(\mathbf{r})\cdot\sigma^{-1}(\mathbf{r})\mathbf{J}_j(\mathbf{r})d\mathbf{r} \tag{3}$$

$$L_{ij} \triangleq \frac{\mu_0}{4\pi}\int_{V_c}\int_{V_c} \frac{\mathbf{J}_i(\mathbf{r})\cdot\mathbf{J}_j(\mathbf{r})}{|\mathbf{r}-\mathbf{r}'|}\,d\mathbf{r}\,d\mathbf{r}' \tag{4}$$

$$V_i \triangleq \int_{V_c} \mathbf{J}_i(\mathbf{r})\cdot\mathbf{A}_0(\mathbf{r})d\mathbf{r} \tag{5}$$

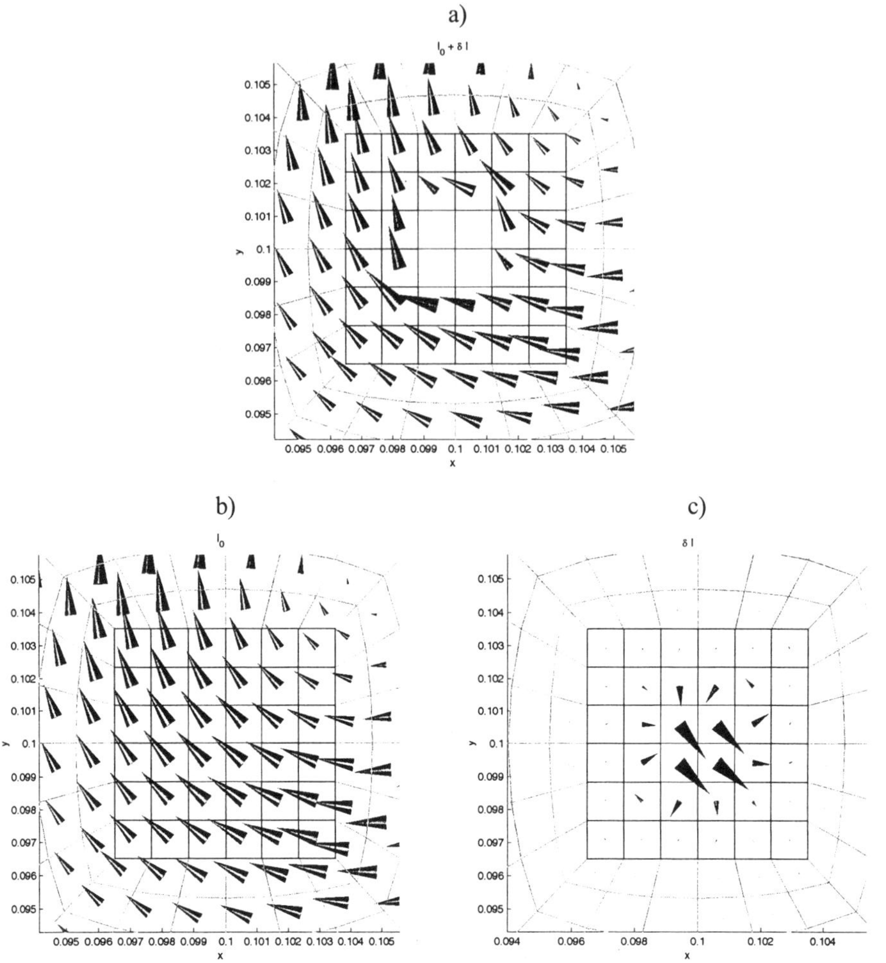

Fig 1 Example of superposition. a) the current density distribution in the presence of a volumetric defect; b) the unperturbed current density distribution $\mathbf{J}_0$; c) the current density perturbation distribution.

where:

$$\mathbf{J}(\mathbf{r}) = \sum_l I_l \mathbf{J}_l(\mathbf{r}) \tag{6}$$

is expressed as a sum of solenoidal basis functions using the electric vector potential **T** ($\mathbf{J}_l = \nabla \times \mathbf{T}_l$), V_c is the volume of the conductor characterized by the conductivity σ, $\mathbf{A}_0$ is the magnetic vector potential due to the sources external to V_c and ω is the angular frequency. In addition, the presence of the perfectly insulating crack in V_0 yields the additional constraint that $\mathbf{J}(\mathbf{r})$ is vanishing in V_0. A suitable discrete form of this constraint consists in requiring that the unknowns I_l's minimize $\int_{V_0} \left\| \mathbf{J}(\mathbf{r}) \right\|^2 \mathrm{d}\mathbf{r}$.

The key point for the obtaining the fast numerical method is to take properly into account the constraint, usually met in practice, that a proper volume V_T, inside which the volumetric defect V_0 is supposed to be located, is a priori known.

Superposition is used in order to solve directly for the current density perturbation due to the presence of the defect, i.e. we assume $\mathbf{J}(\mathbf{r}) = \delta\mathbf{J}(\mathbf{r}) + \mathbf{J}_0(\mathbf{r})$ where $\mathbf{J}_0(\mathbf{r})$ is the unperturbed current density (the current density when the crack is not present). The perturbation $\delta\mathbf{J}$ must compensate the unperturbed current $\mathbf{J}_0(\mathbf{r})$ in V_0, therefore, we require that $\int_{V_0} \left\|\delta\mathbf{J}(\mathbf{r})+\mathbf{J}_0(\mathbf{r})\right\|^2 \mathrm{d}\mathbf{r}$ is minimum. An example of the superposition is shown in fig. 1.

This condition yields to the linear system

$$\underline{\underline{P}}_0 \delta\underline{I} = \underline{w}_0 \tag{7}$$

where $P_{0,ij} = \int_{V_0} \mathbf{J}_i \cdot \mathbf{J}_j \mathrm{d}\mathbf{r}$, $w_{0,i} = -\int_{V_0} \mathbf{J}_i \cdot \mathbf{J}_0 \mathrm{d}\mathbf{r}$.

Let $\underline{\underline{P}}$ and $\underline{w}$ be the matrix and vector appearing in (7), respectively, but referred to V_T. In addition, let us assume that $\underline{\underline{P}}$ is a $m\times n$ full rank matrix with $m<n$.

Then, we make the following change of variables:

$$\delta\underline{I} = \underline{\underline{K}}\delta\underline{X} + \underline{\underline{R}}^T \delta\underline{Y} \tag{8}$$

where the columns of the matrix $\underline{\underline{K}}$ span the null space of $\underline{\underline{P}}$ and the rows of $\underline{\underline{R}}$ are an othonormal basis for the space spanned by the rows of $\underline{\underline{P}}$.

Thus, replacing (8) in (7) we obtain

$$\underline{\underline{P}}_0 \underline{\underline{R}}^T \delta\underline{Y} = \underline{w}_0 \, . \tag{9}$$

It is worth noting that $\underline{\underline{P}}_0 \underline{\underline{K}} = \underline{\underline{0}}$ since the degrees of freedom $\delta\underline{X}$'s are related to current densities external to V_T and, consequently, to V_0.

Representation (8) allows to identify the (new) degrees of freedom accounting for current densities flowing externally to V_T and current densities flowing in V_T (and in $V_c \backslash V_T$).

This is equivalent to replace the original set of shape functions $\mathbf{J}_k$ with a new one. Specifically, by replacing (8) in (6) it results:

$$\delta\mathbf{J}(\mathbf{r}) = \sum_m \delta X_m \tilde{\mathbf{J}}_m(\mathbf{r}) + \mathbf{J}_u(\mathbf{r}) \tag{10}$$

with

$$\tilde{\mathbf{J}}_m(\mathbf{r}) \triangleq \sum_l K_{lm} \mathbf{J}_l(\mathbf{r}) \tag{11}$$

$$\mathbf{J}_u(\mathbf{r}) \triangleq \sum_l (\underline{\underline{R}}^T \delta\underline{Y})_l \mathbf{J}_l(\mathbf{r}) \, . \tag{12}$$

The new basis functions $\tilde{\mathbf{J}}_m(\mathbf{r})$ satisfy the property:

$$\tilde{\mathbf{J}}_m(\mathbf{r}) = 0 \quad \text{in } V_T \tag{13}$$

This follows from:

$$\begin{aligned} \int_{V_T} \tilde{\mathbf{J}}_m(\mathbf{r}) \cdot \tilde{\mathbf{J}}_m(\mathbf{r}) \mathrm{d}\mathbf{r} &= \sum_l K_{lm} \sum_s K_{sm} \int_{V_T} \mathbf{J}_l(\mathbf{r}) \cdot \mathbf{J}_s(\mathbf{r}) \mathrm{d}\mathbf{r} \\ &= \sum_s \sum_l K_{lm} P_{ls} K_{sm} = \left(\underline{\underline{K}}^T \underline{\underline{P}}\,\underline{\underline{K}}\right)_{mm} = 0 \end{aligned} \tag{14}$$

Equation (9), that is constraint (7) expressed only in terms of $\delta\underline{Y}$, is the key point for developing an efficient numerical method. Specifically, we represent the solution of (9) as:

$$\delta\underline{Y} = \underline{\underline{\tilde{K}}}_0 \underline{\alpha} + \underline{y}_0 \tag{15}$$

where the columns of the matrix $\underline{\underline{\tilde{K}}}$ span the null space of $\underline{\underline{P}}_0 \underline{\underline{R}}^T$ and $\underline{y}_0$ is a particular solution of (9).

By replacing (15) in (8) we have the correct change of variables accounting for an impressed source in V_0:

$$\delta\underline{I} = \underline{\underline{K}}\delta\underline{X} + \underline{\underline{R}}^T \underline{\underline{\tilde{K}}}_0 \underline{\alpha} + \underline{\underline{R}}^T \underline{y}_0 \tag{16}$$

where the unknowns are $\delta\underline{X}$ and $\underline{\alpha}$.

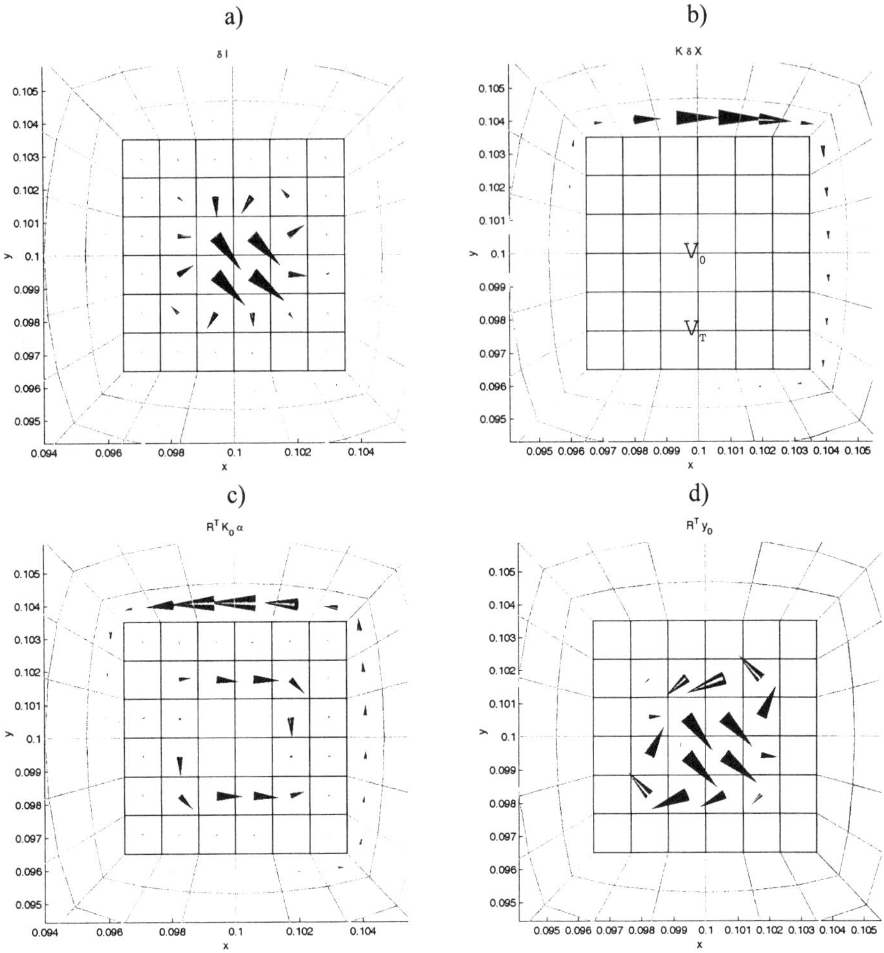

Fig 2 Example of the change of variable (16). a) The current density perturbation distribution in the presence of a volumetric defect; b) the contribution due to $\underline{\underline{K}}\delta\underline{X}$; c) the contribution due to $\underline{\underline{R}}^T\underline{\underline{\tilde{K}}}_0\underline{\alpha}$; d) the contribution due to $\underline{\underline{R}}^T\underline{y}_0$.

An example of the current density distribution associated to this change of variables is shown in fig. 2.
The numerical formulation arising from representation (16) and (1) (related to $V_c \backslash V_0$) is:

$$\underline{\underline{Z}}_{xx}\delta\underline{X}+\underline{\underline{Z}}_{xy}\underline{\underline{\tilde{K}}}_0\underline{\alpha}=-\underline{\underline{Z}}_{xy}\underline{y}_0 \tag{17}$$

$$\underline{\underline{\tilde{K}}}_0^T\underline{\underline{Z}}_{yx}\delta\underline{X}+\underline{\underline{\tilde{K}}}_0^T\underline{\underline{Z}}_{yy}\underline{\underline{\tilde{K}}}_0\underline{\alpha}=-\underline{\underline{\tilde{K}}}_0^T\underline{\underline{Z}}_{yy}\underline{y}_0, \tag{18}$$

with

$$\underline{\underline{Z}}_{xx}\triangleq\underline{\underline{K}}^T\underline{\underline{ZK}} \tag{19}$$

$$\underline{\underline{Z}}_{xy}=\underline{\underline{Z}}_{yx}\triangleq\underline{\underline{K}}^T\underline{\underline{ZR}} \tag{20}$$

$$\underline{\underline{Z}}_{yy}\triangleq\underline{\underline{R}}^T\underline{\underline{ZR}} \tag{21}$$

Solving (17) with respect to $\underline{x}$ and substituting in (18) leads to:

$$\delta\underline{X}=-\underline{\underline{Z}}_{xx}^{-1}\underline{\underline{Z}}_{xy}\left(\underline{\underline{\tilde{K}}}_0\underline{\alpha}-\underline{y}_0\right) \tag{22}$$

$$\left(\underline{\underline{\tilde{K}}}_0^T \underline{\underline{Z}}_{yy} \underline{\underline{\tilde{K}}}_0 - \underline{\underline{\tilde{K}}}_0^T \underline{\underline{Z}}_{yx} \underline{\underline{Z}}_{xx}^{-1} \underline{\underline{Z}}_{xy} \underline{\underline{\tilde{K}}}_0\right)\underline{\alpha} = \underline{\underline{\tilde{K}}}_0^T \left(\underline{\underline{Z}}_{yx} \underline{\underline{Z}}_{xx}^{-1} \underline{\underline{Z}}_{xy} - \underline{\underline{Z}}_{yy}\right)\underline{y}_0, \tag{23}$$

It is worth noting that matrices $\underline{\underline{Z}}_{xx}$, $\underline{\underline{Z}}_{xy}$, $\underline{\underline{Z}}_{yy}$ and, consequently, $\underline{\underline{Z}}_{yx}\underline{\underline{Z}}_{xx}^{-1}\underline{\underline{Z}}_{xy}$ can be computed once for all.

3. The inverse problem

The features of the direct problem solver are well suited for the application of 0-th order methods of inversion. Indeed, for each mesh element of V_T we must decide whether it is part of the defect or not, i.e. we must take a binary decision. The related solution of the direct problem implies the inversion of a system of equations of the order of the unknowns located in V_0. Hence, the application of a genetic-like algorithm seems straight-forward, as described in [7] for planar cracks. Each tentative configuration is evaluated comparing the predicted and expected perturbation of the magnetic field in the region where the probes are located.

Contrary to planar cracks, the number of binary unknowns may grow to unacceptably high levels even for small search regions. Hence, suitable assumptions must be envisaged in order to make the problem affordable. Focusing the attention to the case of one defect, a possibility could be to describe the defect in terms of a few number of global parameters, identifying the position and the dimensions for a given class of shapes.

4. Numerical results

The problem that we solve is to find a volumetric defect in a 20cm×20cm×0.4cm conducting plate. The source is due to a circular coil of radius a= 7.5 mm placed at x=10.5cm, y=10.5cm, z=0.5mm over the plate. A set of 20×20 'simulated' field measurements over the plate is obtained with a fine mesh in which a volume crack of cylindrical shape (radius r=2.4mm, height h=2mm) is located just in the middle of the conducting plate (see fig. 3). In fig. 4 we show the distribution of the real part of the x component of the magnetic field produced by the current perturbation associated to the defect, normalized to a unitary source current.

To solve the inverse problem, we describe the possible defects in terms of the following global parameters: position in the x-y plane, position along z, dimensions along x, dimensions along y, dimensions along z. Each of this parameters can assume only a discrete set of values related to the number of voxels along each direction in the discretization. As a consequence of this choice of parameters, we force the shape of the defects to be a parallelepiped.

The mesh used for solving the inverse problem is made of 1680 elements, 2165 nodes, leading to 2893 unknowns. The search volume V_T is made of 144 (6×6×4) elements and 301 unknowns.

The genetic algorithm evolves using the standard rules: crossover and mutation. At each generation, the values of the various parameters are either modified randomly (mutation) or combined between different individuals (crossover). In the cases presented in this paper we have used 50 individuals with a probability of mutation of 0.7. The crossover is made swapping the dimensions of the crack and its position among the best 40 individuals. The objective function is given by:

$$\Psi(p) = \frac{\left\|\Delta B_x(p) - \Delta B_{x,meas}\right\|}{\left\|\Delta B_{x,meas}\right\|} + \frac{\left\|\Delta B_y(p) - \Delta B_{y,meas}\right\|}{\left\|\Delta B_{y,meas}\right\|} + \frac{\left\|\Delta B_z(p) - \Delta B_{z,meas}\right\|}{\left\|\Delta B_{z,meas}\right\|}, \tag{24}$$

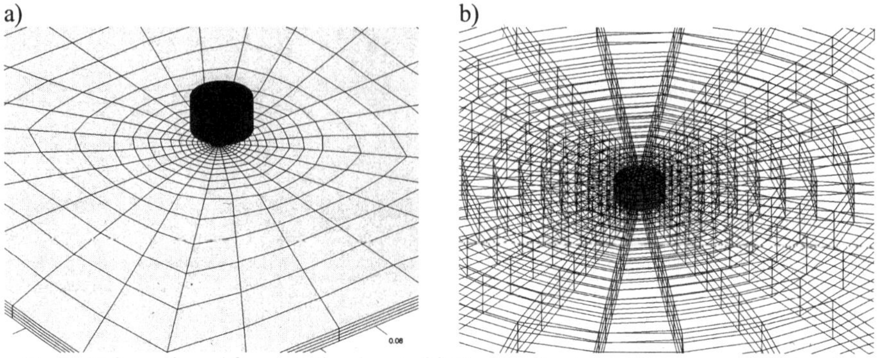

Fig 3 a) The mesh used for computing the set of fields data. The shape of the source coil is also shown. b) the shape of the buried defect

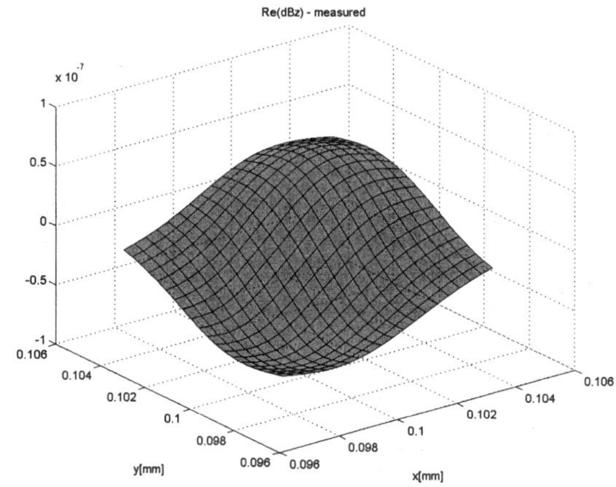

Fig 4 Real part of the x component of the magnetic field produced by the current perturbation associated to the defect, normalized to a unitary source current, at the sensor locations.

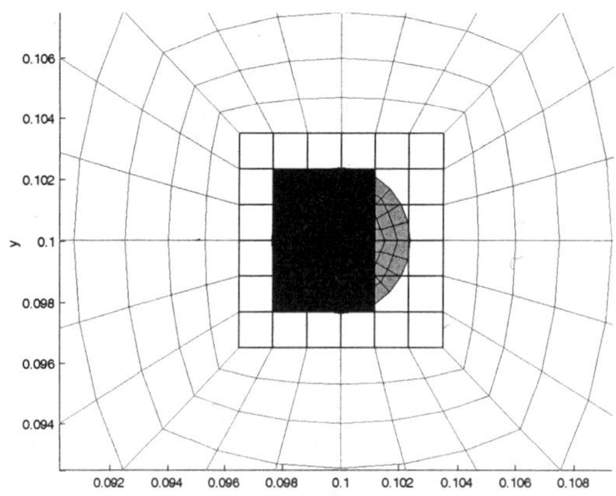

Fig 5 The shape of the defect reconstructed by the genetic algorithm. Light gray: the cylindrical defect and its finite element mesh. Dark grey: the reconstructed defect. The defect is approximated onto a 6×6×4 regular grid (black).

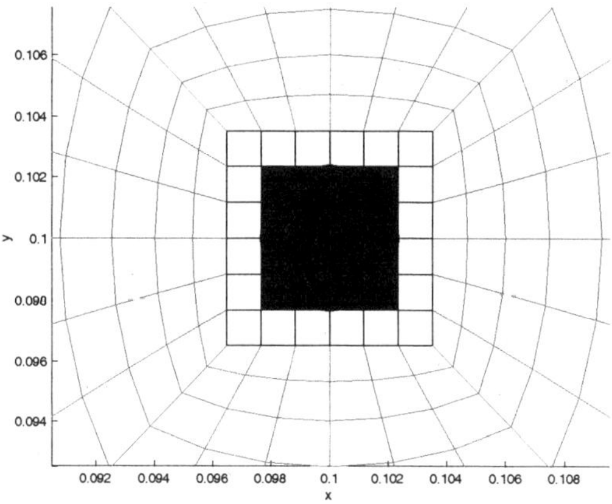

Fig 6 Dark gray: cross section of the smallest parallelepiped enclosing the cylindrical defect.

where p is the array of parameters describing the crack geometry, $\Delta B_{x,meas}$ (ΔB_x) is the measured (numerically computed) x-component of the anomalous of the magnetic flux density and, similarly for the other components.

The shape of the crack obtained after 2000 solutions of the direct problem is shown in fig. 5 and is compared with the true shape. The field perturbation errors associated with this shape are the following: $\varepsilon_x = \dfrac{\|\Delta B_x - \Delta B_{x,meas}\|}{\|\Delta B_{x,meas}\|} = 0.23$, $\varepsilon_y = 0.13$, $\varepsilon_z = 0.21$. They can also be compared with the values $\varepsilon_x = 0.22$ $\varepsilon_y = 0.22$ $\varepsilon_z = 0.20$, obtained in case of the shape of the defect shown in fig. 6. The execution time is reported in Table I and refers to a Pentium IV, 1.7Ghz - 1GB RAM - 20 GB HDD Personal computer. We notice that the formation and inversion of the impedance matrix is instead requested if a conventional formulation is adopted, showing the main advantage of the proposed approach.

5. Conclusions

We presented a numerical method for detecting volume cracks that combines an efficient procedure for solving the direct problem iteratively changing the defect configuration, presented in [5], with a 0-th order inversion algorithm, which benefits from the features of the direct problem solver.

The proposed approach, based on an efficient volume integral formulation, takes advantages from a change of variables which allows to separate the contribution of the unknowns outside the search volume that are not directly affected by the shape of the defects. Therefore, for each crack configuration, only the inversion of a system of the order of the unknowns located in the crack volume is required, leading to a very limited request of computational time. Differently from other integral formulations, the present approach does not have any geometrical limitation related to the computation of the Green function.

Table I - Execution time

Operation	T [s]
Assembling Inductance Matrix	1488
Assembling Resistance Matrix	59
Impedance Matrix Factorisation	119
Precomputations	179
Solution of each direct problem	5
Solution of the inverse problem	14000

Acknowledgements

This work was supported in part by the Italian MIUR and EURATOM/ENEA/CREATE.

References

[1] D. Dos Reis, M. Lambert, D. Lesselier, Eddy-current evaluation of three-dimensional defects in a metal plate, *Inverse Problems*, Vol. 18, pp. 1857-1871, 2002
[2] D. Premel, A. Baussard, Eddy-current evaluation of threedimensional flaws in flat conductive materials using a Bayesian approach, *Inverse Problems*, Vol. 18, pp. 1873-1889, 2002
[3] Z. Badics, J. Pavo, H. Komatsu, S. Kojima, Y. Matsumoto, K. Aoki, fast flaw reconstruction from 3D eddy current data, *IEEE Trans. on Magnetics*, vol. 34, pp. 2823-2828, 1998
[4] J. R. Bowler, S. A. Jenkins, L.D. Sabbagh and H.A. Sabbagh, Eddy-current probe impedance due to a volumetric flaw, *J. Appl. Phys.*, Vol. 70, No. 3, pp. 1107-1114, 1991
[5] G. Rubinacci, A. Tamburrino, S. Ventre, F. Villone, Numerical modelling of volumetric cracks, presented at ISEM Conference, 2003
[6] R. Albanese, G. Rubinacci, Finite element methods for the solution of 3D eddy current problems, *Advances in Imaging and Electron Physics*, vol. 102, pp. 1-86, 1998
[7] R. Albanese, G. Rubinacci, F. Villone, An Integral Computational Model for Crack Simulation and Detection via Eddy Currents, *J. Comp. Phys.*, Vol. 152, 736-755, 1999

Electromagnetic Nondestructive Evaluation (VIII)
T. Sollier et al. (Eds.)
IOS Press, 2004

Identification of Material Degradation for Nonlinear Electromagnetic Problem

Fumio KOJIMA[1], Katsuyuki ARA[2], and Seiki TAKAHASHI[2]
[1]*Graduate School of Science and Technology, Kobe University*
1-1, Rokkodai, Nada-ku, Kobe 657-8501 Japan
[2]*Iwate University, Ueda, Morioka, Iwate 020-8551 Japan*

Abstract. A new computational method for evaluating a level of material degradation is presented. A magnetic reluctivity is approximated by a series of natural cubic spline functions with knot sequence based on $B - H$ curve. An inspection model is then described by a nonlinear elliptic equation with the unknown degradation parameter. The output least square problem is formulated for identifying a material degradation parameter under magnetic density measurements.

1 Introduction

The evaluation of material degradation in pressure vessels in nuclear reactor is quite important issue in the the lifetime elongation and safe operation. During the operation these structural materials suffer from high stress, heat and neutron radiation. The changes of microstructure cause the change of mechanical properties, such as hardness, yield stress, etc. Due to these effects, magnetic properties are affected and the assessment of the associated magnetic parameters provide us the feasible information of degradation level before generating macroscopic crack. The mechanism caused by the strain field due to dislocations gives an influence on the magnetization curve through magneto-elastic coupling. It is well-known that the hysteresis of magnetic materials is characterized by

$$B(H) = \mu_0 H + M(H) \tag{1}$$

where B, H, and $M(H)$ denote the magnetic flux density, the magnetic field, and the magnetic moment of a sample, respectively. Recently, some interesting experimental findings have been explored for the relation between the structure-sensitive magnetic properties and dislocations [2]. Takahashi et *al* have proposed a new simple hysteresis model:

$$M(H) = M_s(1 - \frac{c}{2H^2}) \tag{2}$$

and have showed that the so-called "c" parameter is a structure sensitive parameter with respect to magnetic materials [2]. They have demonstrated experimentally that the increase of c parameter value implies the increase of dislocation rate of the sample material. Taking into account these, our inverse problem is to identify the c parameter for evaluating material degradation.

2 Inspection Model

Let us consider Maxwell's equations for magnetization procedures:

$$\nabla \times \mathbf{H} = \mathbf{J} \tag{3}$$
$$\mathbf{B} = \mu \mathbf{H} \tag{4}$$
$$\nabla \cdot \mathbf{B} = 0. \tag{5}$$

The constitution law (4) is linear in air, where we have $\mu = \mu_0 = 4\pi \times 10^{-7}$, while it becomes nonlinear in magnetic material. In other words, the magnetic permeability μ is a function of $\mathbf{H}(\mathbf{B})$. With divergence free condition (5) we introduce the magnetic potential vector $\mathbf{A}$ such that

$$\mathbf{B} = \nabla \times \mathbf{A}.$$

From this relation, the system (3-5) can be reduced to

$$\nabla \times (\nu \nabla \times \mathbf{A}) = \mathbf{J} \tag{6}$$

where ν is the magnetic reluctivity given by $\nu = \mu^{-1}$. To restrict our idea into two-dimensional problem, the potential vector $\mathbf{A}$ and current density vector $\mathbf{J}$ have a particular structure:

$$\mathbf{A} = \{\, 0,\ 0,\ A \,\} \quad \text{and} \quad \mathbf{J} = \{\, 0.\ 0,\ J \,\},$$

where A and J depend only on x_1 and x_2. In this case, the system (6) can be written by the following nonlinear Euler's equation:

$$-\nabla \cdot \nu \nabla A = J\ . \tag{7}$$

The system (7) is theoretically defined on the whole two-dimensional Euclidean space, but in practice for computer implementation we take a boundary domain. Let Ω be an inspection environment in two spatial dimensions defined by

$$\Omega = \{\mathbf{x} = (x_1, x_2) \mid w_- \le x_1 \le w_+,\ h_- \le x_2 \le h_+\}$$

and let Ω_m be a sectional plan of magnetic sample plate given by

$$\Omega_m = \{\mathbf{x} = (x_1, x_2) \mid w_- \le x_1 \le w_+,\ 0 \le x_2 \le h_t\} \subset \Omega$$

where h_t implies the thickness of the test sample. Figure 1 illustrates the overall configuration of the inspection procedures. While the sample is attached and magnetized to the test sample, magnetic measurements are made from two pick-up coils at both side of the yoke. If we set the boundary of the virtual window Ω as an air region which is sufficiently far from the inspection area, we may set the following boundary condition:

$$A = 0 \ \text{ on } \partial\Omega \tag{8}$$

where $\partial\Omega$ denotes the boundary of Ω. In the two-dimensional case, we note that

$$|\mathbf{B}| = |\nabla \times \mathbf{B}| = |\nabla \mathbf{B}|.$$

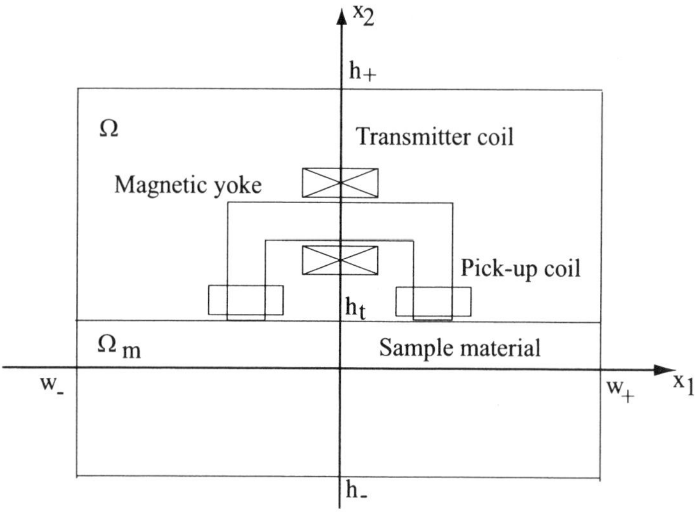

Figure 1: Inspection procedures

From (1) and (2), the reluctivity of the magnetic sample is given by

$$\nu(c, H) = \frac{H}{\mu_0 H + M_s(1 - \frac{c}{2H^2})} \tag{9}$$

$$B^2(c, H) = \left\{\mu_0 H + M_s(1 - \frac{c}{2H^2})\right\}^2 \tag{10}$$

where $B = |\mathbf{B}| = |\nabla A|$. In the sequel, without loss of generality, the magnetic reluctivity is represented as

$$\nu(c, |\nabla A|^2) = \begin{cases} \nu(c, B^2) & \text{for } x \in \Omega_m \\ \mu_0^{-1} & \text{Otherwise} \end{cases} . \tag{11}$$

Thus the nondestructive evaluation system can be rewritten by

$$-\nabla \cdot \nu(c, |\nabla A|^2)\nabla A = \chi_C J \quad \text{in } \Omega \tag{12}$$

$$A = 0 \quad \text{on } \partial\Omega \tag{13}$$

where J denotes the applied current density in the magnetic yoke, and where χ_C is the characteristic function on the transmitter coil of the yoke, respectively. Thus the output model is described by

$$Y(c) \propto \left\{ C_l\left(-\frac{\partial A}{\partial x_1}\right), C_r\left(-\frac{\partial A}{\partial x_1}\right) \right\} \tag{14}$$

where the operator $C_.$ implies the interpolation operator defined on the effective region of two pick-up coils.

3 Approximation of Magnetic Reluctivity

From $\nu = H/B$, it is crucial that the reluctivity of the magnetic material becomes divergent as $H \to +H_c$ and as $H \to +\infty$. To assure the existence and uniqueness of the solution of the forward system (12) with (13), we must restrict ν function into the finite range. To this end, we assume that the magnetic field is defined on the finite interval

$$I_H = \{ H \mid 0 < \bar{H}_- \leq H \leq \bar{H}_+ < \infty \}.$$

With the interval, ν function is approximated by the series of natural cubic splines with respect to B^2 [3]. Let the interval I_H be divided into

$$\bar{H}_- = \bar{H}_0^N \leq \bar{H}_1^N \leq \cdots \leq \bar{H}_N^N = \bar{H}_+$$

and let

$$\Delta(c, B^2) = \{B^2(c, \bar{H}_i^N)\}_{i=0}^{N}$$

be the knot sequence of cubic splines where

$$B^2(c, \bar{H}_i^N) = \{\mu_0 \bar{H}_i^N + M(\bar{H}_i^N)\}^2 \quad \text{with} \quad \bar{B}_- = B^2(c, \bar{H}_0^N),\ \bar{B}_+ = B^2(c, \bar{H}_N^N).$$

Then on the corresponding interval $I_B(c) = [\bar{B}_-(c), \bar{B}_+(c)]$, the reluctivity is approximated by

$$\begin{aligned} \nu^N(c, B^2) &= \alpha_i^N(c)(B^2 - B^2(c, \bar{H}_i^N))^3 + \beta_i^N(c)(B^2 - B^2(c, \bar{H}_i^N))^2 \\ &\quad + \gamma_i^N(c)(B^2 - B^2(c, \bar{H}_i^N)) + \delta_i^N(c) \\ &\quad \text{for} \quad B^2(c, \bar{H}_i^N) \leq B^2 \leq B^2(c, \bar{H}_{i+1}^N), \quad i = 0, 1, \cdots, N-1, \end{aligned}$$

with properties

$$\nu^N(B^2(c, \bar{H}_i^N)) = \nu(B^2(c, \bar{H}_i^N)) \quad \text{for} \quad i = 0, 1, 2, \cdots, N.$$

Furthermore, to assure the existence of the solution of (12), the reluctivity on $[0, \bar{B}_-^2(c))$ and $(\bar{B}_+^2(c), \infty)$ is approximated as

$$\nu^N(c, B^2) = m_-(c)(B^2 - \bar{H}_-) + l_-(c) \quad \text{for } 0 \leq B^2 \leq \bar{B}_-^2(c) \tag{15}$$

$$\nu^N(c, B^2) = m_+(c)(B^2 - \bar{H}_+) + l_+(c) \quad \text{for } \bar{B}_+(c) \leq B^2 < \infty, \tag{16}$$

where the coefficients m_-, l_-, m_+, and l_+ are determined by the differential and functional values at the both ends of the natural cubic spline function. Figures 2 and 3 demonstrate B-H curve and the corresponding approximated ν function using the natural cubic splines.

4 Forward Analysis

The forward problem is to compute the model output (14) with the constraint (12) given the appropriate "c" parameter. To implement those, we use the nonlinear finite element analysis. Let $\mathcal{T}_h$ be a set of triangles T such that

$$\Omega_h = \bigcup_{\mathcal{T}_h} T \subset \Omega,$$

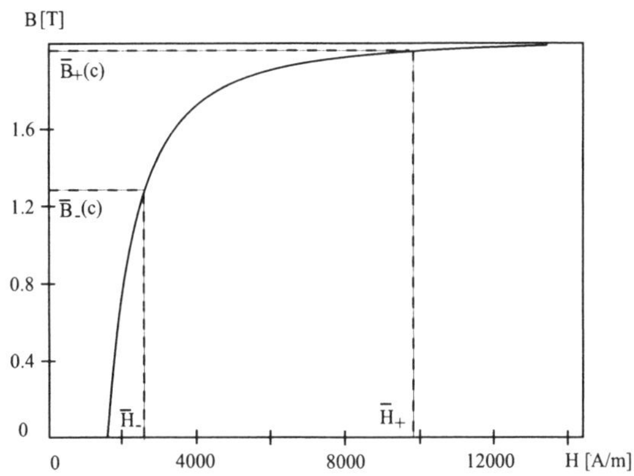

Figure 2: B-H curve and the setting of the interval I_H

where h is the length of the largest edges of $\mathcal{T}_h$. Figure 4 demonstrates the finite element decomposition of the virtual window Ω. With the triangulation used, the interfaces between air and magnetic material and between air and magnetic yoke never cross a triangle and are unions of edges of triangles belonging to $\mathcal{T}_h$. Let θ_h be the nodes of the triangulation $\mathcal{T}_h$ such that

$$\theta_h = \{\mathbf{x}_k = (x_1^k, x_2^k) \mid \mathbf{x}_k \text{ is node of } \mathcal{T}_h \}.$$

Suppose that the nodes which belongs to θ_h can be numbered form 1 to N_h, where $N(h)$ is the dimension of the finite element space. Let us define the set of continuous function ω_j by

$$\omega_j^h(\mathbf{x}_i) = \delta_{ij} \quad \text{for } i, j = 1, 2, \cdots, N(h)$$

with the property

$$\omega_j^h = 0 \quad \text{on } \partial\Omega_h \bigcap \partial\Omega.$$

Then the discrete potential vector A_h is defined as

$$A_h(\mathbf{x}) = \sum_{i=1}^{N(h)} \bar{A}_i^h \omega_i^h(\mathbf{x}). \tag{17}$$

The energy functional of the finite element analysis is given by

$$\mathcal{F}(A_h) = \int_{\Omega_h} \left[\frac{\nu(c, |\nabla A_h|^2)}{2} \left\{ \left(\frac{\partial A_h}{\partial x_1} \right)^2 + \left(\frac{\partial A_h}{\partial x_1} \right)^2 \right\} - JA_h \right] d\mathbf{x}. \tag{18}$$

Then the finite element solution is to seek the optimal coefficient vector $\{\bar{A}_i\}_{i=1}^{N(h)}$,

$$\mathcal{F}(A_i^*) = \min_{A_i} \mathcal{F}(A_h). \tag{19}$$

Since the reluctivity is state dependent, we need iterative procedures for the above optimization. A iterative solver using Newton-Raphson method is described as follows:

$$\{\bar{A}^{(n+1)}\} = \{\bar{A}^{(n)}\} + \{\delta A\} \quad n = 0, 1, 2, \cdots. \tag{20}$$

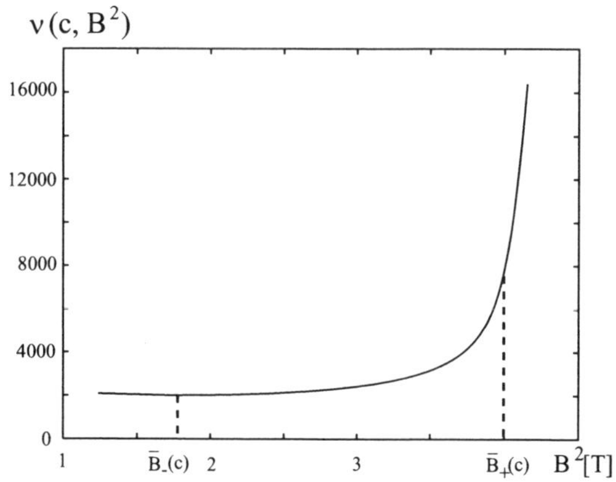

Figure 3: The approximated ν function

The update vector $\{\delta A\}$ is the solution of

$$[K]\{\delta A\} = \{f\} \tag{21}$$

where the element matrix and vector are respectively given by

$$[K]_{ij} = \frac{\partial^2 \mathcal{F}}{\partial A_i \partial A_j} \quad \text{for } i, j = 1, 2, \cdots, N(h) \tag{22}$$

$$\{f\}_j = -\frac{\partial \mathcal{F}}{\partial A_j} \quad \text{for } j = 1, 2, \cdots, N(h). \tag{23}$$

In the computational experiments, we assume that the reluctivity ν and the current density J takes its constant values in each finite element. In that case, the element matrix and vector given by (22) and (23) can be easily evaluated. Consequently, the model output can be computed as

$$\{Y_h(c)\} = [\mathcal{C}_h]\{\bar{A}^*(c)\} \tag{24}$$

where $\{\bar{A}^*(c)\}$ denotes the numerical solution of the iterative solver (20). Figure 5 depicts the magnetic flux density vector nearby the magnetic sample. In this experiments, the number of finite elements and the number of nodes are set as $dim(\mathcal{T}_h) = 2880$ and $N(h) = 1517$, respectively. The c-parameter in the numerical experiment was preassigned as $c = 1.5 \times 10^6[Oe]$.

5 Inverse Analysis

Numerical treatments of (24) can be implemented by the finite element method with triangular piecewise linear elements as stated in the previous section. Our inverse problem is stated as follows:

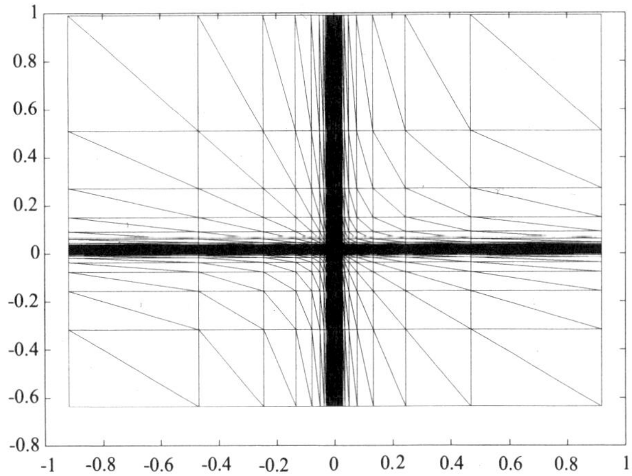

Figure 4: Finite element decomposition nearby sample

Inverse Problem : Given measurement data $\{\tilde{Y}_i\}_{i=1}^{N_p}$ excited through applied current densities $\{J_i\}_{i=1}^{N_p}$, find the optimal parameter $c = c^*$ which is the solution of

$$E[c^*] = \min_c \sum_{i=1}^{N_p} |Y(J_i, c) - \tilde{Y}_i|^2. \tag{25}$$

In the numerical experiments, the lower bound $\bar{H}_-$ was taken as the intersection between the line

$$B = \frac{dB}{dH}(H_c)(H - H_c)$$

and the line $B = M_s$. The upper bound $\bar{H}_+$ was set as the intersection between the function $B(H)$ given by (1) and the line $B = M_s$. Consequently, we set the lower and upper bound of the magnetic field were chosen as

$$H_- = \frac{M_s}{B'(H_c)} + H_c, \quad H_+ = \left(\frac{cM_s}{2\mu_0}\right)^{\frac{1}{3}}$$

respectively. Figure 6 (a) shows the distribution of the c-parameter to be identified and the estimated result is depicted in Figure 6 (b). The unknown function was given by the Gaussian density

$$c(\mathbf{x}) = \sum_{i=1}^{2} \exp\left(\frac{|\mathbf{x} - \mathbf{x}_i|^2}{\sigma_i^2}\right).$$

In the inverse problem, the simulation data with 5% additive noise were tested. We assume that ,in the each search area, the unknown c-parameter takes a constant value. The optimization routine was implemented using the golden section approach ([4]) at each mesh area as shown in Figure 6.

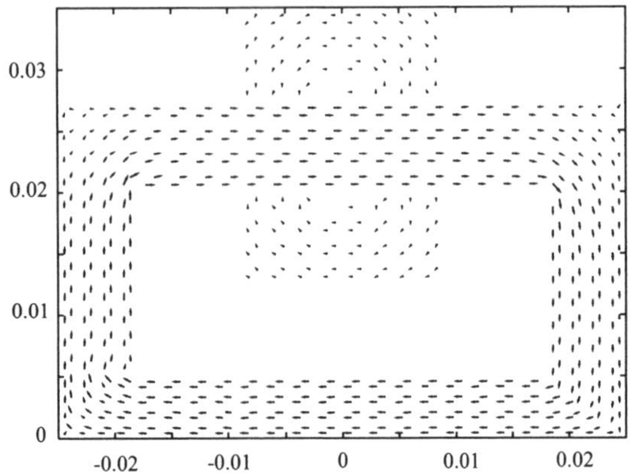

Figure 5: Magnetic flux density vector

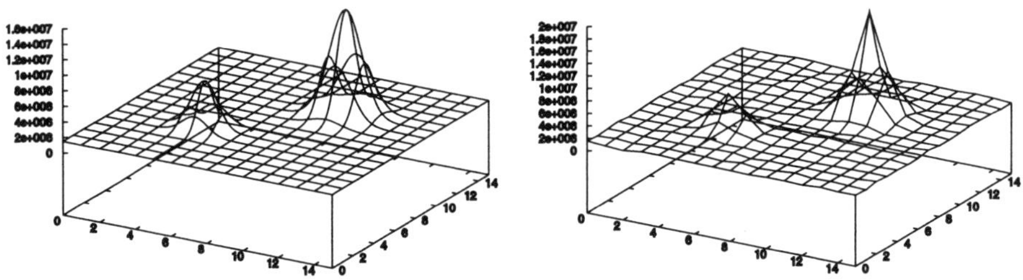

(a) distribution of c-parameter

(b) estimated function

Figure 6: Inverse analysis in the simulation experiments

Acknowledgements

This study was carried out in part while the first author was visiting the Statistical and Applied Mathematical Sciences Institute (SAMSI) NC U.S.A.

References

[1] F. Kojima, Inverse problems related to electromagnetic nondestructive evaluation, *Research Directions in Distributed Parameter Systems* (eds. R.C. Smith and M. Demetriou), Frontiers in Applied Mathematics, (2003), pp.225-241, SIAM Publisher.

[2] S. Takahashi, J. Echigoya, Z. Motoki, Magnetization curves of plastically deformed Fe metals and alloys, Journal of Applied Physics 87 (2000), 805-813.

[3] de Boor, Practical Guide to Splines, Academic Press, New York 1975.

[4] W.H. Press, B.P. Flannery, S.A. Teukolsky, W.T. Vetterling, *Numerical Recipes in C: The Art of Scientific Computation*, Second Edition, (1993), pp.397-401, Cambridge University Press.

Inverse Problem and Signal Processing

Electromagnetic Nondestructive Evaluation (VIII)
T. Sollier et al. (Eds.)
IOS Press, 2004

Signal Processing Techniques for Detecting Strut Fractures in Prosthetic Heart Valves

N V Nair, S Ramakrishnan, S T Thomas, R Ahn and S S Udpa

Department of Electrical and Computer Engineering, Michigan State University, East Lansing, MI, USA

Abstract. The Bjork-Shiley Convexo-Concave (BSCC) prosthetic heart valve employs a carbon disc, which is held in place by two struts, as an occluder to block the flow of blood in one direction. It has been demonstrated earlier that differences in the acoustic resonant frequencies associated with intact and fractured struts can be exploited as a basis for non-invasively detecting fractured struts. The challenge lies in detecting the acoustic resonance signal from among a variety of noise sources present in the human body. We propose to address the problem using blind source separation techniques. This paper suggests a cascade of two existing approaches for separating signals necessary for identifying strut failures. Preliminary results obtained with a human body as a transfer medium are presented.

1. Introduction

Prosthetic heart valves are usually implanted when valve disease leading to stenosis or incompetence is indicated. One type of mechanical valve called the Bjork-Shiley Convexo-Concave (BSCC) heart valve was widely implanted between 1979 and 1986. These valves have a carbon occluder disc held in place by two metallic struts of which, the outlet strut is welded to the suture ring, while the inlet strut is integral to the valve suture ring. Occasionally the weld at one end of the outlet strut fails. Although the patient does not experience any discomfort at this point, this may be a precursor to fracture at the other leg of the strut causing loss of the disc, which leads to fatality. As a result, there is considerable interest in detecting single leg separation (SLS) of the outlet strut in the heart valve [1].

One of the recent techniques [1] proposed for detecting SLS in BSCC heart valves involves excitation of the strut resonance modes using electromagnetic methods. The response of the strut to the excitation is monitored by measuring the acoustic field using an array of sensors placed on the chest of the patient. Finite element studies [2] have revealed that SLS failures alter the vibrational modes of the outlet strut, supporting the possibility of exploiting this phenomenon to diagnose the structural integrity of BSCC valves. These studies were later verified experimentally on a set of IOS and SLS valves. The fundamental frequency of the intact outlet strut was found to be a bending oscillation at 7960 Hz (that varied minimally with diameter and length of the valve), while that of the single-leg separated strut was found to be significantly lower at 2180 Hz. A valve with an SLS of the outlet strut and a residual contact at the separation (termed as 'kissing' contact) produced an intermediate frequency of around 4290 Hz. [2]. Thus, a possible strategy for detecting SLS failures is to estimate the strut's resonant frequency through acoustic measurements. The merit of using the strut resonant frequency as the basis of a BSCC heart valve classifier depends significantly on the frequency remaining constant, over time and under a wide range of patient physiological conditions. Studies in this area have shown that the IOS resonant frequency is not a function of patients' blood pressure, heart rate, and respiration rate, and hence these parameters need not be measured and included in the classification algorithms.

Also, the resonant frequency remains constant over time, thereby making the above technique viable [3].

While analyzing the acoustic signals, the question of whether to evaluate the valve closing or opening sounds was considered. Although both events lead to emission of similar acoustic frequencies, the closing sound had the advantage of greater intensity. Also, the single leg separated strut would be freer to vibrate when the valve is closed than when the valve is open [4]. On the other hand, studies indicate that the opening beats possess better frequency information about potential outlet strut fracture since the disk radiation frequencies that dominate the closing beats are only weakly coupled during the opening. We chose to analyze the closing sounds to take advantage of the higher amplitudes of the closing sounds.

Prior techniques of analyzing closing sounds range from the use of discriminant analysis techniques in the frequency domain that reduce each segment of the frequency spectrum into a discriminant score [4] to classifiers based on few characteristic features of the first resonant peak of the IOS present in the time-windowed, ensemble averaged spectrum of the valve closing sounds [5,6]. Scarborough and Inderbitzen [7] use the Short-Time Fourier Transform to perform a time-frequency analysis of the closing sounds.

One of the challenges in detecting the signal lies in the fact that the signal measured by the transducers placed on the chest of the patient is distorted as it travel through the intervening tissue. Furthermore, it is corrupted by acoustic signals that are generated in the thoracic cavity. The human thoracic cavity is a rich source of acoustic signals most of which cover the same spectral range (for instance, the respiratory tract sounds occur in the region of 600Hz to 2.5 KHz [12]). Additional noise is introduced by the instrumentation employed to capture the data. In addition to low signal-to-noise ratios, the issue is further complicated by the fact that most of the interfering signals lie in the same frequency range as the resonating strut signals precluding the use of traditional frequency domain filtering techniques to de-noise the signals of interest. Therefore, it is useful to employ signal-processing techniques that employ a combination of spatial as well as frequency localization. In this paper we cast this problem as one of blind source separation. We utilize two existing methods, namely the sub-optimal algorithm based on the Maximum Likelihood Principle by Wax [13] to first locate the positions of the sources and then use the Linearly Constrained Minimum Variance beamforming technique to estimate the desired source signal.

A brief description of the method employed for exciting the resonant modes of the strut is given before the signal processing algorithms for enhancing the signal-to-noise ratio is presented.

2. EMAT Technique for Exciting the Strut Resonant Modes

We employ, what is often referred to as, the electromagnetic acoustic transduction (EMAT) technique for exciting the resonant modes of the strut. We immerse the heart valve in a static magnetic field (~1.5 T) that is established using a pair of magnets. The static magnetic field is oriented in a direction parallel to the plane containing the suture ring. Additionally, a time varying field is established using a coil that is excited by a tone burst. This field is oriented in a direction that is parallel to the plane containing the suture ring but orthogonal to the direction of the static field. The time varying tone-burst field induces eddy currents in the suture ring. These eddy currents interact with the static magnetic field to produce a Lorentz force, which acts on the suture ring. The amplitude of the force is proportional to the static flux density and the current induced in the strut. Also, the direction of the force is orthogonal to the direction of the induced current and the static field. The tone frequency is chosen to match the modal frequency associated with either an intact or fractured outlet strut. The application of the tone burst causes the valve to vibrate and generate an acoustic field. Transthoracic recordings representing samples of the acoustic field are made using an array of electronic stethoscopes placed on the chest of the patient. The recordings are made during a short period shortly after the excitation tone burst is switched off. If the strut is intact, its displacement reaches a maximum when the tone frequency is close to one of the resonant modes associated with an intact strut. Since the resonant modes of an intact strut differ substantially from those of a fractured strut, it is possible to diagnose the state of the valve simply by determining the tone frequency at which the displacement amplitude reaches a maximum [1].

In practice, the IOS and SLS valve resonant frequencies are known to vary within a range of frequencies around 7.5 kHz and 2.2 kHz. Hence, the tone frequency is swept over a range covering the expected values of modal frequencies, and acoustic measurements are made.

3. Blind Source Separation Technique

The problem of detection and estimation of superimposed signals can be formally stated as follows. Consider a set of p sensors receiving signals from q sources (assume p>q). The signal received at the sensors can then be written in the form of a complex valued process $x(.)$

$$\bar{x}(t) = \sum_{k=1}^{q} \bar{a}_k(f,r,\theta,\phi)s_k(t) + \bar{n}(t) \tag{1}$$

where $s_k(.)$ denotes the kth complex narrowband signal vector, $\bar{n}$ (.) is the noise vector, $\bar{a}_k$ denotes the p x 1 complex valued propagation or steering vector of the signal that is based on a near-field propagation model, which is parameterized by the frequency and the location of the kth source signal and is defined as follows:

$$\bar{a}_k(f,r,\theta,\phi) = [\alpha_1 e^{-j2\pi f\tau_1} \ldots \alpha_i e^{-j2\pi f\tau_i} \ldots \alpha_N e^{-j2\pi f\tau_N}]^T \tag{2}$$

where $\alpha_i = \alpha_i(r,\theta,\phi)$ is the attenuation factor, or the ratio of the distance traveled by the wavefront to the reference sensor to that traveled to the ith sensor and $\tau_i = \tau_i(r,\theta,\phi)$ is the difference in the time taken by the wavefront to reach sensor i with respect to the reference sensor. In our case, the acoustic signal from the resonating heart valve is treated as the desired source located amidst a few other (q-1) interfering body noise sources. The problem can further be more compactly posed in matrix form as

$$\tilde{x}(t) = \mathrm{A}(f,r,\theta,\phi)\tilde{s}(t) + \tilde{n}(t) \tag{3}$$

where

$$A(f,r,\theta,\phi) = \left[\overline{a_1}, \overline{a_2}, \ldots, \overline{a_p}\right] \tag{4}$$

$$\tilde{s}(t) = \left[\tilde{s}_1(t), \tilde{s}_2(t), \ldots, \tilde{s}_p(t)\right]^T \tag{5}$$

are matrix of source parameters and the vector consisting of the source signals respectively. Wax [14] has shown that it is possible to cast this problem as a model selection problem under the assumption that the noise vectors are gaussian random processes with zero mean and have a covariance matrix of the form $\sigma^2 I$. This model selection can be further solved using the minimum description length method. This method is optimal but is computationally very intensive. Wax has suggested an alternative suboptimal approach that is computationally less intensive. The estimation of the source parameters in the EMAT problem is done using the sub-optimal technique.

3.1 *Sub-optimal algorithm*

The sub-optimal solution suggested by Wax [13] is used to obtain an accurate estimate of the source parameters, namely the location and frequency, given the number of sources 'q' (usually available based on knowledge of human physiology), and an initial estimate of these parameters. It has been shown [13] that the solution to the parameter vector is the solution to the maximization of the pseudo likelihood function given by

$$L(f,r,\theta,\phi)=\sum_{i=1}^{q}(l_i-\sigma^2)\frac{|\bar{a}(f,r,\theta,\phi)^H u_i|^2}{|\bar{a}(f,r,\theta,\phi)^H \bar{a}(f,r,\theta,\phi)|} \tag{6}$$

where l_i are the eigen values of the sensor-data covariance matrix **R**, u_i the corresponding eigen vectors associated with the input (sensor) signals, and σ^2 is given by:

$$\sigma^2=\frac{1}{p-q}\sum_{i=q+1}^{p}l_i \tag{7}$$

The q optimum solutions (f,r,θ,ϕ) obtained through an exhaustive search yield the parameters associated with the q sources.

3.2 *Beamforming*

Once the parameters of the sources (the locations and frequencies) are determined, we use a beamformer to separate out the signals that may overlap in the frequency domain but originate from different physical locations. If the weights of the beamformer are denoted by w, then the output of the beamformer [14] can be written as $\bar{y}(\varsigma)=w^H * \bar{x}(\varsigma)$. The *Linearly Constrained Minimum Variance* (LCMV) beamformer has been used here to design an optimal weight matrix (w) that minimizes the output variance under the constraint that the signal from the desired direction is retained, and those from the direction of the interfering sources are nullified. The LCMV problem is thus stated as:

$$\min_w w^H R w \text{ subject to } C^H w = g \tag{8}$$

where C is the constraint matrix, and g is a constant, that may be defined as follows to enhance a desired signal while nullifying an interferer.

$$C=\left[d_{des}(f,r,\theta,\phi)\quad d_{\text{int}}(f,r,\theta,\phi)\right],\ g=\begin{bmatrix}1\\0\end{bmatrix} \tag{9}$$

The optimal solution to the above problem can be obtained as [14]

$$w_{opt}=R^{-1}C[C^H R^{-1}C]^{-1}g \tag{10}$$

The spectrum of the desired heart valve signal is then computed to determine if the dominant component is near either 2.2 or 7.5 KHz (SLS or IOS case respectively).

4. Experiments

The above described signal processing algorithms were validated by conducting three experiments. The first one dealt with simulated source locations, frequencies and transfer parameters; the second one was an acoustic experiment that used air as the transfer medium while the third experiment used the human body as the transfer medium. In the above acoustic experiments, the source locations and frequencies were known approximately.

For simulation tests, an experimental setup as shown in Figure 1 was assumed. Three sources at frequencies 7.5 kHz, 7 kHz and 3 kHz were considered. An array of five sensors measured a mixture of these three source signals. Figure 2 shows the frequency spectra of the sensor signals. In order to test the suboptimal algorithm, we begin with the knowledge of the approximate locations and frequencies of the sources. The algorithm estimated the true locations and frequencies with a maximum relative error of 0.05% (in distance form the origin) for position estimate and 0.16% for the frequency estimate. The LCMV beamforming algorithm subsequently was used to obtain estimates of the three sources. Figure 3 shows the frequency spectra of the estimated source signals.

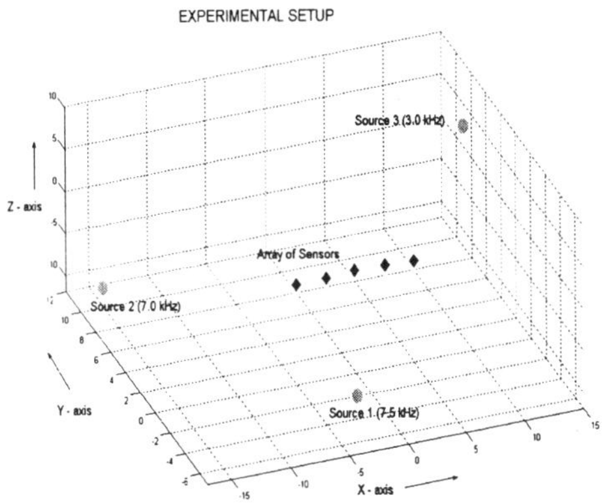

Figure 1: Experimental setup for the simulation test (units for axes in cm)

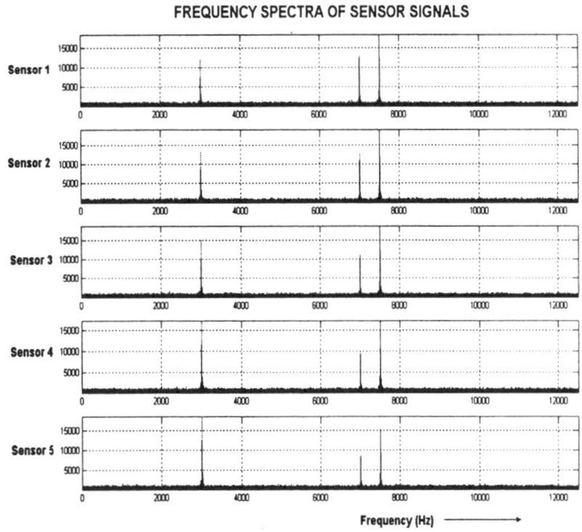

Figure 2: Frequency spectra of sensor signals

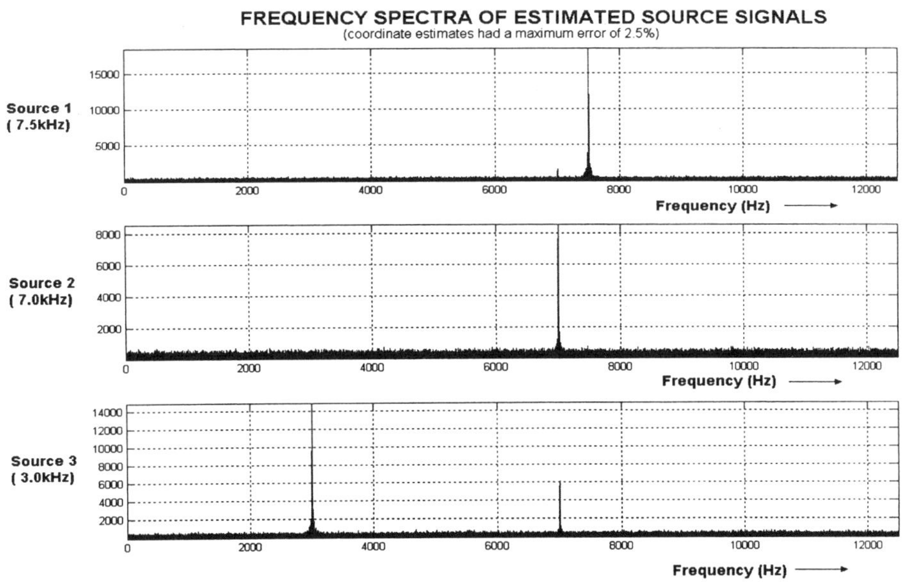

Figure 3: Frequency spectra of the estimated source signals (Experiment 1)

The second experiment was carried out with 4 microphones and 3 computer speakers in a closed room. The positions of the sensors and sources were as shown in Figure 4. This test was performed to validate the algorithm performance in a real acoustic transfer medium. The sources were pure tones at frequencies of 5 kHz, 6 kHz and 3 kHz. As in the previous case, approximate locations and frequencies were provided to the sub-optimal algorithm. The maximum relative error for position (as a % of distance from the center of the circle) and frequency estimates of the sources were 0.5% and 3% respectively. The beamformed estimates of the three sources are shown in Figure 5.

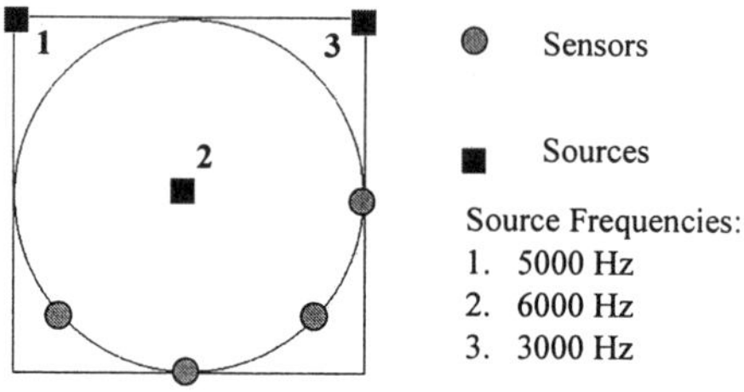

Figure 4: Setup for the second experiment that used air as the transfer medium

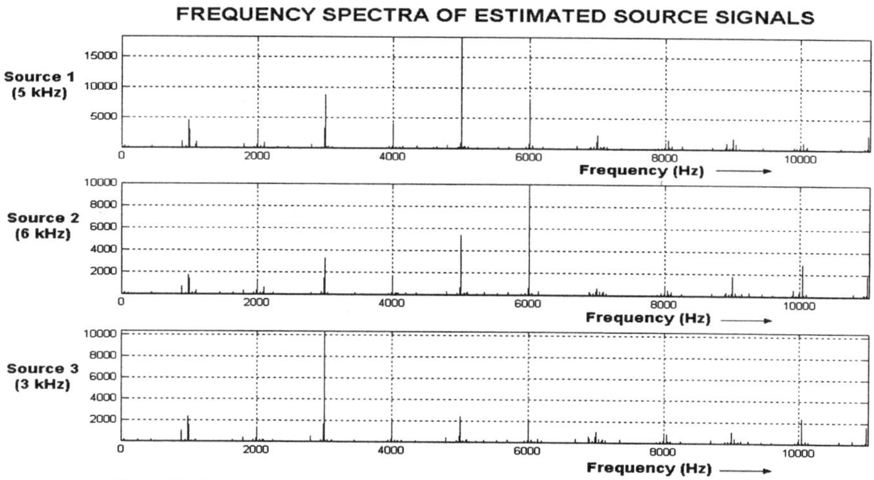

Figure 5: Frequency spectra of the estimated source signals (Experiment 2)

The third experiment was performed on a human body with 5 sensors placed on the chest and two interfering sources on the back. The interfering sources were generated by placing two small tweeters on the back of the subject (Figure 6).

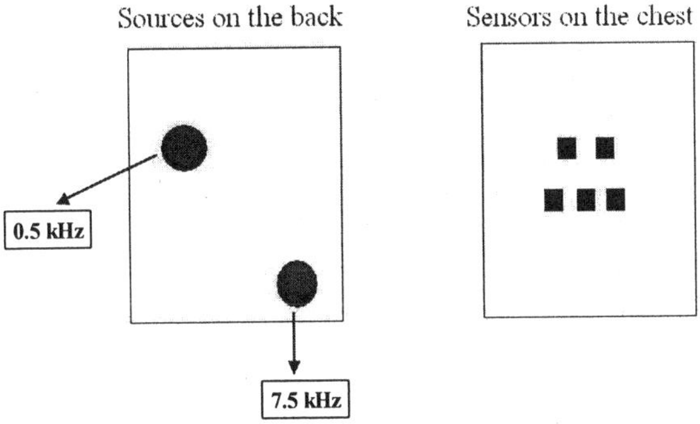

Figure 6: Experimental setup for experiment 3 with human body as the transfer medium

The source frequencies used were 500 Hz and 7500 Hz. The frequency spectra of the sensor signals recorded are shown in Figure 7. Table 1 shows the errors in the estimate of location and frequency of the source signals in the first step (sub-optimal algorithm). The spectrum of the reconstructed estimate for the 7.5 kHz signal is shown in Figure 8.

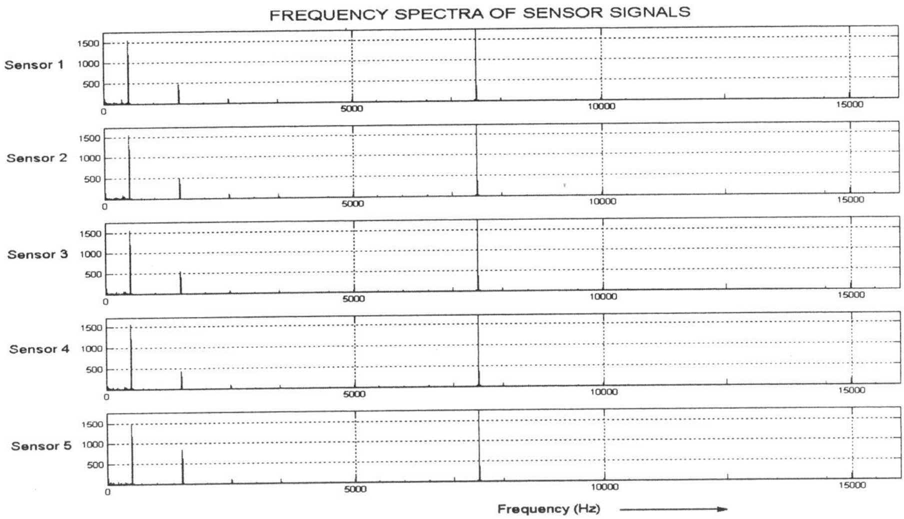

Figure 7: Frequency spectra of the sensor signals

Table 1: Error in parameters (Actual reference dimension was 50 cm)

Parameter	Source 1 (0.5 kHz)	Source 2 (7.5 kHz)
Relative error in X (%)	0.008	0.01
Relative error in Y (%)	0.01	0.01
Relative error in Z (%)	0.01	0.01
Absolute error in frequency (Hz)	50	40

In calculating the error values in the above experiment, it has been assumed that the body (especially the chest wall where the sensors are placed) is stationary. It is however still worthwhile to note that the errors obtained in the algorithm under this assumption are much smaller than the errors that would be caused by the movement of the body during the experiment. Taking into account the various assumptions made in the problem, the idea in designing the search interval is to make it as small as possible, thus minimizing the errors in the algorithm implementation. This in turn leads to very low values of relative errors in position (as low as 0.008%) that is much smaller than the error caused by movement of the thoracic wall.

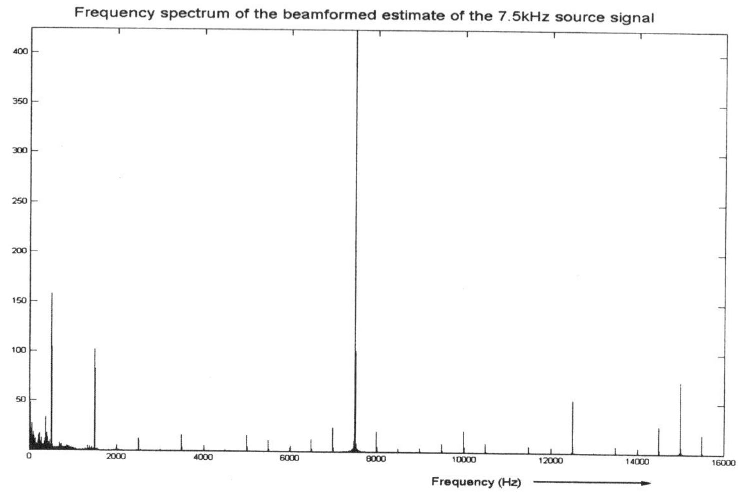

Figure 8: Frequency spectrum of the reconstructed estimate of the 7.5 kHz source signal

5. Conclusions

The initial results show considerable promise. Results show that the approach is capable of distinguishing the strut signals reasonably well and it is possible to estimate the dominant frequency.

Acknowledgement

The work reported in this paper was supported by the Bowling-Pfizer Settlement Fund. The support is gratefully acknowledged.

References

[1] S. Udpa, "New Electromagnetic Methods For The Evaluation Of Prosthetic Heart Valves", *Journal of Applied Physics*, Vol. 91, No. 10, pp.1-5, 2002.

[2] R. Chia, "Finite element analysis of vibrations of the Bjork-Shiley convexo-concave heart valve," *7th IEEE Symp. Computer-based Med Sys*, pp. 48-50, 1994

[3] T.D. Plemons, S.G. Schreck, and R.S. Inderbitzen, "Stability of BSCC heart valve outlet strut resonant frequency under changing physiological conditions and observation times," *7th IEEE Symp. Computer-based Med Sys*, pp. 119-124, 1994

[4] D.K. Walker and L.N. Scotten, "Discrimination *in vitro* between the acoustic emissions from Bjork-Shiley convexo concave valves with and without a broken minor strut," *Medical and Biological Engineering and Computing*, vol. 29, pp. 457-464, 1991

[5] J.J. Dow, T.D. Plemons, K. Scarbrough, H. Reeder, M. Hovenga, D.W. Wieting, and J.G. Chandler, "Acoustic assessment of the physical integrity of Bjork-Shiley convexo-concave heart valves," *Circulation*, vol. 95, pp. 905-909, 1997

[6] A.C. Eberhardt, C.E. Chassaing, R.S. Inderbitzen, and D.W. Wieting, "Acoustic Evaluation of Progressive Failure in BSCC Heart Valves," *7th IEEE Symp. Computer-based Med Sys*, pp. 112-118, 1994

[7] K. Scarbrough and R.S. Inderbitzen, "Application of time-frequency analysis to the classification of Bjork-Shiley convexo-concave heart valve condition," 7th *IEEE Symp. Computer-based Med Sys*, pp. 131-138, 1999
[8] J.V. Candy and H.E. Jones, "Classification of prosthetic heart valve sounds: A parametric approach," *Journal of Acoustical Society of America*, vol. 97, pp. 3675-3687, 1995
[9] J.V. Candy and H.E. Jones, "Processing of prosthetic heart valve sounds for single leg separation classification," *Journal of Acoustical Society of America*, vol. 97, pp. 3663-3673, 1995
[10] L.G. Durand, P.D. Stein, M.C. Grenier, J.W. Henry, R.S. Inderbitzen, and D.W. Wieting, "In Vitro and In Vivo low frequency acoustic analysis of Bjork-Shiley convexo-concave heart valve opening sounds," 7th *IEEE Symp. Computer-based Med Sys*, pp. 61-66, 1994
[11] L.G. Durand, M.C. Grenier, R.S. Inderbitzen, D.W. Wieting, and P.D. Stein, "Low frequency analysis of opening sound for detection of single leg separation of Bjork-Shiley convexo-concave heart valves," *Journal of Heart Valve Disease*, vol. 4, supp. I, pp. S32-S37, 1995
[12] P. Harper, S. S. Kraman, H. Pasterkamp and G. R. Wodicka, "An Acoustic Model Of The Respiratory Tract", *IEEE Transactions on Biomedical Engineering,* Vol. 48, No. 5, May 2001.
[13] M. Wax, "Detection And Estimation Of Superimposed Signals", *PhD dissertation*, Stanford University (1985).
[14] I. A. McCowan, D. C. Moore, S. Sridharan, "Near Field Adaptive Beamformer for Robust Speech Recognition", *Digital Signal Processing*, Vol. 12, No. 1, pp. 87-106, Jan 2002.
[15] B. D. Van Veen and K. M. Buckley, "Beamforming: A Versatile Approach to Spatial Filtering", *IEEE ASSP Magazine*, pp 4–24, April 1988.

Electromagnetic Nondestructive Evaluation (VIII)
T. Sollier et al. (Eds.)
IOS Press, 2004

Eddy Current Testing of Conductive Materials Using Time-of-Flight Measurements

Antonello TAMBURRINO
Ass. EURATOM/ENEA/CREATE, Lab. of Computational Electromagnetics and Electromagnetic Nondestructive Evaluation, DAEIMI, University of Cassino, v. G. di Biasio, 43, Cassino-03043, ITALY

Yong TIAN, Satish UDPA
Department of Electrical and Computer Engineering, Michigan State University, 2120 Engineering Building, East Lansing, MI 48832, USA

Abstract. The aim of this work is to develop methods detecting crack in conductive materials within the framework of time-of-flight based (TOF) approaches. The TOF can properly be described in the context of diffusive phenomenon through the Q-transform, which is a one-to-one transform relating solutions of wave propagation problems to solutions of diffusion problems. We assume that the conductive material that contains the crack has abrupt discontinuities. The abrupt discontinuity could, for instance, be the interface between the specimen boundary and air, as in the case of many practical situations. These material discontinuities complicate the extraction of the TOF from the measurements considerably because of the interaction of the field with the interfaces. The main contribution of the paper lies in the introduction of a linear and stable mapping that is capable of "removing" the interfaces, thereby leading the way for solving of the inverse problem using the TOF approach.

1. Introduction

This paper focuses on crack detection in conductive materials by using eddy current testing measurements. This problem, which is cast as a problem of determining the position of a proper source current density occupying the region of the crack, is solved by morphing the concept of time-of-flight (TOF), which is traditionally associated with wave based phenomena, to diffusion phenomena. The presence of air-metal interfaces pose severe problems for introducing the TOF for diffusion phenomena. The main contribution of the paper consists in introducing a linear (integral) and stable transformation to overcome this difficulty. An explicit form for this transformation is given for a canonical problem involving an infinite homogeneous conducting cylinder embedded in the air and TM fields.

The concept of TOF for diffusive problem can, perhaps, best be introduced using the Q-transform that is the operator $Q: u(\mathbf{x},q) \rightarrow v(\mathbf{x},t) = \left[2\sqrt{\pi t^3}\right]^{-1} \int_0^{+\infty} u(\mathbf{x},q)\, q \exp\left(-q^2/4t\right) \mathrm{dq}$. The operator Q maps a solution u of a proper initial value wave propagation problem, the so-called fictitious problem, into the solution v of an initial value diffusion problem [1]. Here, $[q] = T^{1/2}$.

The Q-transform establishes a one to one correspondence between wave propagation and diffusion problems. Therefore, it is natural to define the TOF for a solution of the diffusion problem as the TOF for the solution of the associated fictitious wave propagation problem.

The evaluation of the TOF for diffusion problems using the Q-transform has been considered in [2-9]. Specifically, in [2, 3] the TOF is extracted from the fictitious waveform obtained by applying the inverse Q-transform to diffusive data. The principal problem associated with this approach is that the inversion of the Q-transform is an ill-posed problem and regularization algorithms are required. In [4, 5], the TOF is obtained by assuming a parametric approximation of the impulse response for the fictitious problem containing the TOF. Then, by applying the Q-transform, it has been possible to get a corresponding parametric approximation for the diffusive data in frequency domain and the TOF can be associated with diffusive measurements. In [6, 7] we have exploited the degree of freedom given by the excitation waveform. Specifically, we showed that the peak of the time domain response of a diffusion problem excited by a suitable source waveform is related to the TOF of the fictitious problem. This approach has been extended to 3D vector problems in [8].

For scalar diffusion problems, as in the case of problems treated in this paper, assuming linear, non-dispersive and non-magnetic materials, we have that the wave "velocity" for the fictitious problem is $c(\mathbf{x}) = 1/\sqrt{\mu_0 \sigma(\mathbf{x})}$ where μ_0 is the magnetic permeability of the free space and σ is the conductivity of the material. Notice that $\left[c(\mathbf{x})\right] \doteq LT^{-1/2}$.

The presence of air-metal interfaces, which occur in many practical problems, pose severe problems for introducing the TOF for diffusion phenomena because the air can be considered a perfect insulating material, i.e. its conductivity is zero. Indeed, on the one hand the TOF associated with the passage of the energy in air is zero since the fictitious field wave velocity $c(\mathbf{x})$ is infinite and, therefore, the TOF cannot be used to estimate distance in air. On the other hand, the multiple reflections of the fictitious wave at the interface, complicate the structure of frequency response of the measured quantity and, consequently the extraction of TOF information from the diffusive field becomes difficult. The main contribution of this paper lies in the treatment of this class of problems and is based on [9]. Specifically, we introduce a proper linear and stable transformation which allows us to replace the air with an equivalent conductive material. This modified problem does not have interfaces and, as will be clarified in section 3.2, the TOF information can be easily extracted. We notice that this problem has not been addressed in previous work on the subject [2-8].

2. Mathematical model of the reference problem

In this section a simple illustrative example consisting of a 2D scalar diffusion problem in the frequency domain is introduced. Specifically, the example refers to the problem of retrieving the position of a localized source current density embedded in a conductor. In practical ECT problems this source represents the equivalent source producing the anomalous field, i.e. the equivalent source current density modeling the presence of the crack.

It is worth noting that the solution of this problem is an important step to address the solution of the inverse problem of crack detection and localization within the framework of TOF approach for diffusive problems.

The reference geometry of the problem consists of an infinite conducting cylinder embedded in a homogeneous non-conducting space. The axis of the cylinder is the *z*-axis. The source is an infinite current line embedded inside the cylinder and parallel to the *z*-axis. We assume that the conducting cylinder is a linear, isotropic, non-dispersive and non-magnetic material. In addition, both the conductivity and the source current density are *z*-independent. Consequently, the electric field **E** has the z-component only and is independent of the *z* coordinate, i.e. $\mathbf{E} = \mathbf{i}_z E(r,\theta; j\omega)$, assuming time harmonic field ($e^{j\omega t}$ time variation). By neglecting the displacement current, the electric field E satisfies the following 2D problem (see geometry in Figure 1):

$$\begin{cases} \nabla^2 E - j\omega\mu_0\sigma_0 E = j\omega\mu_0 J_0 & 0 < r < R \\ \nabla^2 E = 0 & r > R \\ E(R^-,\theta;j\omega) = E(R^+,\theta;j\omega) & \\ \partial_r E(R^-,\theta;j\omega) = \partial_r E(R^+,\theta;j\omega) & \\ \lim_{r\to+\infty} E(r,\theta;j\omega) = 0 & \end{cases} \tag{1}$$

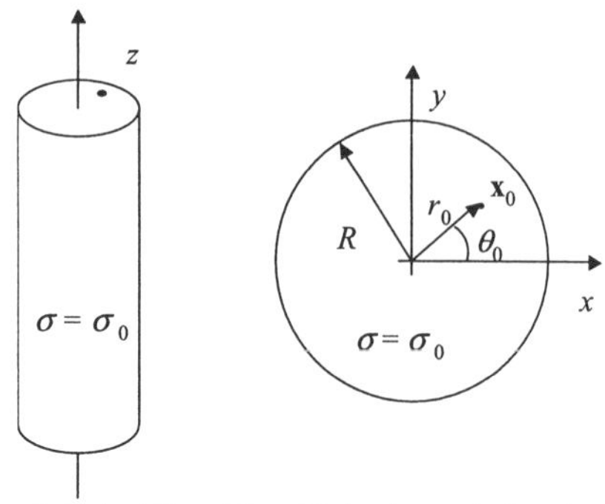

Figure 1. The reference geometry. Left the 3D problem. Right: the equivalent 2D schematization. A point source of strength I_0 is located at $\mathbf{x}_0$.

where ω is the angular frequency, μ_0 is the permeability of the free space, σ_0 is the conductivity of the cylinder, $J_0(\mathbf{x}; j\omega) = I_0(j\omega)\delta(\mathbf{x} - \mathbf{x}_0)$ is the current density of the line source located at $\mathbf{x}_0$ ((r_0, θ_0) in polar coordinates) and $I_0(j\omega)$ is the source strength.

Problem (1) is a time-harmonic problem. In this case, it is convenient to examine the fictitious wave problem in the Laplace domain (or, equivalently, in the frequency domain). In the Laplace domain, the action of the Q-transform can be expressed in a particularly "simple" form. Specifically $Q: U(\mathbf{x}, s) \rightarrow V(\mathbf{x}, s) = U\left(\mathbf{x}, \sqrt{s}\right)$ where U is the Laplace transform of u (w.r.t. q) and V is the Laplace transform of v (w.r.t. t) [10]. Therefore, the fictitious wave propagation field satisfies:

$$\begin{cases} \nabla^2 \tilde{E} + \tilde{k}^2 \tilde{E} = j\tilde{\omega}\mu_0 \tilde{J}_0 & 0 < r < R \\ \nabla^2 \tilde{E} = 0 & r > R \\ \tilde{E}\left(R^-, \theta; j\tilde{\omega}\right) = \tilde{E}\left(R^+, \theta; j\tilde{\omega}\right) & \\ \partial_r \tilde{E}\left(R^-, \theta; j\tilde{\omega}\right) = \partial_r \tilde{E}\left(R^+, \theta; j\tilde{\omega}\right) & \\ \lim_{r \to +\infty} \tilde{E}\left(r, \theta; j\tilde{\omega}\right) = 0 & \end{cases} \tag{2}$$

where $\tilde{k} = \tilde{\omega}/c$ and $c = 1/\sqrt{\mu_0 \sigma_0}$ is the fictitious wave speed inside the cylinder. The connection between $\tilde{E}$ and E in the frequency domain is given by

$$E(\mathbf{x}; j\omega) = \tilde{E}\left(\mathbf{x}; \sqrt{j\omega}\right) \tag{3}$$

under the condition $\sqrt{j\omega} J_0(\mathbf{x}; j\omega) = \tilde{J}_0\left(\mathbf{x}; \sqrt{j\omega}\right)$.

Problem (2) can be solved analytically using the method of separation of variables (see Appendix). The solution is:

$$\tilde{E}(r, \theta; j\tilde{\omega}) = \begin{cases} -\dfrac{\mu_0 \tilde{\omega} \tilde{I}_0}{4} H_0^{(2)}\left(\tilde{k}\left|\mathbf{x} - \mathbf{x}_0\right|\right) + \sum\limits_{n=-\infty}^{+\infty} \tilde{E}_n^s(r; j\tilde{\omega}) e^{jn\theta}, r \le R \\ \sum\limits_{n=-\infty}^{+\infty} \tilde{E}_n(r; j\tilde{\omega}) e^{jn\theta}, r \ge R \end{cases} \tag{4}$$

where the angular Fourier coefficients $\tilde{E}_n^s(r; j\tilde{\omega})$ and $\tilde{E}_n(r; j\tilde{\omega})$ are

$$\tilde{E}_n^s(r; j\tilde{\omega}) = \frac{\mu_0 \tilde{\omega}}{4} J_{|n|}\left(\tilde{k} r\right) H_{|n|-1}^{(2)}\left(\tilde{k} R\right) \Big/ J_{|n|-1}\left(\tilde{k} R\right)\left[\tilde{I}_0 J_{|n|}\left(\tilde{k} r_0\right) e^{-jn\theta_0}\right] \tag{5}$$

$$\widetilde{E}_n(r;j\widetilde{\omega}) = -j\frac{\mu_0 c}{2\pi R}\left(\frac{R}{r}\right)^{|n|}\frac{1}{J_{|n|-1}(\widetilde{k}R)}\left[\widetilde{I}_0 J_{|n|}(\widetilde{k}r_0)e^{-jn\theta_0}\right] \tag{6}$$

Here $J_n(\cdot)$ is the n-th order Bessel function of the first kind and $H_n^{(2)}(\cdot)$ is the n-th order Hankel function of the second kind. We notice that $-0.25\mu_0\tilde{\omega}\tilde{I}_0 H_0^{(2)}(\tilde{\omega}|\mathbf{x}-\mathbf{x}_0|/c)$ is the free space radiation due to a point source [11]. $\widetilde{E}_n^s(r;j\widetilde{\omega})$ and $\widetilde{E}_n(r;j\widetilde{\omega})$ are due to the presence of the material interface at $r=R$.

3. Interface Removal

In this section we introduce a linear and stable integral operator that is capable of "removing" the effect of the material interface. This operator maps the solution of problem (2) into the solution of a new problem where the point source is embedded in an infinite homogenous material. Then, by using the Q-transform, we derive the equivalent operator but for the solution of (1). Finally, we solve the inverse problem of locating the point source by means of the TOF approach applied to the (easier) new problem where the interface has been removed.

3.1 Interface Removal for Fictitious Wave Problems

To construct the interface removal scheme for our example (see [9] for details), we start from the field $\widetilde{V}(r,\theta;j\widetilde{\omega})$ which is a fictitious field related to the problem consisting of the same point source but embedded in a infinite homogeneous media with conductivity σ_0 (the interface has been removed in this problem). As is well known, $\tilde{V}$ is given by

$$\tilde{V}(r,\theta;j\tilde{\omega}) = -0.25\mu_0\tilde{\omega}\tilde{I}_0 H_0^{(2)}(\tilde{\omega}|\mathbf{x}-\mathbf{x}_0|/c). \tag{7}$$

We notice that $\tilde{V}$ depends explicitly on the TOF term $|\mathbf{x}-\mathbf{x}_0|/c$. Therefore, starting from $\tilde{V}$ we can easily retrieve the TOF that can be used to estimate the distance $|\mathbf{x}-\mathbf{x}_0|$ between the source $\mathbf{x}_0$ and a given observation point $\mathbf{x}$. On the other hand, it can be proved (see [9]) that the fictitious field $\tilde{E}$ depends on TOF's of type $(R-|\mathbf{x}_0|+2Rm)/c$ and $(R+|\mathbf{x}_0|+2Rm)/c$ where m is a non-negative integer. Consequently, $\tilde{E}$ is not suitable for extracting the TOF $|\mathbf{x}-\mathbf{x}_0|/c$.

The field $\tilde{E}$ can be "converted" into the field $\tilde{V}$ by a linear integral operator that does not depend on the source position and strength that, in an inverse problem, are unknowns. To determine this integral operator, we notice that:

$$\widetilde{V}(r,\theta;j\widetilde{\omega}) = \sum_{n=-\infty}^{\infty}\widetilde{V}_n(r;j\widetilde{\omega})e^{jn\theta} \tag{8}$$

where (see, for instance, [12]) $\tilde{V}_n(r;j\tilde{\omega}) = -0.25\mu_0\tilde{\omega}H_{|n|}^{(2)}(\tilde{k}r)\left[\tilde{I}_0 J_{|n|}(\tilde{k}r_0)e^{-jn\theta_0}\right]$. Therefore (see (6)) $\tilde{V}_n(r;j\tilde{\omega})$ is proportional to $\tilde{E}_n(r';j\tilde{\omega})$ and their ratio does not depend on the source term $\tilde{I}_0 J_{|n|}(\tilde{k}r_0)e^{-jn\theta_0}$:

$$\tilde{V}_n(r;j\tilde{\omega}) = -\frac{j\tilde{k}R}{4}\left(\frac{r'}{R}\right)^{|n|} H_{|n|}^{(2)}(\tilde{k}r) J_{|n|-1}(\tilde{k}R) 2\pi\tilde{E}_n(r';j\tilde{\omega}),\ r'\geq R \tag{9}$$

Replacing (9) in (8) and taking into account that $\tilde{E}_n(r';j\tilde{\omega}) = (2\pi)^{-1}\int_0^{2\pi}\tilde{E}(r',\theta';j\tilde{\omega})\,\mathrm{d}\theta'$ (see (4)) we have:

$$\tilde{V}(r,\theta;j\tilde{\omega}) = -\frac{j\tilde{k}R}{4}\int_0^{2\pi}\left[\sum_{n=-\infty}^{+\infty}\left(\frac{r'}{R}\right)^{|n|} H_{|n|}^{(2)}\left(\tilde{k}r\right)J_{|n|-1}\left(\tilde{k}R\right)e^{jn(\theta-\theta')}\right]\cdot \tilde{E}(r',\theta';j\tilde{\omega})d\theta'. \tag{10}$$

Equation (10) indicates that $\tilde{E}$ measured onto a circle of radius r' can be converted into $\tilde{V}$ at a circle of radius r by a proper integral operator. Moreover, it can be proved that, for $r'<r$, this integral operator is a compact and, therefore, continuous operator (see [9]).

Finally, we notice that (10) has been obtained by considering a point source located at (r_0,θ_0). However, thanks to the linearity of the problem, (10) is also valid for sources having support (in the *xy* plane) that is not a point but a line or a surface. Therefore, (10) can be applied to remove the air-metal interface also in the presence of more realistic cracks of finite extension.

3.2 Inversion Method for Diffusion Problems

By using the Q-transform in the frequency domain (replace $\tilde{\omega}$ with $\sqrt{\omega/j}$), we obtain the equivalent of (10) but for the diffusion problem. Specifically, we have:

$$V(r,\theta;j\omega) = -\frac{R\sqrt{j\omega}}{4c}\int_0^{2\pi}\left[\sum_{n=-\infty}^{+\infty}\left(\frac{r'}{R}\right)^{|n|} H_{|n|}^{(2)}\left(\frac{\sqrt{\omega}}{\sqrt{jc}}r\right)J_{|n|-1}\left(\frac{\sqrt{\omega}}{\sqrt{jc}}R\right)e^{jn(\theta-\theta')}\right]\cdot E(r',\theta';j\omega)d\theta' \tag{11}$$

where V is the diffusive field produced by the source J_0 placed in an homogenous infinite conductor of conductivity σ_0.

Since V is the Q-transform of $\tilde{V}$ we also have that:

$$V(r,\theta;j\omega) = -0.25\mu_0\omega I_0(j\omega)H_0^{(2)}\left(q_{TOF}\sqrt{-j\omega}\right) \tag{12}$$

where $q_{TOF} = |\mathbf{x}-\mathbf{x}_0|/c$ is the TOF that appears clearly in (12). In practice, the measure of $E(r',\theta';j\omega)$, $0\le\theta'<2\pi$ is affected by noise. Therefore, by applying the integral operator defined in (11) to the measured data $E_{meas} = E+\Delta E$ (ΔE is the noise term), we obtain $V_{meas} = V+\Delta V$, a noisy "version" of V. As discussed in section 3.1, for $r'<r$ the operator defined in (11) is compact, thus the noise affecting E is not amplified. The estimation of the TOF q_{TOF} from V_{meas} can be achieved by solving the following one parameter minimization problem:

$$\text{find } q^* \text{ s.t. } \psi(q) \triangleq \left|-0.25\mu_0\omega I_0(j\omega)H_0^{(2)}\left(q\sqrt{-j\omega}\right)-V_{meas}\left(r^*,\theta^*;j\omega\right)\right| \text{ is minimum} \tag{13}$$

where (r^*,θ^*) is a given observation point and q^* is the related estimate of the TOF. Finally, once the distance between the point source and the observation point (r^*,θ^*) has been estimated for at least two observation points, we can obtain the position of the point source.

The inversion algorithm can be summarized as follows:

1. measure the data $E(r',\theta';j\omega)$ for $r'=const.$ and $0\le\theta<2\pi$;
2. convert $E(r',\theta';j\omega)$ into $V(r_k,\theta_k;j\omega)$ where (r_k,θ_k) is the *k*-th observation point. Repeat for *k*=1,.., N_T, where N_T is the number of observation points;
3. estimate q_k^* by solving the minimization problem (13) for $(r^*,\theta^*)=(r_k,\theta_k)$. Repeat for *k*=1,.., N_T.
4. combine the distances $d_k = cq_k^*$ from the observation point (r_k,θ_k) to locate the source.

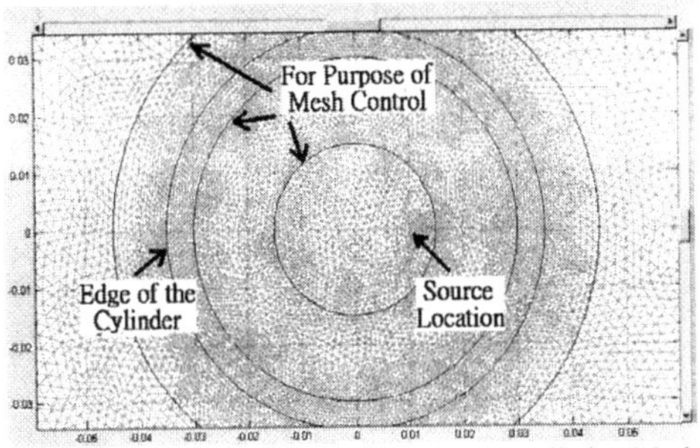

Figure 2. Particulars of the finite element mesh in a neighborhood of the cylinder. Mesh refinement has been enforced introducing artificial boundaries (the circles for mesh control purpose).

It is worth noting that problem (13) requires the knowledge of $I_0(j\omega)$ that in a crack identification problem is unknown. The simplest strategy to solve this problem consists in introducing the value of $I_0(j\omega)$ as additional unknown in (13) yielding to a two parameters minimization problem.

4. Numerical Examples

In this section, we present a numerical example to validate the inversion method. With reference to Figure 1, the basic parameters for numerical simulations are $R = 3\text{cm}$, $r_0 = 1\text{cm}$ and $\theta_0 = 0°$, $\sigma_0 = 3.54\times10^7$ S/m (copper), the radius of circles where E is measured is $r' = 3.2\text{mm}$, the radius of the circle where V is computed is $r = 3.5\text{mm}$. The excitation frequency is 75Hz which corresponds to a skin depth of 9.8 mm.

The electric field $E(r,\theta;j\omega)$ has been numerically computed using the commercial finite element package FEMLAB®. The solution domain is a circular domain of radius 1m. A homogenous Dirichlet boundary condition is imposed. The mesh consists of 24497 nodes and 48800 triangular elements of the Lagrange Quadratic type. The mesh density in the region of interest, i.e. the region adjacent to the source and to the evaluation circles (Figure 2) is high. The source strength, $I_0(j\omega)=1$.

The numerical computation of the integral operator appearing in (11) has been implemented by means of the Fast Fourier Transform algorithm (its kernel depends on $\theta-\theta'$). Specifically, we use a uniform sampling scheme with 64 points located at (r',θ_k), where $\theta_k = 2\pi k/64$, to numerically compute V on the circle of radius r.

To choose the region for locating the observation points (the point where we compute V by (11)), we analyze the plot of $|V(r,\theta;j\omega)|$ (not reported here). Specifically, we choose the observation points in the region where $|V(r,\theta;j\omega)|$ presents the largest values. In the present example, this region is in a neighborhood of θ=0.

In the example we choose only one observation point (N_T=1 in the inversion algorithm of section 3.2) to focus our attention on the solution of (13), that is key to locating the source. As highlighted in section 3.2, the location of the source requires at least two observation points.

Figure 3 shows the error functional ψ as a function of $d = cq$ and with reference to the observation point(r,θ_4). The relative error affecting the estimation of the distance between the point source and the observation point is 0.16% (left) for a noise-free E (E is affected by the numerical error only) and 1.36% for a 10% noisy data.

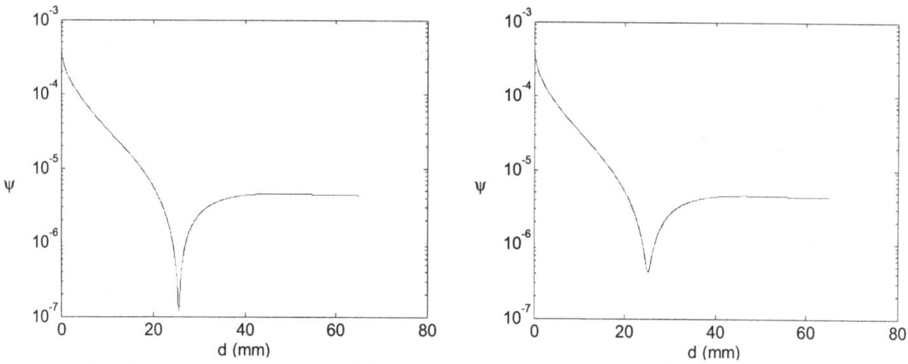

Figure. 3. The discrepancy ψ as a function of $d = cq$ assuming as observation point (r, θ_4) for two different noise levels. Left: 0% (only numerical errors) noise level. Right: 10% noise level.

The noise has been modeled as an additive noise uniformly distributed in $(-\Delta, \Delta)$ where Δ is 10% of the peak value of $\max_{\theta}\left|E(r', \theta; j\omega)\right|$. It is possible to show that the noise is strongly filtered by the compact integral operator defined in (11). This explains the moderate reconstruction error (1.6%) corresponding to the 10% noise level.

5. Conclusions

In this paper we solve a simple canonical problem consisting of locating a line source in an infinitely long conductive cylinder. This problem has been solved by introducing the concept of time-of-flight for diffusive problems through the Q-transform. The main contribution of the paper is the definition of a linear integral operator that is capable of mapping the field produced by the line source embedded in the conducting cylinder into a field produced by the same line source but embedded in a infinite homogenous conductor. This mapping allows us to overcome the complications associated with introducing the TOF for diffusive problem due to the presence of the material interface at the boundary of the conductive cylinder. We stress that the Q-transform has been applied, not to process the measured data, but, as a tool to treat a diffusion problem in terms of an equivalent wave propagation problem.

Acknowledgements

This work is supported by the Italian MIUR and EURATOM/ENEA/CREATE.

Appendix: Method of Separation of Variables

The solution of problem (2) can be obtained analytically by using the separation of variables in polar coordinate.

We start from representation (4) where we observe that the two series $\sum_n \tilde{E}_n(r; j\tilde{\omega})e^{jn\theta}$ and $\sum_n \tilde{E}_n^s(r; j\tilde{\omega})e^{jn\theta}$ are bounded solutions of the Laplace and homogeneous Helmholtz equations, respectively. Consequently $\tilde{E}_n^s(r;j\tilde{\omega}) = C_n^s J_{|n|}(\tilde{k}r)$ and $\tilde{E}_n(r;j\tilde{\omega}) = C_n / r^{|n|}$, where C_n^s and C_n are proper constant, $J_n(\cdot)$ is the Bessel function of the first kind of order n and we recall that $\tilde{k} = \tilde{\omega}/c$. Coefficients C_n^s and C_n can be found by imposing the interface conditions $\tilde{E}(R^-,\theta;j\tilde{\omega}) = \tilde{E}(R^+,\theta;j\tilde{\omega})$, $\partial_r \tilde{E}(R^-,\theta;j\tilde{\omega}) = \partial_r \tilde{E}(R^+,\theta;j\tilde{\omega})$ (see (2)) and by using the addition theorem for Bessel functions (see [12], for instance) $H_0^{(2)}(\tilde{k}|\mathbf{x}-\mathbf{x}_0|) = \sum_n H_{|n|}^{(2)}(\tilde{k}r) J_{|n|}(\tilde{k}r_0) e^{jn(\theta-\theta_0)}$ $r_0 < r$, where (r,θ) is the representation of $\mathbf{x}$ in polar coordinates and $H_0^{(2)}(\cdot)$ is Hankel function of the second kind of order zero. Specifically, it turns out that

$$\begin{cases} C_n^s J_{|n|}(\tilde{k}R) - C_n / R^{|n|} = \dfrac{\mu_0 \tilde{\omega}}{4} H_{|n|}^{(2)}(\tilde{k}R) J_{|n|}(\tilde{k}r_0) \tilde{I}_0 e^{-jn\theta_0} \\ C_n^s \tilde{k} \dot{J}_{|n|}(\tilde{k}R) + |n| C_n / R^{|n|+1} = \dfrac{\mu_0 \tilde{\omega}}{4} \tilde{k} \dot{H}_{|n|}^{(2)}(\tilde{k}R) J_{|n|}(\tilde{k}r_0) \tilde{I}_0 e^{-jn\theta_0} \end{cases} \quad (A.1)$$

Finally, (5) and (6) can be obtained by solving (A.1) and using the following properties of the Bessel functions:

$$Z_{-n}(z) = (-1)^n Z_n(z), \qquad \dot{Z}_n(z) = \frac{1}{2}\left[Z_{n-1}(z) - Z_{n+1}(z)\right]$$

$$Z_{n-1}(z) = \dot{Z}_n(z) + \frac{n}{z} Z_n(z), \qquad -\frac{2j}{\pi z} = J_n(z)\dot{H}_n^{(2)}(z) - \dot{J}_n(z) H_n^{(2)}(z)$$

where $Z_n(z)$ can be any of $J_n(z)$ and $H_n^{(2)}(z)$.

References

[1] L. R. Bragg and J. W. Dettman, An operator calculus for related partial differential equations, J. Math. Analysis and Applic. **22** (1968) 261-271.

[2] K. H. Lee and G. A. Xie, A new approach to imaging with low frequency electromagnetic fields, Geophysics **58** (1993) 780-96.

[3] S. Ross, M. Lusk, W. Lord, Application of a diffusion-to-wave transformation for inverting eddy current nondestructive evaluation data IEEE Trans. on Magnetic **32** (1996) 535-546

[4] D. Gibert, B. Tournerie and J. Virieux High-resolution electromagnetic imaging of the conductive Earth interior Inverse Problems **10** (1994) 341-51.

[5] T. J. Lee, J. H. Suh, H. J. Kim, Y. Song, K. H. Lee, Electromagnetic traveltime tomography using an approximate wavefield transform Geophysics **67** (2002) 68-76

[6] Y. Tian, A. Tamburrino, S. S. Udpa and L. Udpa, Time-of-flight Measurements from Eddy Current Tests. In: D.O. Thompson and D.E. Chimenti (Eds.), Review of Progress in Quantitative Nondestructive Evaluation, American Institute of Physics, 2003

[7] A. Tamburrino and S. S. Udpa, Solution of inverse problems for scalar parabolic equations using a hyperbolic to parabolic transformation: Time Domain Analysis, submitted for publication

[8] A. Tamburrino, R. Fresa, S. S. Udpa, Y. Tian, Three-dimensional defect localization from time-of-flight/eddy current testing data, submitted at COMPUMAG 2003 Conference (Saratoga Spring, USA), July 2003.

[9] A. Tamburrino, "TOF in the presence of material interfaces: a scalar axisymmetric geometry", Internal report, Department of Electrical Engineering, Michigan State University, 2003.

[10] K. H. Lee and H. F. Morrison, "A new approach to modeling the electromagnetic response of conductive media", Geophysics, vol. 54, pp1180-1192, 1989.

[11] L. B. Felsen and N. Marcuvitz, "Radiation and Scattering of Waves", Prentice-Hall Press, 1973.

[12] D. Colton and R. Kress, "Inverse Acoustic and Electromagnetic Scattering Theory", Applied Mathematical Sciences, vol **93**, Springer-Verlag Press, 1991.

Electromagnetic Nondestructive Evaluation (VIII)
T. Sollier et al. (Eds.)
IOS Press, 2004

Blind Source Separation for the Detection and Classification of Rail Surface Defects

Mohamed BENTOUMI[1,2], Gérard BLOCH[1], Patrice AKNIN[2], Gilles MILLERIOUX[1]
[1] : *Centre de Recherche en Automatique de Nancy (CRAN, UMR CNRS 7039)*
ESSTIN, Rue Jean Lamour, 54519 Vandoeuvre-les-Nancy Cedex, France
gerard.bloch@esstin.uhp-nancy.fr
[2] : *Institut National de Recherche sur les Transports et leur Sécurité (INRETS)*
2 avenue du Général Malleret-Joinville, 94114 Arcueil Cedex, France
aknin@inrets.fr

Abstract. The non-destructive evaluation of the rail is crucial to provide a high safety level in railway transportation. To detect and classify on line rail surface defects (splitted rail, shelling...), a specific double-coils double-frequencies eddy current sensor is used, which gives 8 real differential signals. Independent Component Analysis (ICA) for Blind Source Separation (BSS) is applied to dissociate the different classes of defects from the sensory signals. In an initial context comprising more sources than mixtures, typical defect signatures are analyzed to determine relevant signals for separation and estimate the corresponding separation matrices. A hierarchical separation procedure is then proposed and applied to real signals.

1. Introduction

The splitted rail detection is a crucial task for the railway managers to provide a high level of security. The current device providing this function, the track circuit, will disappear in an automatic control context. It is therefore necessary to develop a new system to control the rail in-operating conditions. With that purpose, a specific eddy current (EC) sensor has been developed [1]. This sensor has been designed and optimized according to the following specifications: positioning at *40 mm* height, vertical and horizontal displacements of the sensor due to the bogie dynamics, *100 km/h* maximum speed of the train, strong acceleration levels (up to *10g*), electromagnetic disturbances caused mainly by the traction currents that circulate in the rails.

An EC sensor is sensitive to all modification of the geometry and/or electromagnetic characteristics of the target. Obviously, transverse splits of the rail (Figure 1a) are detected, but minor defects as well (shelling (Figure 1b), welded joints or corrugation for example). Detecting and monitoring such defects allow to set up a predictive maintenance policy.

Figure 1: a) splitted rail (left), b) shelling (right).

The Independent Component Analysis (ICA) for Blind Source Separation (BSS) of rail defects from the EC sensory signals is presented here. The ICA has been already applied to process EC measurements, in similar non-destructive evaluation applications. In [2], the crack detection and recognition are studied. By fusing magnitude and phase signals in different measuring contexts, the ICA is used to extract the edge effect noise and to separate it from the signal related to the crack. In [3], the EC signals affected by strong noise and disturbances are processed to discover the flaws affecting a metallic slab. The availability of multiple measurements allows to perform a linear unsupervised data-fusion, which returns independent latent signals, one of which representing the flaw related signal recovered from the noisy mixture. A one-unit neural ICA system is employed. In [4], the performance of a smart EC sensor is improved by BSS. The sensor is dedicated to the recognition of metal tags buried in the ground, but is also sensitive to conductive elements located close to the target. Some hardware transformations and the use of BSS algorithms allow to restore the tag response.

Section 2 presents the chain of instrumentation and processing and the associated signals delivered by the EC sensor. In the next part, a brief description of Blind Source Separation (BSS) is given. Finally, the results obtained by applying the BSS to the EC signals are presented in the last part.

2. Instrumentation and processing chain

The sensor structure has been optimized and comprises two differential coils with different inner-distance and two control frequencies [1]. Differential measures are insensitive to the sensor/target relative movements, variations of temperature, electromagnetic perturbations... But the differential mode modifies the signal shape and the interpretation becomes more difficult. The sensor gives therefore 4 complex differential channels after demodulation, and 8 real signals after digitization (Figure 2).

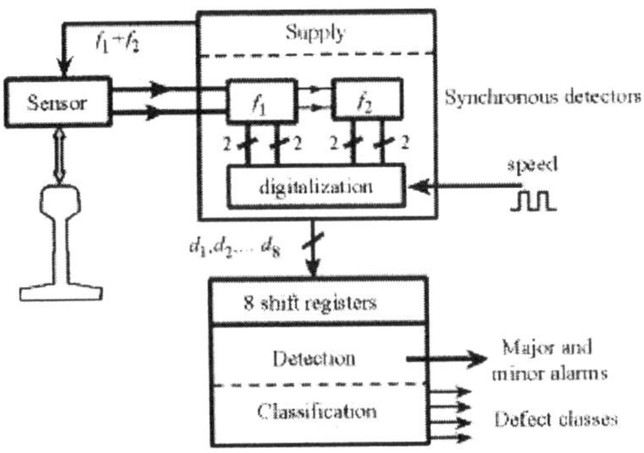

Figure 2: the instrumentation chain.

The digitization is performed using a fixed space step computed from the vehicle speed and fixed to *5 mm* during tests. The detection block generates major and minor alarms according to the kind of detected defect and activates a classification module if needed. Figure 3 shows a sample of the first signal (active part d_1) along *500 m* of rail. Particular points can be located, like switch (Sw), undulatory wear (UW) of the rail, fishplated joint or splitted rail (FJ), welded joint (WJ) and shelling (Sh). After perusal of records, track visits are required, in order to precisely label each particular detected event. A series of tests led to about 600 defects distributed into four classes (Sw, FJ, WJ, Sh). For sake of simplicity, these four types of singular points will be called defect in the following, even if only shelling (Sh) and splitted rail (FJ) constitute strictly speaking defects.

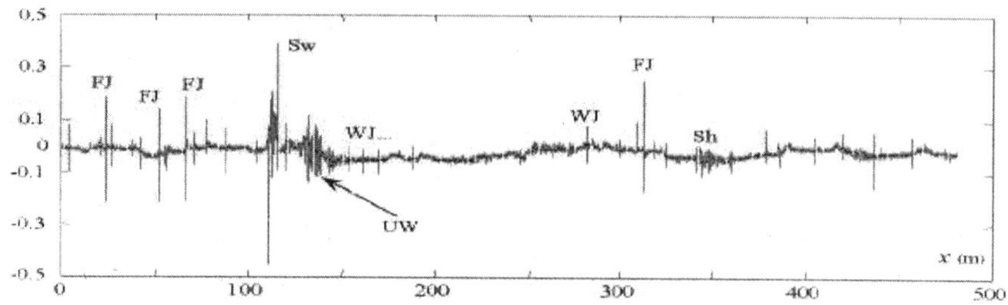

Figure 3: Sample of the EC sensory signal d_l on *500 m* of track.

Several methods have been tested to detect and classify the different defects. The first works were focused on the classification problem [1]. The performances of neural structures have been compared, with special attention to the definition of the signal representation space. Different detection techniques were also tested, among which two are dedicated to the minor defects (shelling and welded joints) [5, 6]. The first one is a time heuristic approach based on priori knowledge of the EC sensor. The second one is based on wavelet basis projections and solves the detection problem in the time-scale plan. These methods can be applied on original EC differential measurements or on absolute signals reconstituted by inverse filtering.

3. Principle of Blind Source Separation

The Blind Source Separation (BSS) or Independent Component Analysis (ICA) has been a very active field of research for a decade, with an impressive associated litterature (see, for instance, the surveys [7, 8, 9]). BSS/ICA consists in recovering unmeasured source signals from observed mixtures of these source signals, without knowing the mixing process. The standard BSS model considers n statistically independent source signals $s_i(t),\ i=1,\cdots,n$, p mixed signals $x_i(t),\ i=1,\cdots,p$, with $p \geq n$, and a linear and instantaneous mixture:

$$x_i(t) = \sum_{j=1}^{n} a_{ij}\, s_j(t), \quad i=1,\cdots,p \tag{1}$$

or, with $S(t) \in R^n$ the source vector, $X(t) \in R^p$ the mixture vector and A the unknown mixing matrix:

$$X(t) = A\, S(t) \tag{2}$$

As depicted in Figure 4, the BSS reconstructs the source vector $S(t)$ from the observed signal vector $X(t)$ by estimating a demixing or separation matrix W:

$$\hat{S}(t) = W\ X(t) \tag{3}$$

$$X(t) = \hat{A}\ \hat{S}(t) \tag{4}$$

and, in the simplest case where $p = n$:

$$\hat{A} = W^{-1} \tag{5}$$

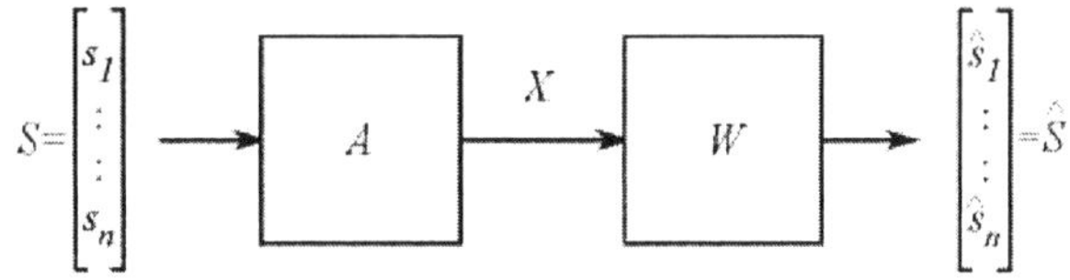

Figure 4: principle of the blind source separation.

There is a basic indeterminacy in (4) because it is possible to construct another solution $\hat{A}'\ \hat{S}'(t)$ satisfying the same condition:

$$X(t) = \hat{A}\,\hat{S}(t) = A\,P^T \Lambda^{-1} \Lambda\, P\, S(t) = \hat{A}'\,\hat{S}'(t)$$

with any permutation matrix P and any diagonal scaling matrix Λ. The solution to an ICA problem is then always defined up to permutation and scaling of each component.

Several extensions of the basic model have been investigated, including the cases with more/less sources than mixtures, noisy observations, ill-conditioned mixtures, more complex models of mixtures (linear convolutive, nonlinear, non-stationary...), use of prior information [10].

The basic assumption for ICA is that the sources are independent, i.e. their probability density functions verify: $p(s_i, s_j) = p(s_i)\, p(s_j)$, and are not Gaussian, at most only one.

The basic BSS/ICA algorithms consist of two stages.

- Data preprocessing to obtain vectors $V(t)$: centering $(V(t) = X(t) - E\{X(t)\})$ and whitening by Principal Component Analysis ($V(t) = V\ X(t)$, with $E\{V(t)V^T(t)\} = I$).
- Iterative optimization of an objective function, often a measure of non-Gaussianity: kurtosis, differential entropy, nengentropy, mutual information.

The algorithm used here is the FastICA algorithm [11].

4. Blind Separation of defects from EC signals

4.1 Determination of the independent sources

There are here $I = 8$ sensory signals d_i, $i = 1, \cdots, I$, and $J = 4$ defects to be detected: fishplated joint (FJ), welded joint (WJ), shelling (Sh), switch joint (Sw). Each particular defect j_0 produces I different signatures and then I source signals. The total number of source signals $s_{ij}(t)$, $i = 1, \cdots, I$, $j = 1, \cdots, J$ is then $I.J$. It is worth noting that, if, for any defect j_0 and for each signal pair (i_1, i_2), $s_{i_1 j_0}(t) \neq s_{i_2 j_0}(t)$ in some sense, the source separation, and then the defect classification, cannot be achieved, as the number of sources is (far) greater than the number of mixtures. Thus, the really *independent* source signals have to be determined. That is achieved by examining the typical defect signatures.

From the actual signals, recorded from the EC sensor during onsite tests, and the labeling of the defects present in the records, 8 typical signatures for each defect have been built by averaging and scaling the real defect signatures (71 signatures for FJ, 348 for WJ, 72 for Sh, 22 for Sw). To be insensitive to the measurement chain tuning and the variable sensor/rail distance, for each defect, the signatures have been scaled between -1 and +1 for the first signal d_1 and the same scaling is applied for the other signals, before averaging.

Typical defect signatures S_{ij}, $i = 1, \cdots, 8$, $j = 1, \cdots, 4$, are therefore available. Figure 5 presents the 8 typical signatures for a) fishplated joint (FJ) and b) welded joint (WJ).

Synthetic source signals $s_{ij}(t)$ for BSS can be then built including these typical signatures at the instants the corresponding defect occurs. In absence of defects, the synthetic sensory signals are simply zero mean noise and the effect of the defects on these signals can be expressed by a mixture, assumed to be linear: $d_i(t) = \sum_{j=1}^{4} a_{ij} s_{ij}(t)$, $i = 1, \cdots, 8$.

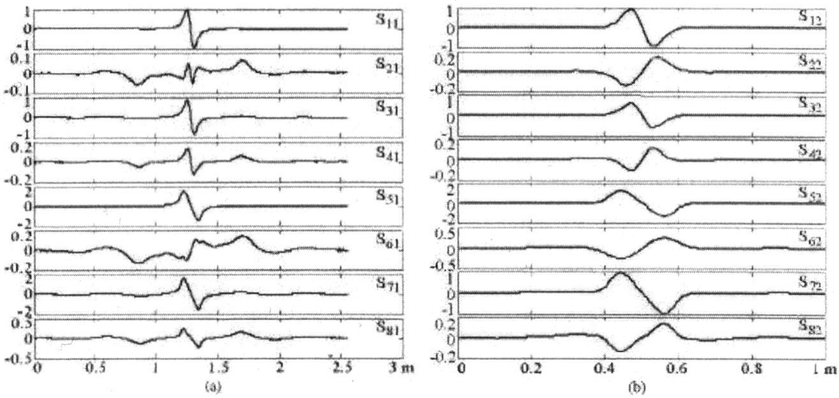

Figure 5: typical signatures for a) FJ (S_{i1}), b) for WJ (S_{i2}), for the 8 signals.

Figure 6 shows, for fishplated joint (FJ) and switch (Sw), the joint probabilities of the typical signatures estimated and represented by simply plotting for each defect the signature for a particular signal with respect to another one. For the example of FJ, the strong linear dependence of the signatures (S_{11}, S_{31}, S_{41}) can be noticed and could be confirmed by checking the corresponding correlation coefficients. For this group, the number of independent sources reduces then from 3 to 1.

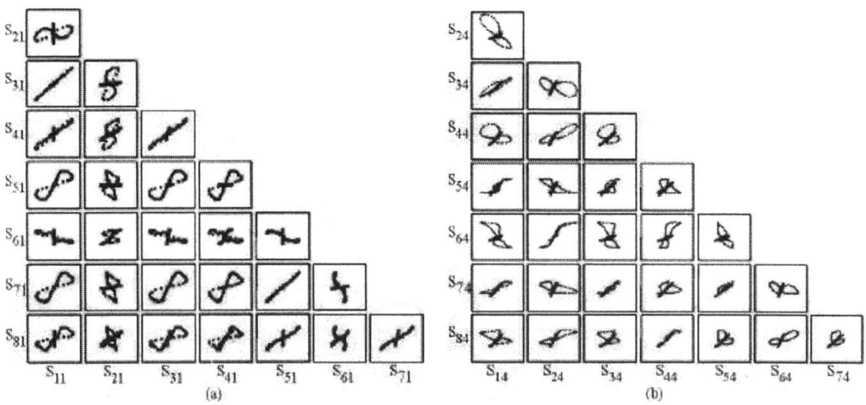

Figure 6: estimated joint probabilities of a) the FJ signatures, b) the Sw signatures.

In the same way, for each defect, the *independent* sources among the different signals have been determined. The results are given in Table 1. In order to separate defects from the mixed signals $d_i(t)$, a group of signals containing only one source for each defect must be selected. For the 3 defect classes FJ, WJ and Sh, such a group exists $((1,3,4)$, for instance). On the contrary, for the 4 defect classes FJ, WJ, Sh and Sw, no group can be found. Moreover correlations exist between signatures of different defects for different signals, complicating the separation.

Table 1: independent sources.
(signals with indices in parentheses correspond to only one source)

Defect # j	Index i for independent sources s_{ij}	Independent sources #
1 (FJ)	$(1,3,4), (2), (5,7,8), (6)$	4
2 (WJ)	$(1,3,4), (2), (5,6,7), (8)$	4
3 (Sh)	$(1,2,3,4), (5,6,7,8)$	2
4 (Sw)	$(1,3), (2,6), (4), (5,7), (8)$	5

4.2 Synthetic signals

Separation matrices obtained from different samples of real sensory signals are always different. Moreover, applying a separation matrix obtained from a particular sample to another one gives poor results. The generalization ability is then weak.

For this reason, synthetic signals build from typical signatures are processed to determine a unique separation matrix. Indeed, the typical signatures sum up the various and noisy real defect signatures recorded during tests and applying the unique separation matrix to real signals gives the best detection and classification results.

On the other hand, the context of the application allows to use a hierarchical detection-classification procedure, where the more energetic a defect signature is, the more priority its processing has. The separation will then be ordered as follows: switch (Sw), then simultaneously fishplated joint (FJ), welded joint (WJ) and shelling (Sh).

According to table 1, to separate switch (Sw) from the other defects first, the BSS has been applied on synthetic mixed signals d_1 and d_3 containing switch (Sw), fishplated joint (FJ), welded joint (WJ) and shelling (Sh) signatures. Figure 7 shows above a part of these signals, obtained by linearly mixing the typical signatures and adding zero mean Gaussian noise with 0.005 standard deviation. The corresponding separated signal is shown below.

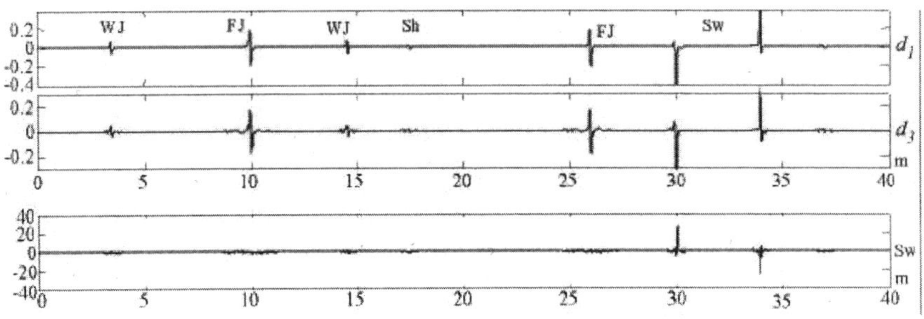

Figure 7: a part of the synthetic signals d_1 and d_3 (above), the separated Sw signal (below).

According to table 1 too, FJ, WJ and Sh defects can be separated from only the linearly mixed signals (d_1, d_3, d_4), which contain only one independent source for each considered defect. Figure 8 presents the 3 mixed signals above and the corresponding separated signals below on the same time period. The quality of the separation can be noticed.

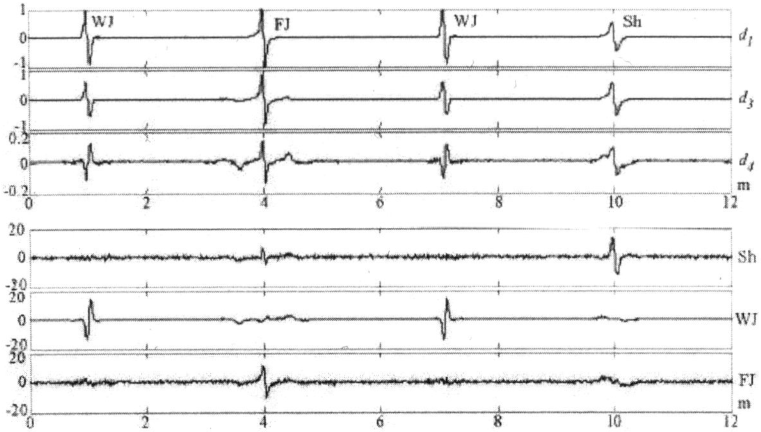

Figure 8: a part of the synthetic mixed signals (above), the separated signals (below).

4.3 Real signals

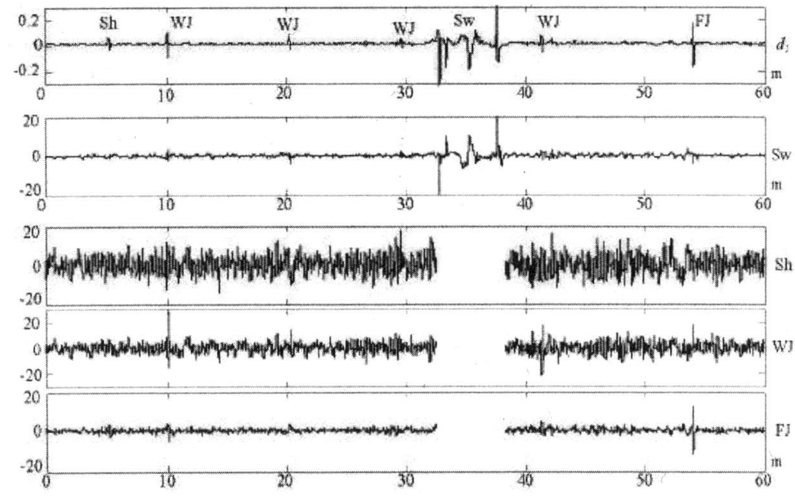

Figure 9: from top to bottom: first sensory signal, separated signals Sw, then Sh, WJ, FJ.

The separating matrices obtained previously from synthetic signals are applied to real sensory signals. The switch (Sw) signal is separated by demixing signals d_1 and d_3. Then fishplated joint (FJ), welded joint (WJ) and shelling (Sh) signals are obtained from d_1, d_3 and d_4. An example of results is shown in Figure 9. The switch is present in the other separated signals, but is ignored by the hierarchic detection. FJ (fishplated joint) and WJ (welded joint) are well separated. On the contrary, Sh (shelling) is buried into noise.

The detection thresholds have been tuned from all the recorded and processed defects, using the Receiver Operating Characteristic (ROC) curves and so simultaneously minimizing the false alarm rate and maximizing the well detection rate. The detection of switches is perfect: 100% of the 22 switches (Sw) are well detected without false alarms

(0%). 93% of the 71 fishplated joints (FJ) are well detected with a false alarm rate of 16%. 58% of the 348 welded joints (WJ) are well detected with a false alarm rate of 47%. The results for shelling (Sh) are not satisfactory.

This can be explained by several reasons. First, we mentioned the various inter- and cross-correlations between the signatures for different defects and different sensory signals. Second, the typical shelling signatures are not very representative of the various real shellings. Third, the real shelling signatures have a very low level variance compared to the other signatures and noise.

5. Conclusion

The BSS has been applied in a non-destructive evaluation application aiming at detecting and classifying on line rail surface defects. The 8 real differential signals provided by a specific eddy current sensor were processed to classify different defects. In an initial context comprising more sources than mixtures, the corresponding typical signatures have been analyzed to determine the independent signatures and relevant sensory signals for separation. Synthetic sensory signals built from these signatures were used to estimate the corresponding separation matrices. A hierarchical separation procedure was then proposed and applied to real signals.

The results for the separation of switches, fishplated joints and welded joints can be considered as satisfactory, with only a subset of the available signals. Nevertheless the correlations between the signatures, for different defects and sensory signals, and the differences of variance level between the signatures for different defects, particularly the low level of shelling signature variance, prevent standard linear BSS from achieving simultaneous complete separation of the 4 defect classes, even if all the signals are used.

References

[1] L. Oukhellou, P. Aknin, J.P. Perrin, "Dedicated sensor and classifier of rail head defects", *Control Engineering Practice*, **7**(1):57–61, 1999.

[2] F.C. Morabito, G. Simone, "Defect Signal Restoration in Eddy Current Measurements by Independent Component Analysis", *AMSE Modelling, Measurement & Control A*, **74**(1):17-36, 2001.

[3] S. Fiori, P. Burrascano, "ECT-data fusion by the independent component analysis for non-destructive evaluation of metallic slabs", *3rd Int. Conf. on independent component analysis and blind signal separation*, San Diego, CA, USA, Dec. 9-12, 2001.

[4] R. Huez, F. Belloir, A. Billat, "Original improvements of a smart eddy current sensor by BSS techniques", *Meas. Sci. Technol.*, **13**:547–555, 2002.

[5] M. Bentoumi, P. Aknin, G. Bloch, "On-line rail defect diagnostics with differential eddy current probes and specific detection processing", *Eur. Phys. J. AP*, **23**:227-233, 2003.

[6] M. Bentoumi, P. Aknin., G. Bloch, S. Talbi, "Improving the detection capacity of differential probes by means of inverse filters delivering estimated absolute signals", in X.P.V. Maldague (Ed.), *Topics in Nondestructive Evaluation* (TONE), **6**:125-130, Advances in Signal Processing for Non Destructive Evaluation of Materials (4th Int. Workshop, Québec, Canada, August 7-10, 2001), ASNT, 2002.

[7] J.F. Cardoso, "Blind Signal Separation: Statistical Principles", *Proceedings of the IEEE*, **9**(10):2009-2025, 1998.

[8] A. Hyvärinen, E. Oja, "Independent component analysis: algorithms and applications", *Neural Networks*, **13**(4-5): 411-430, 2000.

[9] V. David Sanchez A, "Frontiers of research in BSS/ICA", *Neurocomputing*, **49**:7-23, 2002.

[10] A. Parashiv-Ionescu, C. Jutten, A. Ionescu, A. Chovet, A. Rusu, "High performance magnetic field smart sensor arrays with source separation," in *Proc. of the 1st Int. Conf. on Modeling and Simulation of Microsystems*, MSM 98, Santa Clara, CA, USA, April 1998, pp.666–671.

[11] A. Hyvärinen, E. Oja, "A fast fixed-point algorithm for independent component analysis", *Neural Computation*, **9**(7):1483-1492, 1997.

Electromagnetic Nondestructive Evaluation (VIII)
T. Sollier et al. (Eds.)
IOS Press, 2004

The Analog Analysis of Frequency Components in NDT Systems

Ryszard SIKORA, Piotr BANIUKIEWICZ
Chair of Theoretical Electrotechnics and Computer Science, Sikorskiego 37, 70-310 Szczecin, Poland

Abstract. In this paper authors present system of analog active filters, which can improve measurements quality and reduce total time of tests in multifrequency method of cracks detection. Results of crack detection are also presented.

1. Introduction

The multifrequency NDT methods [1] are used to test good and weakly conductive materials such as: ferromagnetic and non-ferromagnetic metals. The right choice of work parameters of testing system such as: testing frequency, type of used sensor, algorithms used to analysis of obtained data, makes use of multifrequency method in testing materials with various electromagnetic proprieties: steels, copper and its alloys, brass, aluminium, Inconel 600, titan and more possible. Various types of discontinuities can be detected, mainly the surface discontinuities and subsurface discontinuities with small dimensions and varied shapes. This is crucial, because all critical tensions, caused by fatigue of material, are concentrated on the surface of material. Multifrequency NDT methods are used in industry for testing half-finished and finished products. These methods use the eddy currents effect. One of the main feature of eddy current is the skin effect, it is associated with the usage of AC sources to energize converters. The skin effect depends on the concentration of currents on and under the surface. Depth of material penetration with eddy currents is a function of conductivity of material, magnetic permeability and frequency. The shape of current's paths depends on the internal structure of material. The output signal will contain the same harmonics as the signal which energized the excitation coil. The output signal produced by eddy currents depends on their amplitude and, as a result, from internal structure of material. Analysis of amplitudes of signal's harmonics, obtained from the sensor, makes detection of discontinuities in material possible. In practice, the FFT [2] algorithm is often used to separate a harmonics of the output signal. The FFT can produce various errors, typical for numerical algorithms, especially when number of samples is small. Theoretically, to obtain a correct Fourier transform of signal, we should analyse an infinite numbers of signal's periods. If not, many unwanted effects can appear in the output signal. Of course, the analysis of the infinite numbers of signal's periods is impossible in real systems, but we should trend to analyse a maximal possible amount of signal's periods. In case of analysing low-frequency signals, this method can drastically increase amount of time required for material testing (long time A/D conversion). For example, while using frequency of 25 Hz during testing the material, we need about 2 sec, to obtain fifty signal's periods. This means, that the sensor must be stopped on each testing point for 2 seconds. Practically hundreds testing points exist on the surface of material. It depends on the step, with which the sensor is moving over the material (typically 0.2 - 1 mm). The long time of A/D conversion is a reason for the use of a step-by-step movement. The NDT system which uses a continuous movement of sensor is difficult to realize in practice. In many cases the technique of digital filtering is used for the analysis of the harmonics of the signal. The digital filters have many features. We are able to design practically any filter with any phase and amplitude characteristic, but these implementation of filters always require a lower or higher computation power of computers.

The presented system uses selective analog filters and can improve measurement speed. It avoids numerical analysis of the signal like the FFT algorithm, and it is not dependent on the computation power of the used computer. The center frequencies of filters are tuned to values suitable for aluminium, but the construction of the system makes the changes of center frequencies possible, when other materials are scanned. In the current version the system can work with fourteen frequencies, but it has modular construction so this range can be improved in easy way.

2. Description of the system

The main constructional principles of the system were: high selectivity and ability to work in real time, with continuous (not step by step) sensor movement over the material. Figure 1 shows a block diagram of the constructed unit.

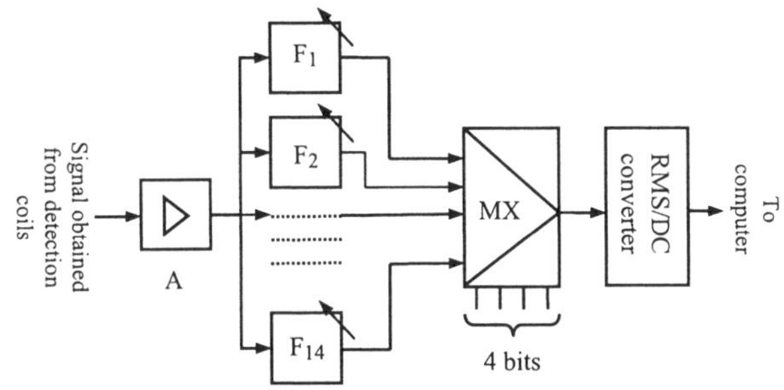

Fig.1. Block diagram of system, A – amplifier, $F_{1\text{-}14}$ – analog filters, MX – multiplexer

The group of fourteen selective analog filters (Fig. 1. blocks $F_1 - F_{14}$) is the main part of the system. The filters make separation and analysis of the individual harmonics of the input signal possible. The monolithic RMS/DC converter calculates RMS value of the sinusoidal component and gives an equivalent dc output voltage. Analysed harmonic is selected by multiplexer MX (analog switch). The multiplexer is controlled by a computer and specialized software.

The filters should have maximally narrow amplitude characteristic to extract only the harmonic with required frequency from the analysed signal. The selective characteristics make use of close frequencies during tests of material possible and guarantee that the neighbouring frequencies don't disturb different channels. Because the testing frequencies depend on the type of material, the filters should offer a possibility to change of their center frequencies in an easy way. To carry out above principles the construction of filters is based on LTC1264 [3] integrated circuits. These circuits are very flexible and make creation of various types of filters with various characteristics possible. The main advantage of LTC1264 is that the center frequency depends on the clock frequency. The change of clock will cause, in an easy way, retuning of filters.

The center frequencies of filters are proportional to the clock frequency, with a proportion coefficient b (equation 1).

$$F_0 = \frac{f_c}{b} \tag{1}$$

Each filter is projected with a different center frequency and a unique coefficient b (about 12 – 33). This coefficient is constant and it is determined on the stage of design. It depends on the configuration of external elements required by an integrated circuit. The block of filters is controlled by the same clock signal. Change of the clock frequency extorts changes of center frequencies of filters in wide range, but the mutual ratios of center frequencies between all filters are always constant. The change of only one center frequency is not possible. Always the whole system will be retuned. The use of integrated circuits instead of filters based on the operational amplifiers causes the higher temperature stability of the whole system, better stability of its point of work.

Each integrated circuit LTC1264 contains four filters in its structure. Each of them can be realized as independent, second order, filter. The maximal selectivity will be obtained, when the order of the filter is high. This also depends on the type of the filter. The good compromise between complexity of electronics circuits and parameters of system is reached, when the construction is based on eighth order filters. In this case, one IC will stand for one filter extracting one harmonic from the input signal. In practice, the elliptic functions were used for approximation of the amplitude characteristic of the filters. This approximation gave better results in case of low-order structures than the usage of Butterworth's or Chebyshev's approximation. Because only the RMS values of harmonics are collected and analyzed, the phase of signal can be ignored. In this case, problem of phase shifts of signal in filters is not important.

On the basis of the principles given above fourteen filters were projected with the following center frequencies (in Hz): 20, 60, 100, 120, 140, 160, 180, 200, 220, 260, 300, 340, 380 and 420. The choice of those frequencies results of the type of materials for which the system is constructed, and the limited selectivity of the filters. The width of passband of filters ranges: from 6Hz for lower frequencies to 16Hz for higher frequencies. This divergence is caused by technological limitations of LTC1264. Thus the two problems appear. First, the design of filters with any Q value greater than 50 is not viable. Second, the values of the external elements should be standard and widely available. Figure 2 shows amplitude characteristics of two filters.

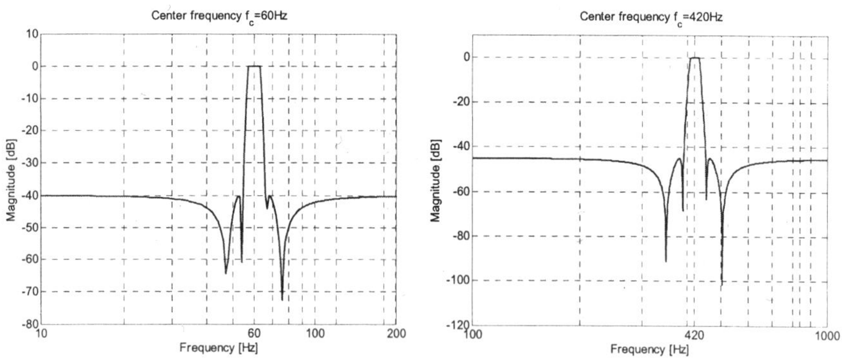

Fig.2. Amplitude characteristic of two filters with center frequencies: 60Hz on the left picture, 420Hz on the right picture.

The minimal value of stopband attenuation is about 40dB, but it isn't equal for all filters. If the center frequencies of filters are altered by the change of the clock frequency, the shape of amplitude characteristic will change in small range, but the passband width will be constant. It was checked empirically. In practice, the filters are tunable up to 200kHz; it is caused by high values of coefficient *b*, which extorts a high clock frequency (Equ. 1). When the clock reaches frequency about 5MHz, the system becomes unstable and many distortions appear.

The values of external components of filter were calculated with the use of specific software, created by Linear Technology, but the standard procedure of filters design [4][5] (based on the transmittance of circuit and the given transmittance of analog prototype of filter) is also available. Figure 3 shows the block scheme of one second order, filter. The type of the filter (lowpass, highpass) is determined by the connections between IC pins. The possible modes of work are explained in the documentation of LTC1264.

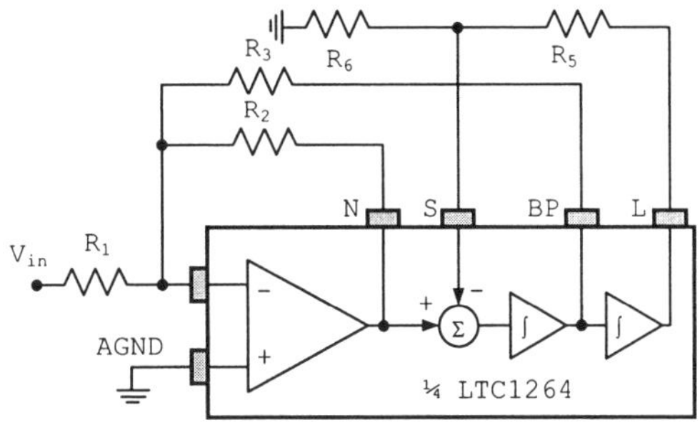

Fig.3. Block scheme of the 2nd order filter (mode 1b), where N, BP, L are: notch, bandpass, lowpass output pins

The approximation of characteristic (elliptic, Butterworth, etc.) is defined by the values of the external resistors. These values can be obtained from equations (2), when the bandpass transfer function *H(s)* is given. The parameter *Q* is the quality factor of the complex pole pair. It is the ratio of center frequency f_0 to the -3dB bandwidth of the 2nd order bandpass function.

$$f_i = \frac{f_{clk}}{20} \qquad Q = \frac{R3}{R2}\sqrt{\frac{R6}{R6+R5}}$$
$$f_0 = f_i\sqrt{\frac{R6}{R6+R5}} \qquad H_{BP} = -\frac{R3}{R1} \tag{2}$$

The next part of the system is the multiplexer. It is used to multiplex signals from filters to RMS converter. The active input is determined by 4-bit binary address line. This part has a significant influence on the total time of analysing the signal, so the fast analog switch ADG407B was used. The main application of ADG407B is audio and video routing. It guarantees a low level of distortions and a high bandpass.

The delay period between switching from one address state to another is about 220μs and it is shorter than the delay periods, caused by other components of the system (Figure 4).

The usage of only one RMS converter and multiplexing the signal obtained from block of filters guarantees that all channels will be measured with the same parameters. This minimize a total cost of the system and increase repeatability of measurements for channels.

Figure 4 shows delay periods caused by multiplexer (left) and total delay of the system (right).

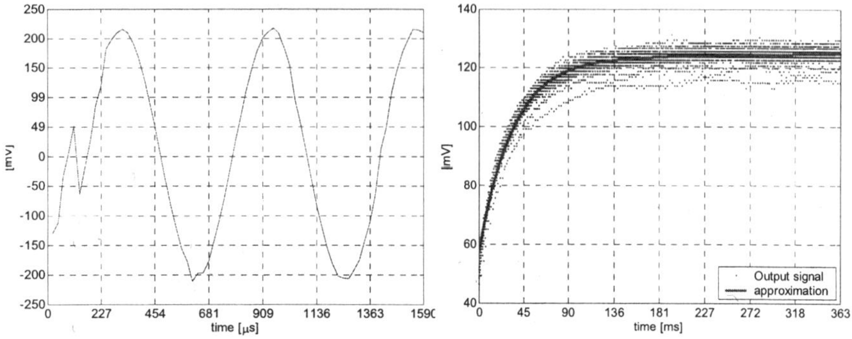

Fig.4. The signal obtained from output of multiplexer (left) and from output of RMS converter (right) after switching the channels.

The integrated circuit AD637 was used as the RMS/DC converter. It is a complete, high accuracy converter that can compute the true RMS value of any complex waveform. It offers high temperature stability and conversion accuracy of ±0.2mV. Because filters work parallel and the switching element is after them (see block diagram Fig.1.), the total time of stabilization only depends on dynamics characteristics of multiplexer and RMS/DC converter. The delay periods caused by multiplexer are short (μs), so the RMS/DC converter has main influence on the rapidity of the system (Fig.4. right). The plots showed above were obtained from A/D converter card. The sampling frequency was 44100Hz.

The short times of conversion and switching mean that the only velocity of the movement of sensor determines the time of testing. The dc output signal is measured by A/C converter. The time of sampling is shorter than in case of measuring AC signals because it doesn't have any requirements of number of periods that the in FFT method has.

To obtain a real characteristic which will be close to a theoretical characteristic, the potentiometers were used as the external resistive elements required by LTC1264. Each of them was tuned to the right value of resistance with tolerance below 1%.

The center frequencies precisely depend on the clock signal. The system uses an external precise generator to warrant the stability of the clock.

3. Results of the measurements

The block diagram of system used to measurements is showed in Figure 5.

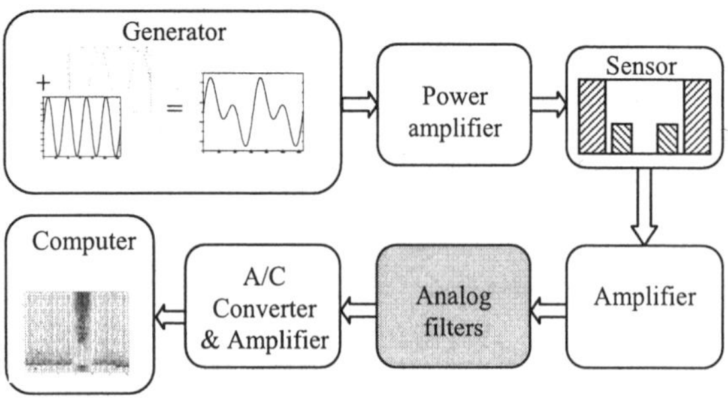

Fig.5. The block diagram of used system during measurements

The system of analog filters described in this paper is marked grey in Figure 5. The sensor [6] contains one excitation coil and two detection coils. The excitation coil has diameter of 7mm, the detection coils have diameter of 2mm and height of 1mm. The excitation coil was energized by AC current of 100mA. The detection coils are sensitive to a normal vector of magnetic flux, generated by eddy currents in plate. The aluminium plate tests were done. The thickness of plate was 10mm. The various configurations of crack locations were tested. The width of the crack was 1mm.

The first example shows results of tests of the plate with depth of crack equal 95%. The crack and sensor were on opposite sides of the plate. The used frequencies depend on coefficient b and the clock signal (Equ. 1.). Theirs mutual ratio is always constant. There are divergences between theoretical coefficients b and its practical realization. It is caused by tolerance of used resistors and thermal unstability of the potentiometers. This is the main problem of the system. Figure 6 shows results of measurements for clock frequency f_c=2000Hz. The equivalent center frequencies of the filters are also showed in Figure 6.

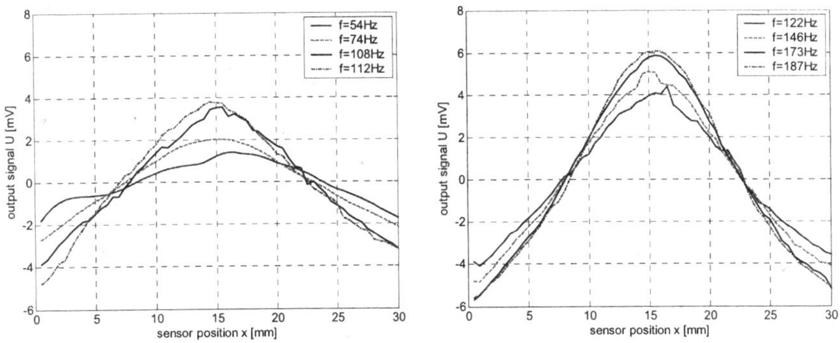

Fig.6. Depth of crack equal 95%, the crack and the sensor were on the opposite sides of the plate

The above characteristics were obtained in one move of the sensor over the plate (eight testing frequencies). The results showed in Figures 7a (left) and 7b (right) were obtained separately.

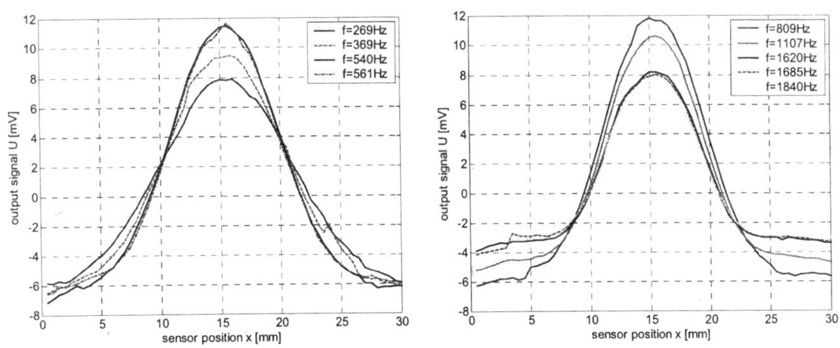

Fig.7. Depth of crack equal 95%, the crack and the sensor were on the opposite sides of the plate (two tests)

Figure 7b shows that output signal caused by crack, decrease with increase of the frequency. The optimal testing frequency for this type of material is about 540Hz. (Fig 7a). The ability to cracks detect under the surface of the plate decrease for high frequencies.

The next figures show results of tests of plate with crack depth of 50% and of 20%. The crack was located on the opposite side of the material. The material's discontinuities which are placed deep under the surface require low testing frequencies. The standard depth of material's penetration by eddy current is larger than dimensions of the specimen. In this case, there is a possibility, that the results of testing are depended on size of the analysed material.

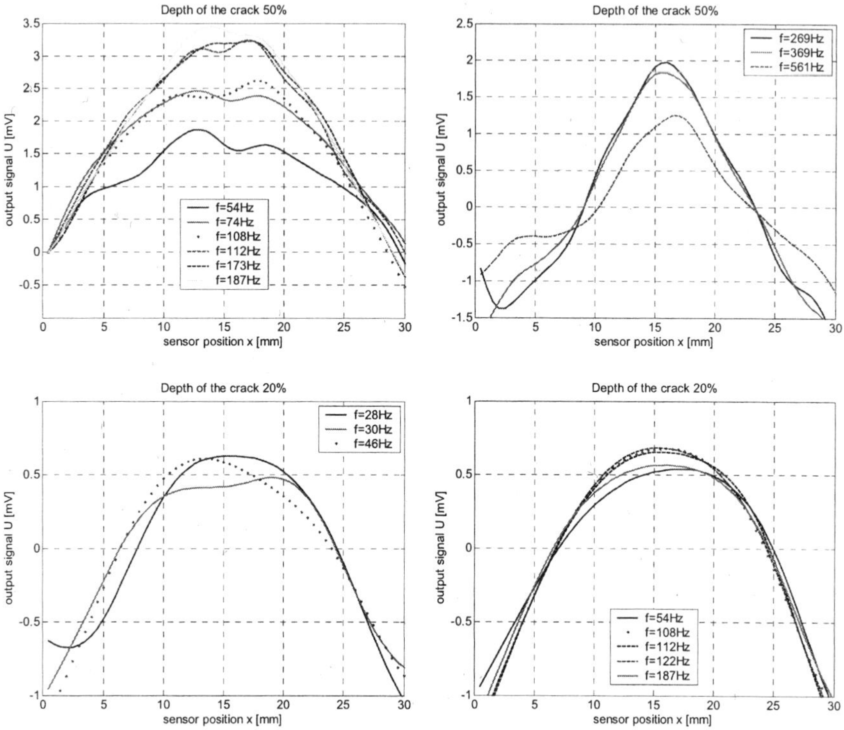

Fig.8. Depth of crack equal 50% and 20%. The crack and the sensor were on the opposite sides of the plate (four tests)

Above plots (Fig. 8) were cleaned by the lowpass filter in postprocessing stage. The next figures show results of crack detection of 20% depth. The crack and the sensor were located on the same side of the plate.

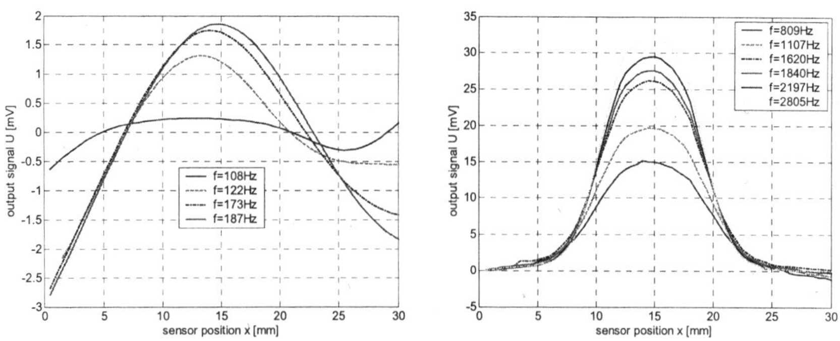

Fig.9. Depth of crack equal 25%. The crack and the sensor were on the same side of the plate

In case of depth of 20%, the signal caused by crack was small and additional amplifier was added to system. The filters add to the signal additional noise, which can distort a low signals.

4. Conclusions

The presented system is one of possible solutions of analysing of the signal in multifrequency NDT testing. It has satisfactory parameters and it can be classified as low-cost system. The previous system with FFT analysis needed about 2 seconds for each testing point to collect data and compute FFT transform for them. Analog filters reduce this time to 1 second (for identical measurement system and computer hardware). In our case the maximal speed of measurement was limited by mechanics of sensor positioning system. The whole system uses a computer only to collect results of the measurements. It doesn't need high-end hardware solutions. The usage of the integrated RMS/DC converter gives lower costs of the system than use of the stationary digital voltmeter. The filters work correctly in case when the amplitude of the input signal is 0.5 - 1V. This extorts usage of the precise amplifier with gain about 100. In case when the input signal is not large, the distance between signal and noise is too small. This causes decrease of sensitivity of the system. The system of analog filters can be used in standalone automatic flaw detectors.

References

[1] T. Chady, M. Enokizono, T. Todaka, Y. Tsuchida, R. Sikora, Flaws Detection and Characterization Using the Multifrequency Excitation and Spectrogram ECT, *Electromagnetic Nondestructive Evaluation (IV)*, S. Udpa et al., IOS Press 2000, p. 87-94
[2] V. K. Madisetti, D. B. Williams, Digital Signal Processing Handbook, *CRCPRESS INC 1999*
[3] LTC1264 Datasheet, Linear Technology
[4] Wai-Kai Chen, The Circuit and Filters Handbook, *CRCPRESS INC 1995.*
[5] Matlab manual
[6] R. Sikora, P. Baniukiewicz, Optimization of Eddy Current Sensors for Cracks Detection in Aluminium Plates, ISTET 2003

Electromagnetic Nondestructive Evaluation (VIII)
T. Sollier et al. (Eds.)
IOS Press, 2004

Data Processing of Corrosion Noise Polluted ECT Signals for Heat Exchanger Tubes of Cu-Ni Alloy

Zhenmao Chen*, Naoki Chigusa**, Hiromu Isaka** and Kenzo Miya*
**International Institute of Universality,*
SB Bldg. 8F, 1-4-6 Nezu, Bunkyo-ku, Tokyo 113-0031, Japan
Phone: +81-3-5814-5350, Fax: +81-3-3827-0682, E-mail: chen@iiu.co.jp
***Nuclear Power Division, The Kansai Electric Power Co., Inc.*
3-3-22,Nakanoshima, Kita-Ku, Osaka, 530-8270, Japan

Abstract. This paper presents a study for data processing of ECT signals in which large corrosion noises are included. Through numerical analysis and experimental investigation, the features of a corrosion noise is clarified at first. Based on these knowledge, a piecewise phase transform algorithm and a pattern matching technique are proposed to extract crack indications from the noise polluted ECT signals. By measuring and processing a mixed signal of both artificial cracks and a seriously corroded zone, the efficiency of the proposed method is confirmed. An OD crack deeper than 30% is possible to be recognized by processing the ECT signal of a Cecco probe with the proposed methods.

1 Introduction

In some kinds of copper alloy tube that are widely used in heat exchangers, material corrosion (here after, Ni corrosion) occurs in the tube surface after years of service under some heat and water conditions. Though the thickness of the corroded layer is small, a big noise is caused in the signal of eddy current testing (ECT) that is routinely applied in the in-service inspection. As the corrosion noise may be even bigger than an inner crack signal some time, it makes a crack initiated in the outside of the tube hardly be recognized from the inspection signals.

In order to solve this problem, numerical simulation and experimental investigation are performed in this paper for investigating the features of the Ni corrosion noise. Based on the obtained knowledges, a piecewise phase optimization algorithm and a pattern matching technique are proposed for recognizing the crack indications. By processing signals due to EDM notches and fatigue cracks fabricated in a tube segment that is cut out from an aged heat exchanger and seriously corroded place by place, it is found that an OD crack deeper than 30% can be recognized from the processed ECT signals.

In this paper, the numerical simulation results of the crack signals, corrosion noise etc. are presented at first. The features of the noise and its difference with the crack signals are clarified in this part, too. Several test-pieces made of the tube segment taken from an aged heat exchanger are fabricated. OD EDM slits of 30%, 40%, and 50% depth, and circumferential OD fatigue cracks are introduced to the test-pieces respectively. The test-pieces are inspected with a Cecco5 ECT probe. The data processing methods are described in section 4. In the last part, a validation using the measured signals is presented.

2 Numerical Analyses

An FEM code based on the reduced vector potential method and the edge element is applied to the simulation of the ECT signals. The validity of the code has been proved by solving benchmark problems and comparing with other experimental data [1],[2],[3]. The numerical model of the simulation is shown in Fig.1 where one of the channels of the Cecco5 probe is illustrated. The sensor consists of one exciter and two pickups of differential output. EDM notches of 0.2mm width are chosen for simulating the signal of the SCC cracks, while a corrosion zone is modeled as a law conductivity region. Based on the observed information of a microscope, the depth of the corrosion is selected as $60\mu m$.

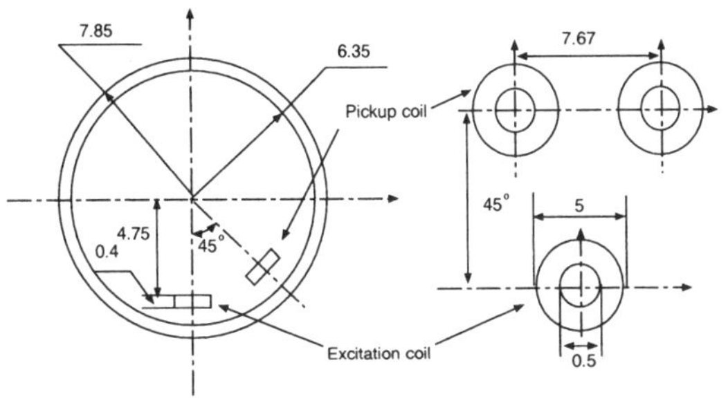

Fig.1 The numerical model of the tube and the probe system

The conductivity of the tube is measured by using a 4 points potential method for used tube segments without any treatment and those after acid washing. It is found that the average conductivity in regions with and without Ni corrosion is not significantly different. In this work, the conductivity of the Cu-Ni alloy is chosen as 3.3×10^6S/m.

As the noise due to a corrosion in the inner side of the tube is much bigger than that from the outside ones, the inner corrosions are considered as major noise source. Signals due to ID cracks, OD cracks of different position and depths are simulated for both corroded and perfect tubes.

Fig.2 shows a comparison of a numerical and an experimental result for cracks in a washed tube segment, which is taken as an example of the validation of the code for present problem. One can find that good agreements are obtained. The differences in the absolute values and the phases come from the unknown amplifier parameters.

Fig.3 shows a simulation result of the crack signal due to an OD axial slit of 60% depth and 5 mm length for frequency 20kHz, 40kHz, 60kHz and 80kHz. The phase changes significantly for different frequency. On the other hand, the signals due to an inner corrosion zone of 2.5mm side length and 60 μm depth are presented in the Fig.4. Though the amplitude of the corrosion noise is about double of the OD60% crack signal, it is possible to distinguish them from a noisy signal as the phase of the corrosion signals are very different with the ones of an OD slit. The other signals, which have not been presented here, show a similar property

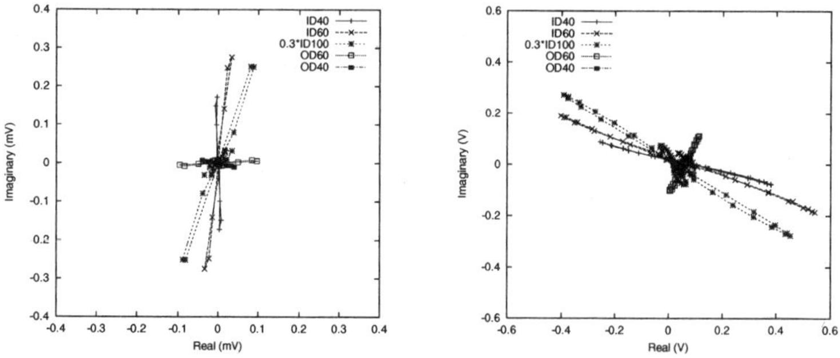

a) Numerical results b) Experimental results

Fig.2 Validation of the numerical code for the Cecco probe and Cu-Ni tube

though the phase of the signals due to different corrosion source may have some differences. This gives a possibility to reduce the noise signal with an optimized phase transformation.

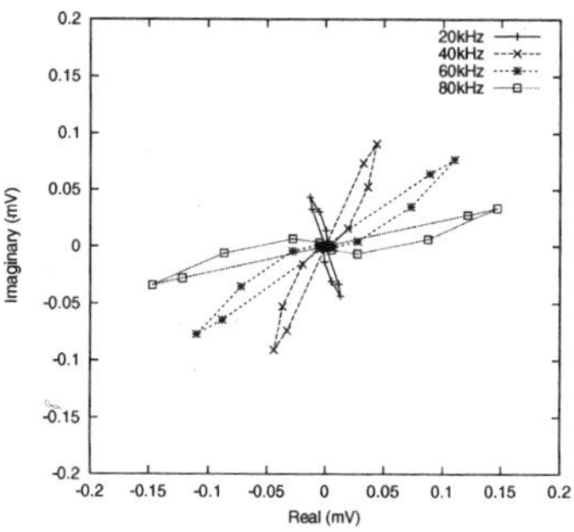

Fig.3 Numerical results for a circumferential OD 60% slit (0.2mm width, 5mm length)

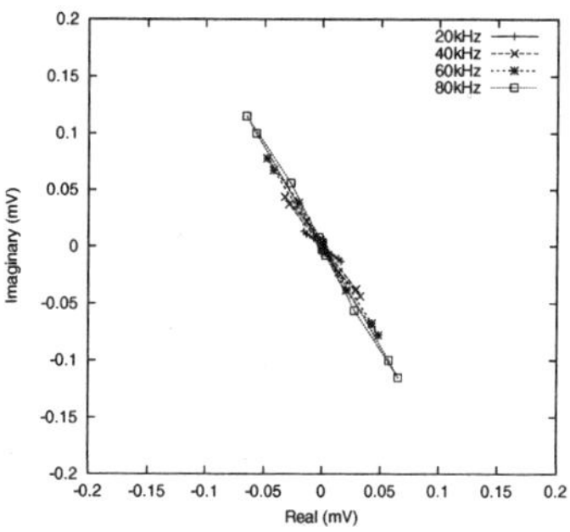

Fig.4 Numerical results for an inner corrosion region (2.5mm side length, 60μm depth)

3 Experimental details

Around 10 test-pieces of tube segment taken from an aged heat exchanger are fabricated. Artificial cracks of slit and fatigue crack are introduced to the test-pieces by the electric discharge machining (EDM) and a rotating fatigue testing. Fig.5 shows a picture of a testpiece with the EDM notches. In this tube segment, the part in the right hand has a heavy corrosion while the other part is not significantly affected. The EDM notches are fabricated in both the region with a large corrosion noise and the normal region. Both axial and circumferential notches are introduced but located 180 degree different in the circumferential direction. The depths of the notches are chosen as 30%, 40% or 50% of the plate thickness respectively.

To simulate the practical application, a SCC should be considered. However, fabrication of an artificial SCC testpiece for the Cu-Ni alloy is difficult as the material is very strong

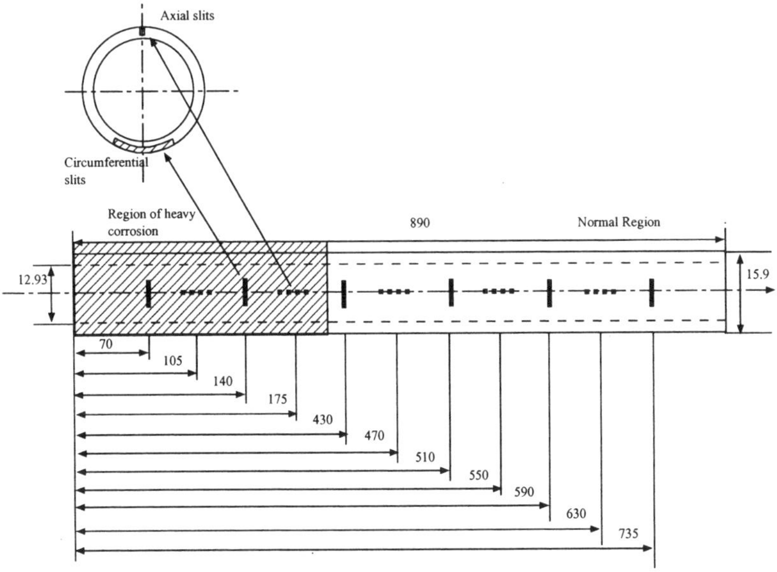

Fig.5 The design of the testpiece with EDM notches

against corrosion. Concerning this fact, circumferential fatigue crack testpieces are fabricated instead. A rotating bending fatigue test machine is employed for creating the fatigue cracks at the central region where a shallow EDM slit (less than 0.1mm in depth) is machined for guiding the crack initiation. Fatigue cracks around 30% depth are taken as the target depth which is controlled by selecting a proper load and loading cycles. Around 5 fatigue testpieces have been successfully made. Excepting one of them, the depth of the cracks are around 30% of the plate thickness.

A Cecco5 ECT probe, which has 16 channels in circumferential direction, is used to inspect the testpieces. The measured signals are transformed to digital values with an A/D converter and stored for data processing. Fig.6 shows a typical testing results of the 13rd channel (axial notches) for a frequency of 60kHz. A large corrosion noise can be observed in the right hand that is much bigger than the signals from an OD50% notch. Therefore the cracks in the corrosion region is impossible to be recognized without a proper data processing.

4 Method for Data Processing

To process the ECT signals obtained by the Cecco5 probe, a scheme including 1) automatic calibration, 2) reduction of the draft noise, 3) separation of crack information from the noise polluted signal, and 4) extraction of crack indications from the separated signals, is developed. For the automatic calibration, the amplitude and the phase of the signal are adjusted to make the output of each channel for a full length circumferential dent or glove to be the same. On the other hand, a polynomial filter [4] is applied to remove the draft noise. As the main work of this paper, algorithms for the signal separation and exaction are originally developed.

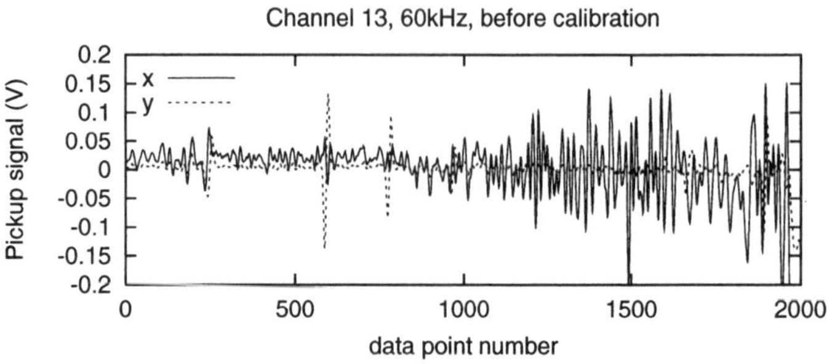

Fig.6 A typical signal with large corrosion noise for EDM notch testpiece (circumferential channel, 60kHz)

The details of the proposed algorithm are described in the following:

4.1 Phase optimization algorithm for crack signal separation

From the numerical results depicted in the previous section, it is observed that the phase property of the crack signal is very different from that of the corrosion noise. From the experiment, it is also found that though the corrosion noise has a similar space frequency with the crack signals, it also shows a very important feature different from the crack signal, say, mass occurrence. In a seriously corroded region, noise signals from a large number of corrosion sources may happen in a long region. However, unlike the signal of supporting plate of the SG inspection, the multi-frequency technique is not applicable for the present problem as the corrosion noise shows some differences for different corrosion zone while the supporting plate gives almost the same signal.

Base on this background, a novel algorithm is proposed in this paper. The basic idea is to find an optimal phase transformation angle that minimizes the y component of the signals, or in other word, transforms the corrosion signal to the x component. In a heavy corrosion region, the optimal transformation angle mainly depends on the corrosion noise because most of the signal indications are from corrosion (mass occurrence property). In this case, the crack signals will leave in the y component after the phase transformation as they have different phase property. On the other hand, concerning the different corrosion status along the axial direction, the phase optimization is necessary to be performed in a piecewise way, i.e. performing the phase optimization in signal segments separately.

However, in a signal segment without significant corrosion noise, the crack signal will also be minimized in the y component. To avoid this problem, the average transformation angle of each signal segment is introduced. If the optimized angle of a segment is quite different with this average value, it is possibly that only crack signals are included in it. In this case, instead of the optimal phase angle, the average one is employed to transform the signal that can avoid the disappearance of the crack information in the y component.

The details of the phase optimization algorithm are as follows:

1) After automatic calibration and reduction of draft noise, input the signal of a selected channel to the phase optimization filter.

2) Subdividing the signal to n signal segments and calculating the optimal phase transform angle for each segment by calculating and comparing $\sum_{i=1}^{m} y_i^2$ for different transform angles, or just using the following formula:

$$\alpha_i = -\frac{1}{2}tan^{-1}\frac{2\sum_{k=1}^{m_i}(x_k - \bar{x}_k)(y_k - \bar{y}_k)}{\sum_{k=1}^{m_i}(y_k - \bar{y}_k)^2 - (x_k - \bar{x}_k)^2}, i = 1, 2, ..., n \quad (1)$$

where m_i is the data point number in the $i - th$ signal segment, x_k, y_k are the $k - th$ data point of the x and y signal, and $\bar{x}_k$,$\bar{y}_k$ are their average values.

3) Calculating the average optimal phase angle by $\bar{\alpha} = \sum_{i=1} n\alpha_i/n$.

4) Taking phase transformation with angle α_i for $i - th$ data segment if $|\alpha_i - \bar{\alpha}| < \alpha_0$, and with angle $\bar{\alpha}$ else. Here α_0 is a threshold value selected on the basis of the average phase different of the crack signal and the corrosion noise. In case considering the inner surface corrosion and the OD crack, it can be selected between $\pi/12$ to $\pi/6$.

4.2 *Extraction of crack signal by using a simple similarity factor*

By using the phase optimization procedure, most corrosion noise can be excluded from the y component of the transformed signal. However, to extract the crack indication from the y signal directly, a special strategy is still necessary as a noise larger than the signal of an OD 30% slit is still possibly remained. A conventional extraction way using the amplitude may give too many mispredictions.

By processing the experimental data with the phase optimization filter, it is found that the noise signal included in the processed y component usually has a different waveform with the crack signals, while the signals due to cracks with different depth and direction do not show a significant difference (if the crack length is not too large). Based on this knowledge, an extraction strategy using a simple similarity factor [4],[5] is proposed as follows:

1) Preparing a standard crack signal $s(x_k), k = 1, 2, ..., m$ (m is the number of data points) by inspecting a washed EDM notch testpiece with the same testing conditions of the practical inspection.

2) Calculating simple similarities between the standard signal and the signal segments of the measured ones (each segment has the same number of data points with the standard one, and shifted point by point) by:

$$\rho(l) = \sum_{k=1}^{m} s(x_k)f(x_{k-l}) / \left(\sum_{k=1}^{m} s(x_k)^2 \sum_{k=1}^{m} f(x_{k-l})^2\right)^{1/2}, l = 1, 2, ..., n - m \quad (2)$$

3) Setting $\rho(l) = 0$ if $\rho(l)$ is less than a threshold value ρ_0 that is usual taken from 0.6 to 0.8 based on a preprocessing.

4) Subdividing the practical signal into n/m signal segments with data points m in each, and taking the maximum similarity $\rho_{max}^j = Max_{l \in R_j}\rho(l)$ in $j - th$ segment as the similarity of all the data point in signal segment R_j, ($j = 1, 2, ..., n/m$). However, if the amplitude of the signal in $j - th$ segment is too small (for instance, less than half) comparing with the signal due to an OD30% slit, the similarities of the data points in the segment R_j will be taken as zero (the detection limit is set as OD30% slit).

5) Setting $\bar{f}(x_k) = f(x_k)\rho_k, k = 1, 2, ..., n$ as the final extracting result.

5 Results of Data Processing

Fig.7 shows an ECT signal for the circumferential slits (13 channel) after automatic calibration and draft reduction. The frequency of the excitation current is 60kHz. From this signal, it is impossible to pickup the crack indications in the right hand of the testpiece where high corrosion noise appeared. After phase optimization, the y component of the signal is shown in Fig.8. One can recognize that there are some crack indications in the right hand of the testpiece.

Fig.9 shows the processing results of the crack indication extraction for the same signal, where a) is the standard signal, b) is the similarities bigger than the threshold value (0.6), c) is the extraction filter obtained and d) is the extracted crack indications. According to the crack information shown in Fig.5, the crack indications are correctly extracted.

Fig.10 shows the x component of ECT signals after phase optimization and the y component of the extracted crack indication for a testpiece with an outside fatigue crack (about 25% in depth and 10 mm in length). Though there is large corrosion noise in the signals, the crack signal can be recognized after phase optimization. By using the extraction filter further, correct crack indication is correctly extracted.

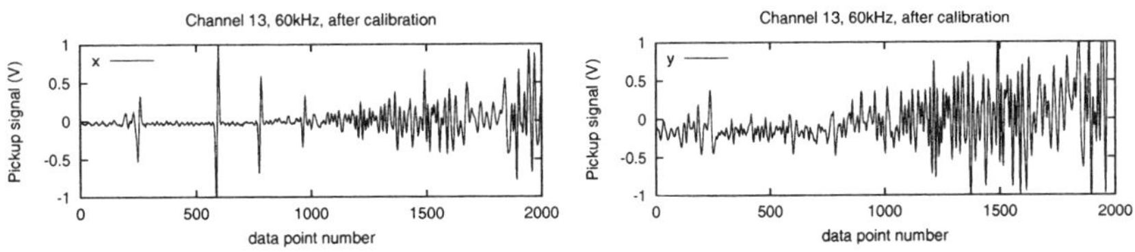

a) x component b) y component

Fig.7 Signal of circumferential slit after calibration (60kHz, channel 13)

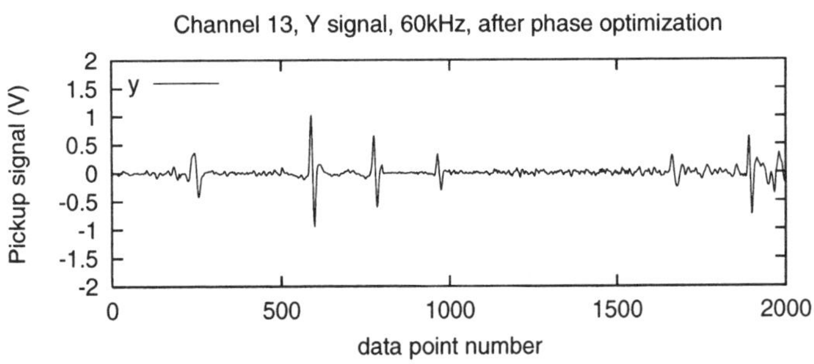

Fig.8 The y signal after phase optimization

6 Conclusions

In this paper, the data processing method is studied for the recognition of the OD cracks from a Ni corrosion noise polluted ECT signal based on numerical simulation results and

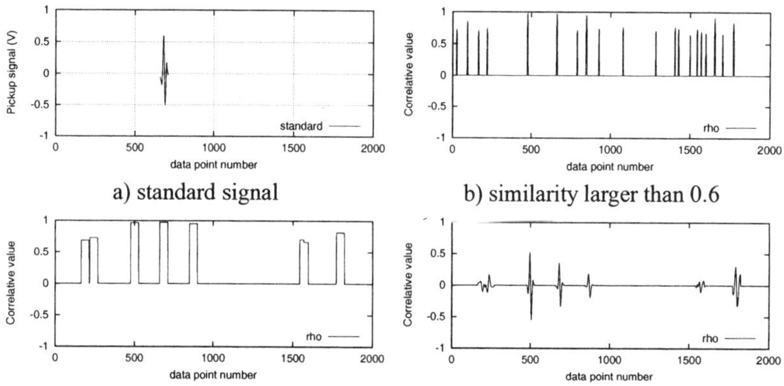

a) standard signal b) similarity larger than 0.6

c) the final extraction function d) the extracted crack signals

Fig.9 Results related to the crack signal extraction filter

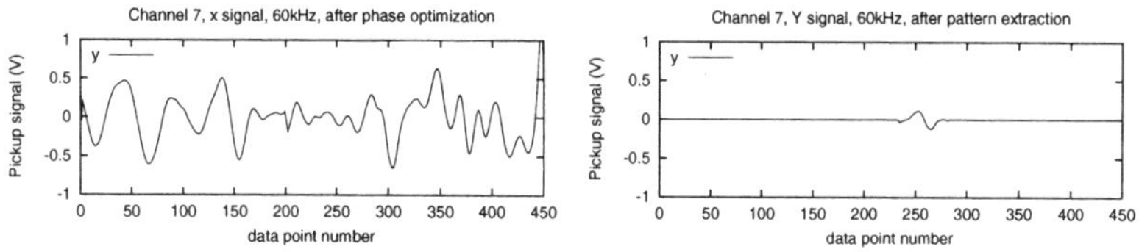

a) x signal after phase extraction b) extracted crack signal

Fig.10 Data processing results for the signal of a fatigue crack testpiece

ECT experiments with testpieces of artificial cracks. Algorithms based on phase optimization and pattern matching technique are proposed for crack signal separation and crack indication extraction. By processing the measured signal with a large corrosion noise, the efficiency of the proposed methods is verified.

References

[1] T.Takagi, and et al., ECT research activities in JSAEM - Benchmark models of eddy current testing for steam generator tube (Part 1), in Stud. Appl. Electromagn. and Mech., Vol.8, 253-264, 1995.

[2] H.Fukutomi and et al., Numerical evaluation of ECT impedance signal due to minute cracks, IEEE Trans.Mag.,33(2), 2123-2126, 1997.

[3] Kameari, Calculation of transient 3-d eddy current using edge element, IEEE Trans. Mag. Vol.26,No.2, 1990.

[4] S-F. Chuang, J.Moulder, J.Basart, Automatic flaw detection system for eddy current inspection of steam generator tubes, in Review. Progr. QNDE, Vol.16, 1997.

[5] P.Xiang and et al., Automated analysis of rotating probe multi-frequency eddy current data form steam generator tubes, Int. J. Appl. Electromagn. Mech., Vol.12, 151-164,2002.

Material Characterization

Electromagnetic Nondestructive Evaluation (VIII)
T. Sollier et al. (Eds.)
IOS Press, 2004

NDE of Pressure Vessel by the Magnetic Method

S. Takahashi, T. Nakano, K. Ara, Y. Kamada, H. Kikuchi, K. Yaegashi and M. Sato
Faculty of Engineering, NDE and Science Research Center, Iwate University, 020-8551 Morioka, Japan

Abstract. The magnetic structure sensitive properties are available parameters to detect lattice defects such as dislocations and grain boundaries in ferromagnetic materials. A comparison of magnetic method with Vickers hardness and Charpy impact test, which is currently employed for the secular degradation of pressure vessels, is investigated and evaluated quantitatively in the cold roll S15C steel. The transition temperature from ductility to brittleness by the Charpy impact test increases from 246 to 280 K by the roll reduction of 40%. The structure sensitive properties the coercive force, H_c and parameter, c, are obtained; the magnetic susceptibility χ_c can be described by a simple relation $\chi_c = c/H^3$ in a limited region of magnetic field H. The mechanical properties, D-B transition temperature and Vickers hardness, agree well with the magnetic structure sensitive properties in the cold rolled steel. We can say conclusively that the magnetic NDE method is an effective test for secular degradation before the crack initiation.

1. Introduction

Dislocations exert a great influence on the magnetic properties. The magnetism due to plastic deformation has been studied theoretically and experimentally since the discovery of roll induced magnetic anisotropy. The mechanism of magnetic change depends on the kind of materials. The basic study makes the application to nondestructive evaluation (NDE) possible; one can obtain the information of the dislocation density, ρ, and its distribution from the magnetic properties. The magnetic NDE method would bring about a great revolution in the NDE for metal fatigue, since the currently employed NDE for metal fatigue is the discovery of micro cracks. The micro cracks are produced by pile of dislocations, where ρ is a maximum locally. By the NDE of dislocations we can make an estimate of the time and place that the micro cracks appear.

One of the important and urgent areas for NDE is the nuclear power station, especially pressure vessels. Most pressure vessels are made of A533B steel which is ferromagnetic. The relationship between magnetism and dislocations was investigated by the Stuttgart group of Max-Planck Institute.[1-4] The strain field around dislocations exerts an influence on the hysteresis curve through the magneto-elastic coupling in ferromagnetic metals such as Ni, Fe, and Co metals. The structure sensitive properties change due to the dislocation density and its distribution. Recently a parameter, χ_c, was found by the present authors which has a simple relation to dislocations and grain boundaries. The magnetic susceptibility, χ_c, above the coercive field can be described by a simple relation to the magnetic field H.[5]

$$\chi_c = \frac{c}{H^3} \qquad (1)$$

The parameter c increases in proportion to ρ and it is sensitive to its distribution.[5] Actually the relation of eq. (1) scatters very widely in the accurate analysis of hysteresis curves.

Table 1 Composition of S15C steel

composition	C	Si	Mn	Fe
wt. %	0.15	0.25	0.45	bal.

Actually the secular degradation of pressure vessels has been examined by Charpy impact test. Charpy impact test is destructive and the test samples in pressure vessels are limited in number. The lifetime of pressure vessels has been prolonged from 40 to 60 years recently in Japan. A new

NDE method in place of Charpy impact test is required in the nuclear power station. If the magnetic structure sensitive properties would be available for NDE of the secular degradation, the magnetic method should be compared with Charpy impact test and the quantitative evaluation should be practiced. The purpose of the present study is the comparison of the magnetic method with Charpy impact test.

The coercive force, H_c, is a good parameter to estimate the value of ρ; H_c increases in proportion to the square root of ρ, that is experimentally confirmed in single crystals with tensile tests.[1,2] Coercive force as well as c-parameter can be obtained in the low field. The other purpose is the evaluation for c-parameter and coercive force from the viewpoint of NDE in pressure vessels

2. Experimental Procedure

We preparedS15C steel plate. The chemical composition of the steel is shown in Table 1. The steel plate was cold-rolled by 0, 5, 10, 30 and 40%. The samples were formed for Charpy impact test; 60 samples were prepared in each roll reduction. Other samples were prepared cutting into picture framed samples by spark cutting for the magnetic measurement.

Charpy impact test was carried out at the temperatures between 200 and 360 K. Absorption energy and the transition temperature from ductility to brittleness (D-B transition temperature) were measured by Charpy impact test. The deterioration by cold roll is estimated by Vickers hardness.

The magnetization curves are measured using a flux-meter at room temperature, the magnetizing and detecting coils on these samples being wounded with 80 and 100 turns, respectively. It was ascertained that the measured results were independent of the number of coil turns.

The broken-out sections by Charpy impact test were observed by a scanning electron microscope (SEM). Dislocations in each roll reduction were observed by a transmission electron microscope (TEM).

3. Experimental Results

The mechanical properties are compared with magnetic structure sensitive properties by use of cold rolled samples. Cold roll was adopted to change the mechanical properties, that introduces dislocations. We observed dislocations in the samples by TEM before the comparison of the mechanical and magnetic properties, since the deterioration is connected with dislocations. Charpy impact test and Vickers hardness are adopted as mechanical test to get the macroscopic information of deterioration in the samples. The magnetic measurement due to the hysteresis loop is also the macroscopic evaluation of deterioration.

Figure 1-a

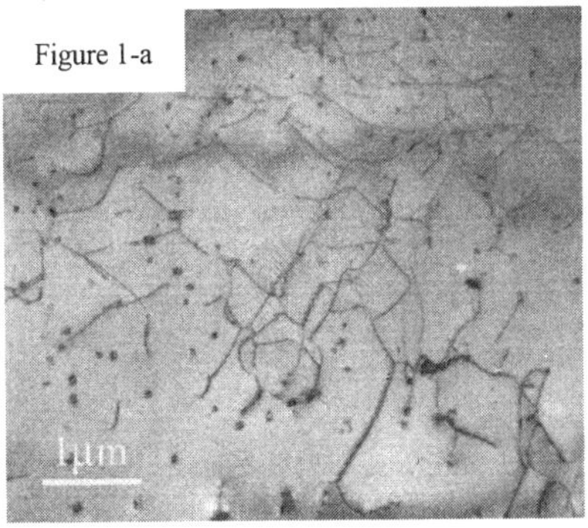

Figure 1-b

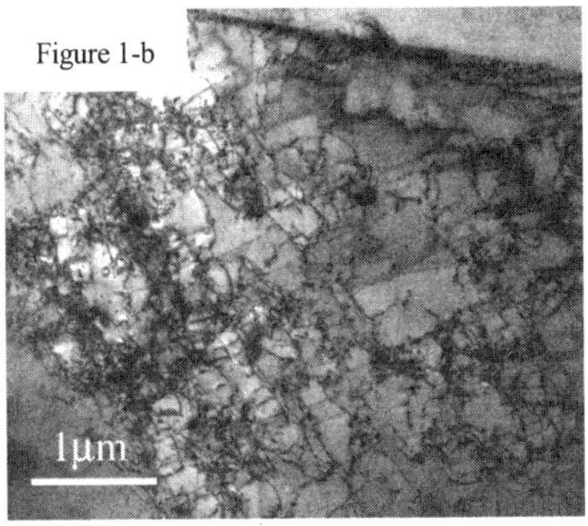

Figure 1-c

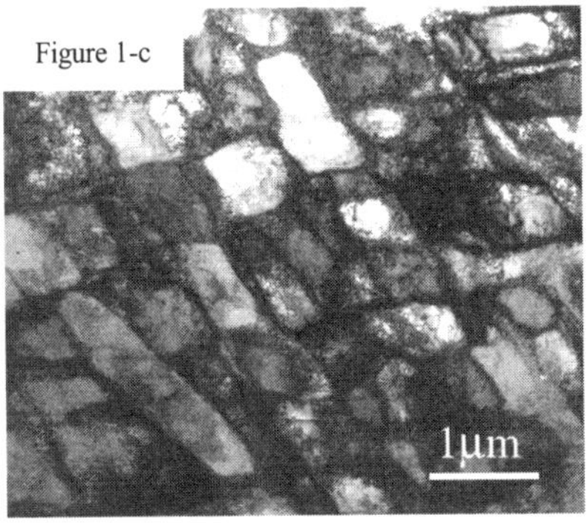

Figures 1-a, 1-b and 1-c Electron micrograph of S15C steel sample before plastic deformation (a), with cold-roll by 5 % (b) and 20 % (c).

3.1 Transmission Electron Microscopy Observation

Dislocations have been observed in many kinds of steels by many investigators. The present observation is similar to the previous ones. Figure 1-a is an electron micrograph of S15C steel sample before plastic deformation. S15C steel sample contains rather high density of dislocations initially, in which dislocations are homogeneously distributed. Figures1-b and 1-c are the electron micrographs of S15C steel samples with cold-roll by 5 and 20 % roll reduction. The dislocation density increases by cold roll and the increasing rate is largest by 5 % roll reduction. The cell structure is made in 20 % roll reduction.

3.2 Charpy Impact Test and Vickers Hardness

Figuere2 shows the relation between the absorption energy and the test temperature obtained by Charpy impact test in cold rolled S15C steel samples. Since Charpy impact test includes a large accidental error, an absorption energy is obtained by five samples; Charpy impact measurement was carried out by use of five samples in the same condition and the average value is adopted removing the highest and lowest values. The absorption energy curve shifts to high temperature as the increase of rolled reductions. The D-B transition temperature can be obtained by the absorption energy curve, and it becomes high with increasing roll reduction. The qualitative tendency comes true to our expectation.

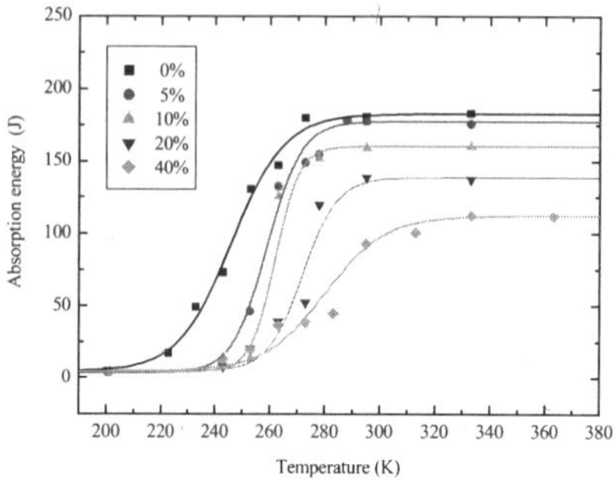

Figure 2 The absorption energy of Charpy impact test versus temperature in cold rolled S15C steel. The rolled reductions are 0, 5, 15, 30 and 40 %.

An absorption energy and /or D-B transition temperature give us the information of the broken mechanism in the impact test; the ductile destruction would be realized above the D-B transition temperature = 243 K or above 94 J in the absorption energy. The broken-out section of samples by Charpy impact test was observed by SEM. Figures 3-a and 3-b show the broken-out sections by Charpy impact test before cold roll at 333 and 243 K, respectively. Their absorbed energies are 183 and 94 J, respectively. These temperatures are above and near the D-B transition temperature as well as the absorption energies being above and equal the critical value. The SEM image indicates that the broken mechanism is the ductile. Figure

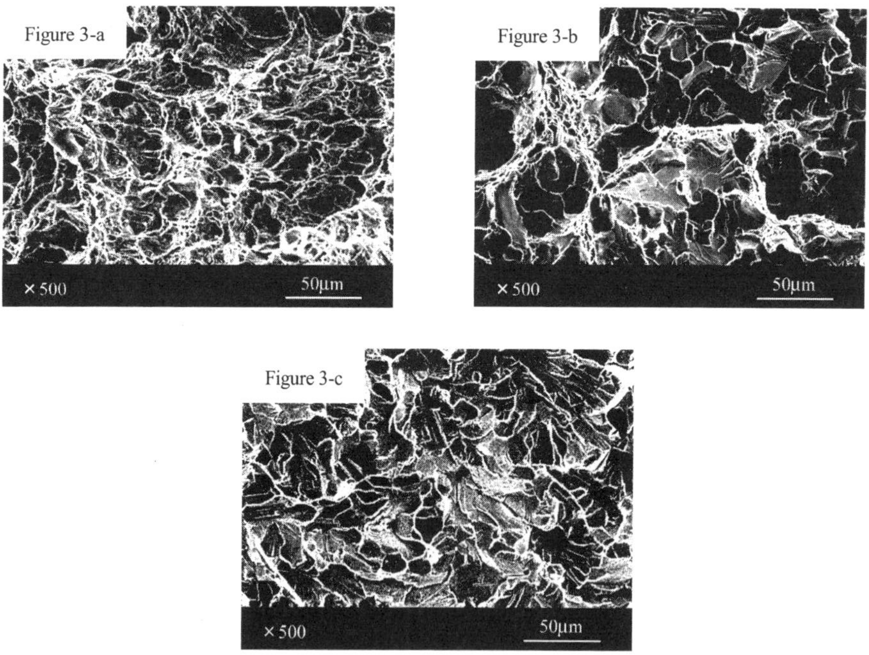

Figures 3-a, 3-b and 3-c The SEM image of the broken-out section by Charpy impact test of S15C steel sample before cold roll. The Charpy impact test is practiced t 333 K (a), at 243 K (b) and 201 K (c). The absorbed energies are 183 J (a), 94 J (b) and 4.5 J (c).

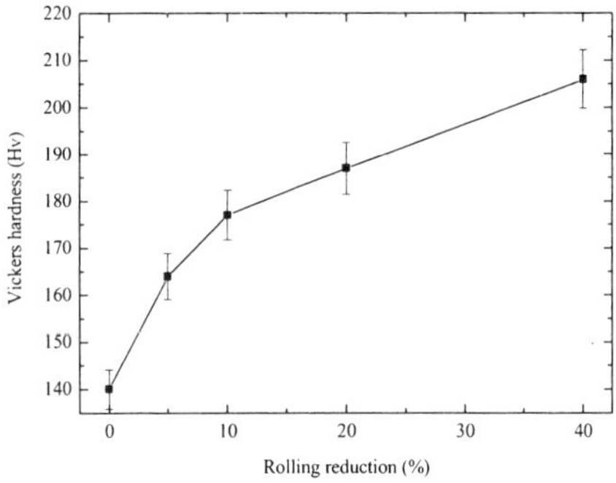

Figure 4 Dependence of Vickers hardness on the roll reduction in S15C steel samples.

3-c shows the broken-out sections by Charpy impact test of S15C steel sample before cold roll at 201.2 K. The dimple structure is observed more in Fig. 3-c than in Fig. 3a. The dimple structure in the SEM image indicates that the destruction mechanism is brittle. Their absorbed energies are 4.5 J. The dimples can be observed in the Charpy impact test at 243 K and there exist ductile and brittle structures in this surface. These SEM images agree well with the results of the absorption energy as well as the D-B transition temperature.

Figure 4 shows the dependence of Vickers hardness on the roll reduction in S15C steel samples. The Vickers hardness corresponds to the D-B transition temperature in cold rolled samples.

3.3 Hysteresis Loops

Hysteresis loops are measured in S15C steel samples with cold roll by 0, 5, 10, 30 and 40 %. A clear change is admitted in hysteresis loops. The values of H_c, and c-parameters are obtained by analyzing these hysteresis loops. Figuere5 shows H_c versus roll reduction in S15C steel samples compared with D-B transition temperature. H_c increases with the roll reduction. D-B transition temperature and H_c show a similar tendency as the parameter indicating degradation.

The c-parameter is obtained by the dependence of the magnetic susceptibility χ_c on the

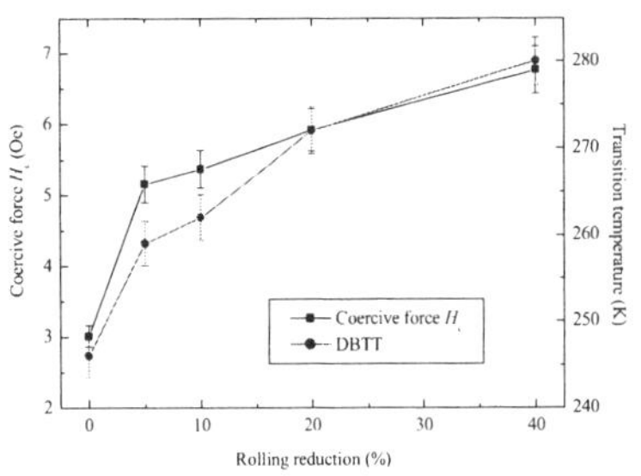

Figure 5 Coercive force versus roll reduction in S15C steel compared with the D-B transition temperature

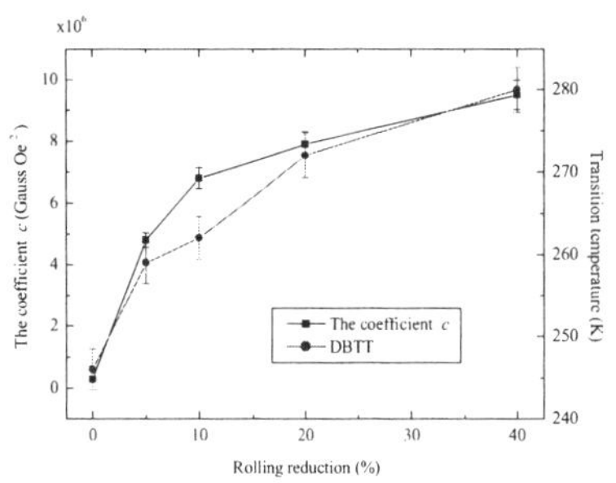

Figure 6 The coefficient of susceptibility, c, versus roll reduction in S15C steel compared with the D-B transition temperature

Table 2 The characters of Charpy impact test and magnetic measurements

Characteristic points	Charpy impact test	Magnetic measurement
Information	Direct	Indirect
Method	Destructive	Nondestructive
Accuracy	Good	Better
Sample number and size	Need many samples and definite size	One sample and indefinite size

magnetic field above H_c. The magnetic susceptibility χ_c and the magnetic field are plotted by the logarithmic scale and the slope of the straight line gives the value of c from Eq. (1). Figure 6 shows c-parameter versus roll reduction in S15C steel compared with the D-B transition temperature. The c-parameter increases with roll reduction and shows a similar tendency with H_c.

4. Discussion

Charpy impact test gives us the direct information of ductility and brittleness. The D-B transition temperature can be explained by the dislocation theory; the D-B transition temperature indicates the delicate affair of macroscopic deformation and the dislocation movement in the high strain rate test. The plastic deformation is supplied by the dislocation movement completely above the D-B transition temperature; the greater part of external energy transfers to the kinetic energy of dislocations in the ductile region. On the contrary, the plastic deformation cannot be supplied by the dislocation movement in the impact test below the temperature and only small part of external energy transfers to the kinetic energy of dislocations. D-B transition temperature depends on the number and size of obstacles for dislocation movements. There are several kinds of obstacles, i.e. dislocations, grain boundaries, copper precipitates and the other lattice defects. Dislocations play an important role of obstacles in themselves. The D-B transition temperature becomes high with the increase of roll reduction.

The magnetic structure sensitive properties depend on ρ and dislocation distribution. H_c, for an example, increases in proportion to the square root of ρ. [2, 3] As for c-parameter it increases in proportion to ρ. [5] We can give an explanation to the experimental fact that these magnetic properties increase with the increase of roll reduction.

Figures 5 and 6 show that magnetic structure sensitive properties and D-B transition temperature are correlative. The correlation of mechanical properties and magnetic structure sensitive properties is clear from dislocations behavior. D-B transition temperature, c-parameter and H_c increase with the increase of roll reduction and the increase rate of them is the largest in 5 % roll reduction. The increasing rate of ρ is also the largest by 5 % roll reduction, which is consistent with the increase of mechanical properties as well as magnetic properties by 5 % roll reduction. The mechanical properties correspond to the magnetic properties in the degradation process. A gap between these properties can be seen at 10 % roll reduction as shown in figures 5 and 6 and it would be caused from the measuring error of Charpy impact test, since the dependence of H_c, c-parameter and Vickers hardness on roll reduction has a similar attitude. The data of Charpy impact test deviate much and the mean value of five data is used in the present paper. Another cause of the gap is the different degree of plastic deformation in the samples. The samples were prepared from a sheet of cold rolled plate. The magnetic measurement would be carried out by the sample with higher ρ than the average value in a sheet. We can conclude that it is possible to estimate the D-B transition temperature by the hysteresis loops nondestructively. And two tests have their own characters as shown in Table 2.

D-B transition temperature is sensitive to ρ. When ρ is high and the mean free path of the dislocation movement is short, dislocations cannot absorb the external energy by Charpy impact test at low temperature. Magnetic structure sensitive properties are also sensitive to lattice defects such as dislocations. H_c corresponds to the highest peak of potential energy for the domain wall movement, which is caused from the internal stress due to dislocations through the magneto-elastic energy.[3] The c-parameter is caused by the deflection of magnetic moments around dislocations.[5] The correlation between D-B transition temperature and magnetic properties is reasonable from the dislocation theory.

Charpy impact test gives us an information on deterioration of materials directly but it is not NDT. Hysteresis loops contains a lot of information on deterioration too. It is possible to measure hysteresis loops nondestructively. The magnetic NDE method has its own merits compared with Charpy impact test. But the magnetic method has a demerit that its information is indirect. The demerit is the essential character in connection with NDE.

References

[1] A. Seeger, H. Kronmuller, H. Rieger and H. Trauble. J. Appl. Phys., 35, 740 (1964)
[2] H., Kronmuller, Canadian J. Phys. 45 631 (1967).
[3] H., Kronmuller, Inter. J. Nondestrctive Testing, 3 315 (1972).
[4] H. Trauble, Magnetism and Meatllurgy, Ed. by A. E. Berkowitz and E. Kneller, Chap. XIII, Acad. Press, 621 (1969).
[5] S. Takahashi, J. Echigoya, and Z. Motoki, J. Appl. Phys. 87, 805 (200).

Electromagnetic Nondestructive Evaluation (VIII)
T. Sollier et al. (Eds.)
IOS Press, 2004

A Probe Using a Magnetic Yoke for NDE of Ferromagnetic Steels

Hiroaki KIKUCHI, Katsuyuki ARA, *Noriya EBINE, **Yoshiaki SAKAI
Akinori TAKAHASHI, Yasuhiro KAMADA, Seiki TAKAHASHI
Faculty of Engineering, NDE & SRC, Iwate University, 020-8551 Morioka, Japan
**Reactor Safety Division, JAERI, 319-1195 Tokai-mura, Japan*
***JFE NKK Engineering Research Center, 514-0393 Tsu, Japan*

Abstract. A probe with a magnetic yoke for nondestructive evaluation of material fatigue has been developed. This probe aims to evaluate magnetic properties of ferromagnetic materials without machining them to a shape of ring core or a frame core. To confirm the feasibility of the probe, the magnetic properties measured by the probe are compared with those of ring sample for steels used in nuclear reactor pressure vessels and stainless steels fabricated at different tempered temperatures. The magnetic properties measured by the probe have a good correlation with those of the ring samples. Additionally, to assess fatigue of ferromagnetic materials, the magnetic properties of specimen applied tensile and cyclic loads have been measured using the probe. Those results show a dependence on the magnitude of stress and the number of cycles. This illustrates the potential of application of the proposed probe to NDE of fatigue in ferromagnetic materials.

1. Introduction

The main causes of degradation in aging structures, such as nuclear reactor pressure vessels (RPVs), elevated motorways and bridges, are based on metal fatigue in a wide sense. The metal fatigue is degradation of the mechanical properties resulting from lattice defects, such as dislocation. Recently the structures for which dozens of years have passed after construction are increasing. Thus, it is necessary to evaluate the extent of metal fatigue before crack initiation to prevent an accident resulting from degradation due to aging. Ultrasonic method [1] and eddy current testing [2] have been used as nondestructive evaluation methods in order to inspect degradation of these structures. However, their purpose is to detect generated micro cracks and those methods cannot assess the degradation before crack initiation. However, it is known that the magnetic properties of the ferromagnetic materials based on Fe used for the structures have a good correlation with the mechanical properties [3-6]. Hence, means to investigate the metal fatigue before crack initiation from changes of the magnetic properties has been proposed and fundamental research carried out previously [7].

The evaluation of the magnetic properties, such as coercive force and susceptibility, which are relative to mechanical properties has been performed until now by using a ring sample or a frame sample. The samples for magnetic properties measurement are usually cut out from the tensile deformed or the fatigued materials and then primary and secondary coils are wound on those samples in order to detect the magnetic changes. However, in machining of installed structure with metal fatigue is impossible. Therefore a technique to evaluate the magnetic properties in situ will be required strongly. Besides, the construction company will need to assess the mechanical properties nondestructively when accepting the materials used for construction. Thus, the method of inspecting the mechanical properties of the materials by using the changes of the magnetic properties is expected.

On the basis of such a background, in this paper, we propose a new probe, which has a winding coil for excitation, a pickup coil to detect magnetic flux and a magnetic yoke, without machining of structural materials, and examine how to evaluate the magnetic properties using the probe. Since it is not easy to identify the hysteresis curve of specimen by directly using the probe, we aims to clarify the correlation between the magnetic properties measured by the proposed probe and that of the ring sample. In addition, we try to assess the magnetic properties of specimen which were plastically deformed by tensile and fatigue testing.

2. Probe Structure

Fig. 1 shows the photographs of the proposed probe. The probe consists of a magnetic yoke, excitation and pickup coils. The magnetic yoke is made of 2%V-Permendur that has high saturation magnetic flux density of 2.2 T, low coercive force of 50 A/m and high maximum permeability of 15000. The size of the yoke part is given in Fig. 2. The magnetic flux pass length of the magnetic yoke part is 80 mm. The excitation coil, made of Cu-wire of 0.37 mm diameter, surrounds the magnetic yoke, and the number of turns is 1800. The pickup coil, made of Cu-wire of 0.1 mm diameter, is wound around the portion of the slot established in the center of magnetic yoke pole as is shown in Fig.1 (b). The number of turns of the pickup coil is 590, and the cross sectional area is 56 mm^2.

3. Experimental Procedure

3.1 Setup for measurements

The schematic representation of a measuring setup with the probe is illustrated in Fig. 3. The probe was set on a plate specimen of ferromagnetic material and the

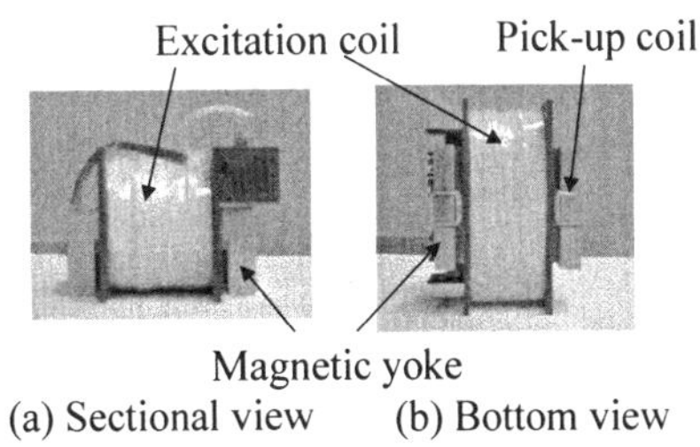

Fig. 1 Photographs of the probe.

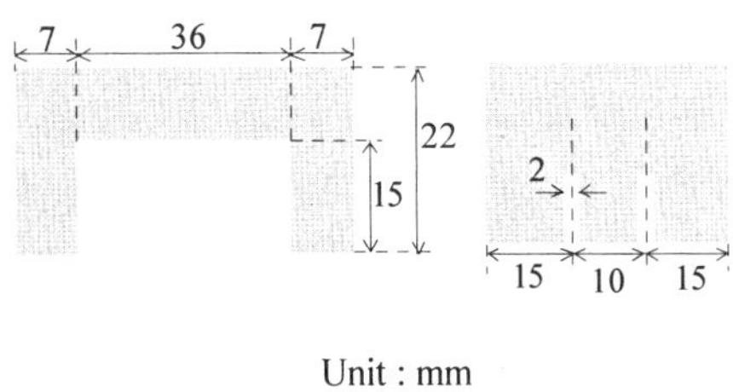

Fig. 2 Dimensions of the magnetic yoke.

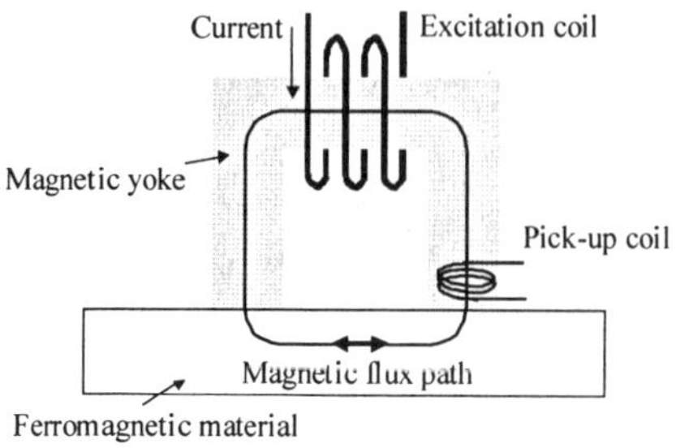

Fig. 3 Schematic representation of the measuring setup.

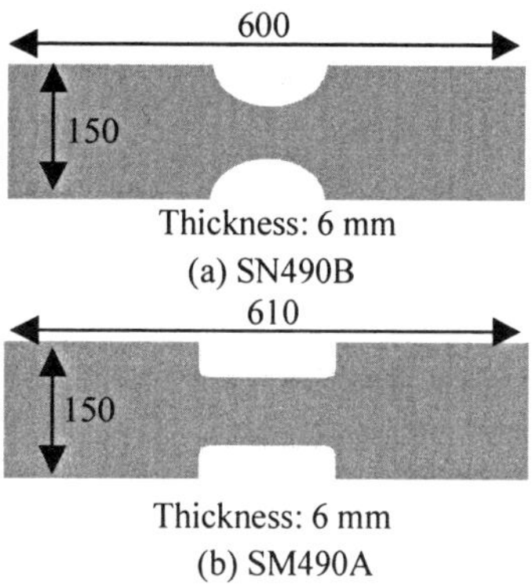

Fig. 4 Dimension of the specimen applied fatigue and tensile test. (unit : mm)

Table 1 Chemical compositions of A533B and SUS410. (wt%)

Material	C	Si	Mn	P	S	Cu	Ni	Cr	Mo
A533B	0.17	0.24	1.37	0.003	0.003	0.02	0.6	0.07	0.46
SUS410	0.06	0.34	0.41	0.029	0.002	-	-	12.66	-

Table 2 Heat treatment of A533B and SUS410.

Material	Quenching	Tempering
A533B	950 °C × 30 min. (watercooling)	620 °C × 1hr. (aircooling) 650 °C × 1hr. (aircooling) 680 °C × 1hr. (aircooling)
SU410	950 °C × 20 min. (watercooling)	600 °C × 1hr. (aircooling) 640 °C × 1hr. (aircooling) 680 °C × 1hr. (aircooling) 720 °C × 1hr. (aircooling)

closed magnetic circuit was arranged. By applying current to the excitation coil, a part of specimen was magnetized. The strength of excitation field, H, was calculated from the flux pass length of the magnetic circuit, the magnitude of the current and the number of turns of the excitation coil. The magnetic flux pass length of specimen was 50 mm. The pickup coil detected the magnetic flux concentrated in the magnetic circuit. The magnetic flux density, B, was calculated by integration of detection signal of the pickup coil. From these results, a B-H curve was obtained. The triangular wave current of 0.02 Hz was applied for the excitation.

3.2 Tempering specimen

Typical low-alloy steel for RPVs, A533B, and a martensitic stainless steel, SUS410, have been used for plate specimen. The chemical composition of each sample is provided in Table 1. Dimensions are 290 mm×290 mm×9 mm. In order to change magnetic properties of these specimen, tempering temperatures have been varied as 620, 650 and 680 °C in A533B steels, and 600, 640, 680 and 720 °C in SUS410 steels shown in Table 2. Each of their microstructures after heat treatment is of tempered martensite. The probe has been placed in the central part of specimen.

3.3 Specimen applied tensile and fatigue test

Fig. 4 shows the shape and size of specimen applied the tensile and the fatigue test. We have been using SN490B steels for the fatigue test, and SM490A steels for the tensile test. Their chemical composition is provided in Table 3. The yield stresses of

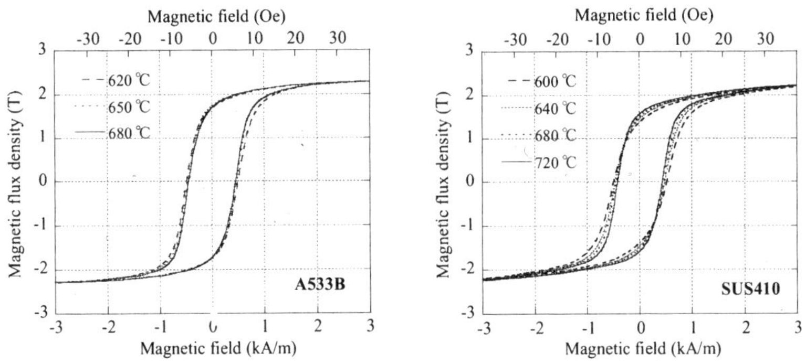

(a) A533B

(b) SUS410

Fig. 5 B-H curves measured by the probe (A533B and SUS410).

Table 3 Chemical composition of SN490B and SM490A. (wt%)

Material	C	Si	Mn	P	S
SN490B	0.09	0.24	1.45	0.013	0.003
SM490A	0.018	0.04	1.19	0.013	0.005

Table 4 Conditions of fatigue and tensile test.

Samp.	Cycles
1	0
2	5×10^4
3	7×10^4

Samp.	Stress (MPa)
1	0
2	299
3	457
4	518

SN490B and SM490A are 365 MPa and 457 MPa respectively and the tensile strength of those are 554 MPa and 580 MPa respectively. Table 4 shows the conditions for the fatigue and the tensile test. In the fatigue test the applied stress was 450 MPa. When carrying the measurements, the probe was placed in those region of specimen with plastic deformation, i.e. at the center part of gauge length, and no plactic deformation, i.e. near the edge.

4. Experimental Results and Discussion

4.1 Evaluation for magnetic properties of tempering specimen

Fig. 5 shows B-H curves of A533B and SUS410 which were measured using the probe. The shape of B-H curve differs with the tempering temperatures of each specimen. All the specimen show magnetic flux density of 2.2 T at the field strength of nearly 3 kA/m. This value is equal to the saturation magnetic flux density of yoke material, 2%V-Permendur, which means that yoke is saturated. Since the saturation magnetic flux density of A533B and SUS410, about 1.9 T, 1.7 T respectively, are smaller than the saturation magnetic flux density of 2%V-Permendur, it seems that magnetic field and magnetic flux are spread around the probe in the specimen than where the probe is actually placed. From this, it is difficult to calculate the magnetization curve of specimen directly. However, the magnetic properties of the yoke have not changed so that differences of measured B and H are the reflection of changes of the magnetic properties of specimen.

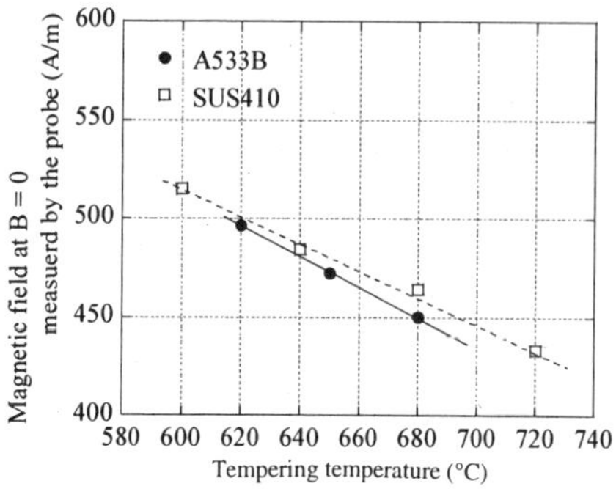

Fig. 6 Dependence of magnetic field strength at B = 0 on tempering temperature.

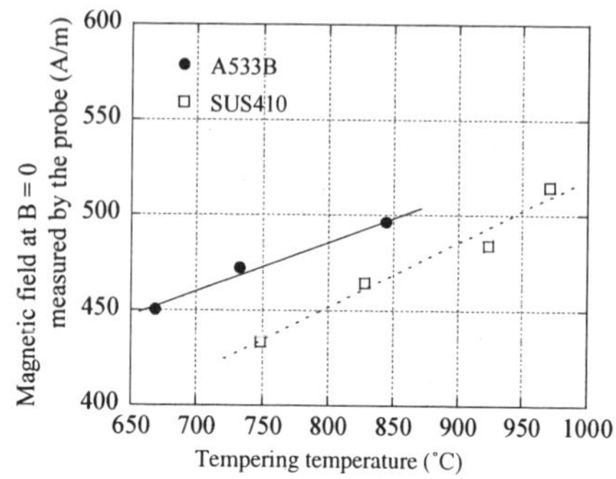

Fig. 8 Comparison between magnetic field at B = 0 measured by the probe and coercive force of the ring.

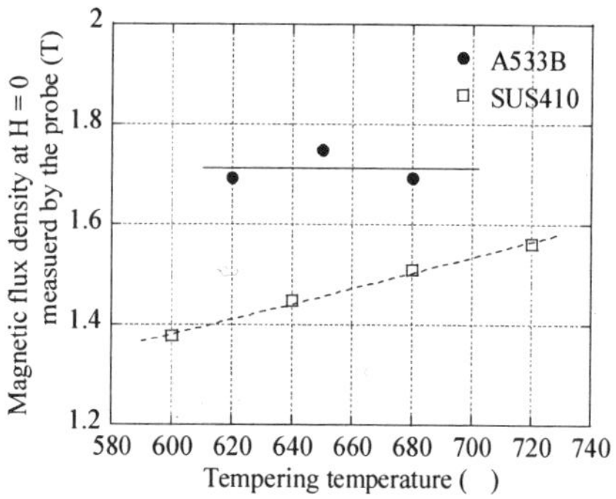

Fig. 7 Dependence of magnetic flux density at H = 0 on tempering temperature.

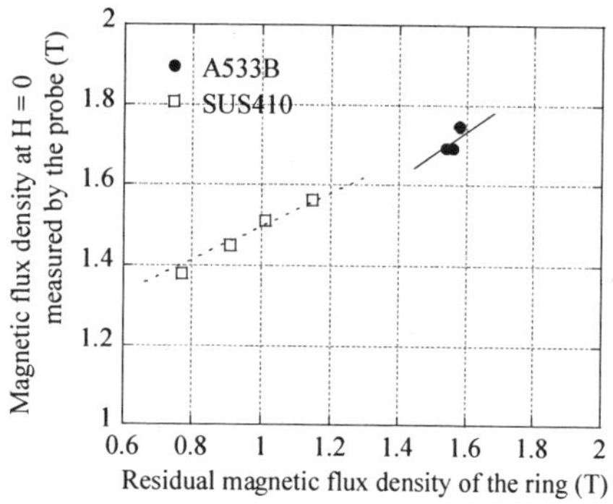

Fig. 9 Comparison between magnetic flux density at H = 0 measured by the probe and residual magnetic flux density of the ring.

The magnetic field strength, H_{cp}, where the magnetic flux density becomes zero, is plotted against the tempering temperatures in Fig. 6. The value of H_{cp} was obtained from B-H curves shown in Fig. 5. The value of H_{cp} decreases with increasing tempering temperature in both A533B and SUS410. The magnetic flux density, B_{rp}, where the magnetic field becomes zero, is plotted against the tempering temperatures in Fig. 7. The value of B_{rp} was obtained from the B-H curve shown in Fig. 5. The value of B_{rp} increases with increasing tempering temperature in SUS410 and the one of B_{rp} was constant with increasing tempering temperature in A533B.

Fig. 8 shows the comparison between the magnetic field strength, H_{cp}, measured by the probe and the coercive force, H_c, measured from the ring samples using B-H tracer. The data of the ring samples are taken from reference [8]. As shown in Fig. 8, for samples A533B and SUS410 the value of H_{cp} is proportional to the value of coercive force by the ring samples. Fig. 9 shows the comparison between the magnetic flux density, B_{rp}, measured by the probe and residual magnetic flux density, B_r, measured from the ring samples. As shown in Fig. 9, for samples A533B and SUS410 the value of B_{rp} is proportional to the value of residual magnetic flux density.

As a result, magnetic properties measured by the probe can be said to have a

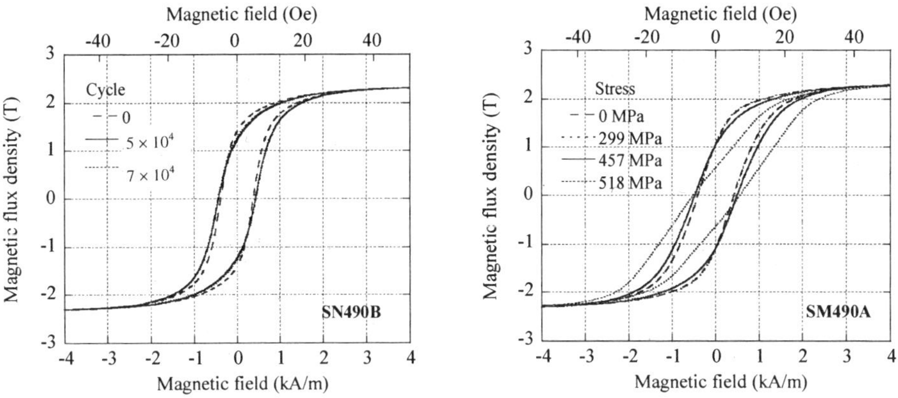

(a) SN490B (b) SM490A

Fig. 10 B-H curves of SN490B and SM490A

good correlation with those of the ring samples. This tells us that the evaluation of magnetic properties by the proposed probe is useful.

4.2 Trial for specimen applied tensile and cyclic test

Fig. 10 shows the B-H curves, measured using the probe, of SN490B subjected to the fatigue tests and those of SM490A subjected to the tensile tests. In both tests, the B-H curves measured at the central part of specimen were shown. As compared with the B-H curve in 0 cycle, the B-H curves in 5×10^4 or 7×10^4 cycles swell and their inclination also changes with a slight decrease. On the other hand, in the tensile test, although any difference is not observed in the B-H curves of 0, 299MPa, that of 457MPa beyond its yield stress swells and inclination also becomes small. In 518MPa, that tendency becomes more remarkable. The B-H curves measured at the edge part of specimen do not change in any conditions.

Fig. 11 shows the value of ΔH_{cp}, which is the difference of H_{cp} measured at the central part and that measured at the edge part; ΔH_{cp} is plotted against the number of cycles in the fatigue test. The value of ΔH_{cp} increases when applying fatigue test, and the increase in ΔH_{cp} may take place in the early stage of fatigue, and then it becomes constant after that. Fig. 12 shows the value of ΔH_{cp} plotted against the applied stress in the tensile test. When the applied stress is less than the yield stress, the value of ΔH_{cp} does not change, but when the value of stress is above the yield stress, ΔH_{cp} increases rapidly.

The changes in the value of ΔH_{cp} as shown in Figs. 11 and 12 are agree well with the results obtained from ring samples after tensile and fatigue test [7,9]. This shows the potential of nondestructive tests with this probe to evaluate the extent of plastic deformation of ferromagnetic materials.

5. Conclusion

We proposed a probe, with magnetic yoke, excitation coil and pickup coil, for

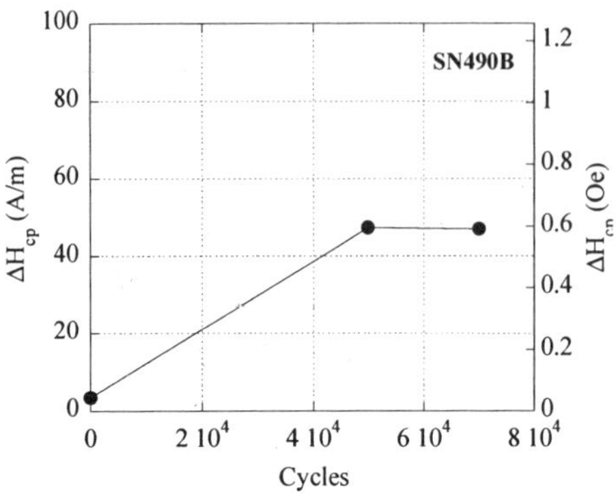

Fig. 11 Dependence of ΔH_{cp} on the number of cycle.

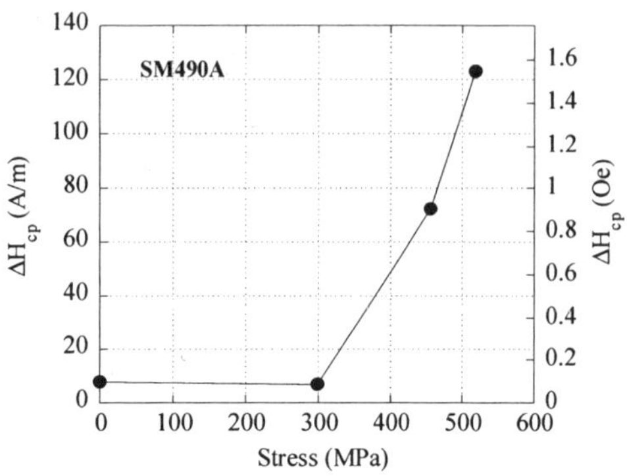

Fig. 12 Dependence of ΔH_{cp} on the stress.

NDE without machining materials. The magnetic properties of A533B and SUS410 measured using the probe have been compared to those of the same materials into the ring shape and is good correlation. This therefore shows the potential of the evaluation for magnetic properties using this probe. Additionally, magnetic properties of specimen practically deformed by tensile testing and cyclic loads were measured using the probe. Those results show dependence on the magnitude of stress and the number of cycles. Hence this probe can be applied to NDE of fatigue in ferromagnetic materials.

References

[1] The Japanese Society for Non-destructive inspection, A New Handbook of Non-destructive Evaluation, 2.6 Ultrasonic inspection test (1992).

[2] H. L. Libby : Introduction to Electromagnetic Nondestructive Test Methods, Wiley (1971).

[3] H. Kronmüller, Magnetic Techniques for the Study of Dislocations in Ferromagnetic materials, Int. J. Nondestruct. Testing, **3** 315-350 (1972).

[4] K. Propp, M. Blochwitz, G. Brauer, E. Hampe, W.D. Leonhardt, and H. P. Schützer, Äuderung meechanischer und physikalischer Eigenschaften von Reaktordruckbehälterstahl bei Neutronbestrahlung und thermischer Ausheilung, Kernenergie, **29**, 22 (1986).

[5] E. H. El-Kalla, F. A. Rahnman, and A. A. Maaty, The effect of neutron irradiation on the magnetic properties of low carbon steels, Applied Radiation and Isotopes, **40**, 159-162 (1989).

[6] J. F. Stubbins, J. G. Williams, and A.M. Ougouag, Measurement of Irradiation Damage in Nuclear Pressure Vessel Steels by Magnetic Properties Change, Proc. Fifth Int. Symp. Environmental Degradation of Materials in Nuclear Power Systems-Water Reactors, 719-727 (American Nuclear Society, 1991).

[7] S. Takahashi, J. Echigoya, Z. Motoki, Magnetization curves of plastically deformed Fe metals and alloys, J. Appl. Phys., **87** 805-813(2000).

[8] N. Ebine, K. Tachida, Gilanyi Attila, K. Ara, M. Uesaka, K. Miya, Correlation between the Magnetic and Mechanical Properties of Ferromagnetic Structual Steels, A533B and SUS410, J. Magn. Soc. Jpn., **20** 665-668(1996).

[9] K. Yaegashi, Y. Kamada, J. Echigoya, K. Ara, Y. Sakai and S. Takahashi, Nondestructive Evaluation of Deformation Substructure in Fatigue SM490YA Steel Using Magnetic Measurement, Tetsu-to-Hagane, **89** 871-876 (2003).

Electromagnetic Nondestructive Evaluation (VIII)
T. Sollier et al. (Eds.)
IOS Press, 2004

Evaluation of Case Hardening Depth by the MFES Method

Yuichiro KAI[1], Yuji TSUCHIDA[1], Masato ENOKIZONO[1], and Shoichiro NAGATA[2]
[1]*Oita University, 700 Dannoharu, Oita 870-1192, Japan*
[2]*Miyazaki University, 1-1 Gakuenkibanadai-nishi, Miyazaki 889-2192, Japan*

Abstract. The hardness of steel can be increased using hardening methods. The evaluation of the depth that is hardened by case hardening is necessary. In this paper, we employ the multi-frequency excitation spectrogram method for estimating the case hardening depth.

1. Introduction

It is well known that Japanese steel swords have been made very hard by repeated heating and quick cooling. Hardening methods are used extensively in many industrial fields. Steel after smelting cannot be used for fabricating components due to residual strain and stress. The residual stain and stress are reduced using annealing methods and the surface of the steel is often hardened using appropriate hardening methods. Heat treatment methods such as annealing and hardening are very important for industrial use of steel. Hardening methods can increase the hardness of steel at the risk of increasing its brittleness. Therefore, the measurement and the evaluation of the hardness of steel are very important.

Mechanical hardness tests like the Vickers hardness test and the Rockwell hardness test are used to estimate hardening depth. The hardening depth is estimated on the basis of mechanical hardness measured by these tests. All of the mechanical hardness tests determine the hardness of the sample by making small holes in the measurement area on the surface of the sample thereby destroying the specimen. Therefore, reference samples are required to estimate the depth of the hardening. The estimation of the hardening depth using such mechanical methods is not convenient in a process setting due to the destructive nature of the test. We propose to evaluate hardening using electromagnetic methods instead of mechanical tests [1]. Electromagnetic methods are nondestructive and fast in comparison with mechanical tests. In this paper, we use eddy current testing methods to estimate the hardening depth. Eddy current testing methods have been used for the detection of surface breaking cracks and defects. However it has been shown that it is difficult to measure material changes inside the sample. We propose to employ the multi-frequency excitation spectrogram (MFES) method to estimate the hardening depth [2]. This method is effective for detecting material changes inside the sample since it uses a wide band of frequencies from low to high frequencies. This paper presents results obtained with the MFES method for measuring case hardness.

2. Principle of measurement

In general, eddy currents are induced in response to a time-varying magnetic field in conductive materials in accordance with the Faraday's law. In this case, the eddy current flows primarily on the surface at high frequencies. This is called the skin effect. On the other hand, the eddy current penetration increases when the excitation frequency is decreased, however the signal measured by the sensor becomes very small. Consequently, the detection of material changes in the interior of the sample is difficult in comparison with the detection of the cracks and the material changes on the surface when the eddy current techniques are employed. It is evident that it is necessary to choose an optimal excitation frequency in eddy current testing.

The relationship between the penetration depth δ and the frequency can be written as,

$$\delta = \frac{1}{\sqrt{\pi f \mu \sigma}} , \qquad (1)$$

where, f is the frequency, μ the permeability and σ the conductivity, respectively. The permeability and the conductivity of the case hardened region are different from those of the material at the recesses of the specimen. Conventional eddy current testing methods are unsuitable for evaluating case hardening since the material change due to the case hardening affects the inside of the specimen. Such methods are not effective since they normally use only a single high frequency excitation signal. In this paper, we use a multi-frequency method to detect changes in materials properties caused by hardening. The MFES method is effective because it uses several frequency components at once.

3. Case Hardening Depth

As mentioned above, the hardness of carbon steel increases with case hardening. This is due to change in the crystal structure that is referred to as martensite phase. The magnetic properties also change with the change in the crystal structure. Fig. 1 shows typical magnetization characteristics before and after case hardening. Fig. 2 shows the relationship between the Vicker's hardness and the case hardening depth. Case hardening causes the surface to harden. The hardness decreases with depth inside the specimen. The hardness can be detected nondestructively using electromagnetic sensors since the magnetic properties are also affected.

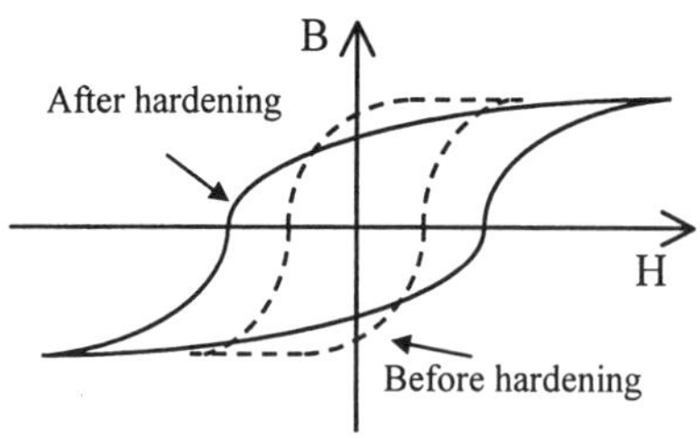

Fig.1 Hysteresis loop

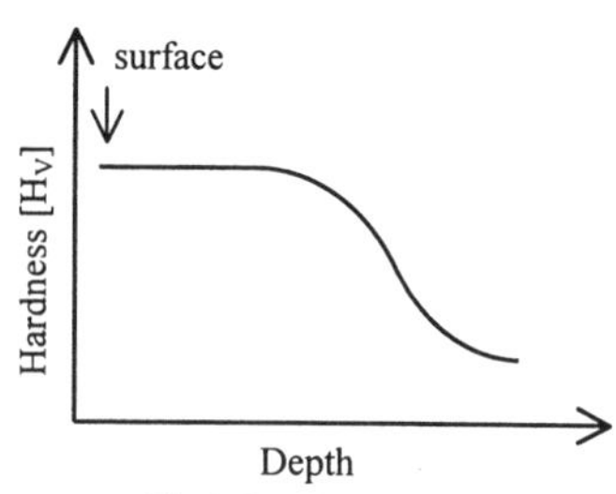

Fig.2 Case hardened depth

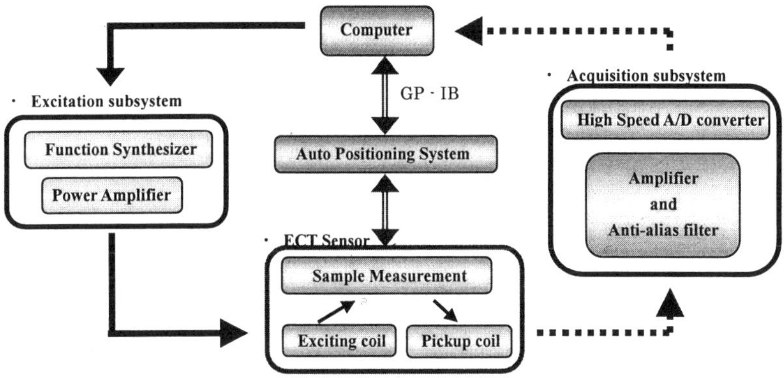

Fig.3 Measurement system

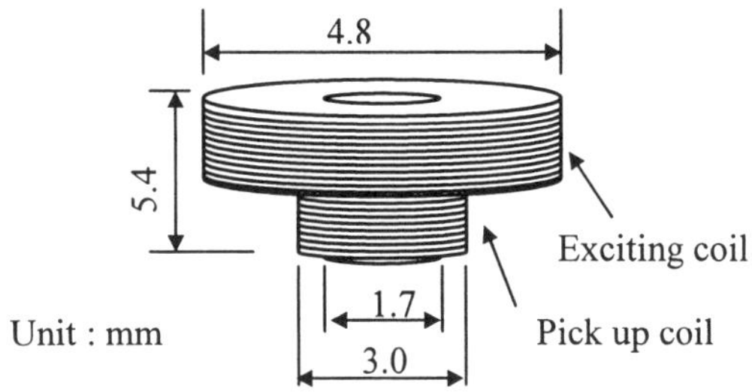

Fig.4 Exciting coil and Pick up coil

4. Measurement System and Sample of Measurement

Fig. 3 shows a block diagram of the measurement system. We begin the measurement process by entering the initial parameters describing the multi-frequency excitation waveform into the personal computer (PC). The multi-frequency excitation current generated by the function synthesizer is applied to the exciting coil through the power amplifier. Signals from the pickup coil are transferred to the PC after digitization by the high speed and high resolution A/D converter. The spectrogram of the signals is computed by the PC and displayed. The horizontal axis in the spectrogram shows the position of the sensor and the vertical axis shows the frequency. The auto positioning system enables the sensor to move on the sample with high accuracy. Fig.4 shows the pancake-type sensor employed to test the specimen.

Fig.5 (a) shows a case hardened test specimen whose shape is cylindrical (100 mm diameter and 200 mm long). The surfaces of the samples are hardened using induction heating methods. The sensor is scanned in the direction shown in Fig. 5(a). The measurements were made within a 100 mm window straddling the center as shown in Fig. 5(b). The signals are measured at 200 points every 0.5 mm in this region. Table 1 shows the hardening depths of the samples, which were measured using the Vickers test. The critical Vickers hardness was selected as 400 Hv. We applied the MFES method to those samples in order to estimate case hardening depth without destroying the specimen in contrast to the Vicker's test.

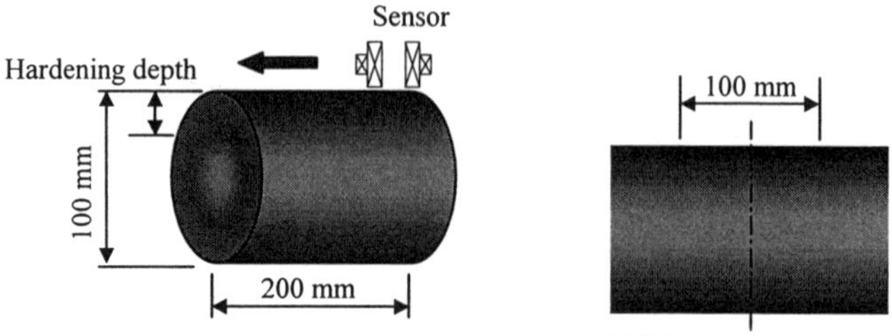

(a) Moving direction (b) Measurement range

Fig.5 Structure of sensor

Table 1 Case hardening depth

Hardening depth [mm]			
0.0	2.5	3.5	4.5

Table 2 Measurement condition.

Exciting current [mA]	130
Exciting frequency [Hz]	100 150 200 250 300 400 450 500 550 600 650 700 750 800 850 900 950 1000

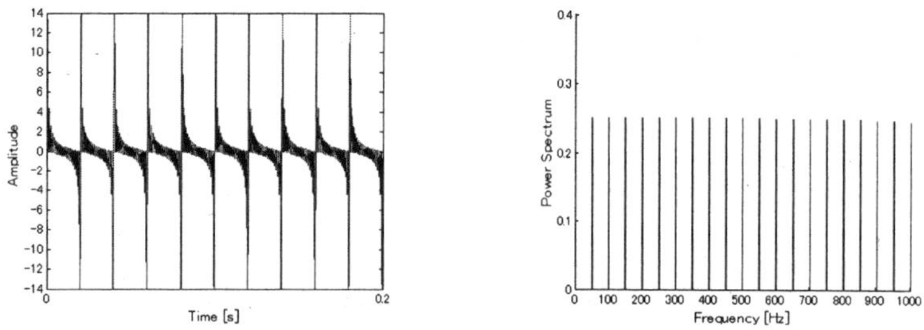

(a) wave form of input voltage (b) power spectrum of input voltage

Fig.6 Wave form and power spectrum of input voltage.

5. Measurement and Result

The frequency and the excitation current were set at the same value for every measurement. The measurements are performed along four arbitrary lines in one sample to determine the differences in signals in the same sample. Table 2 shows the excitation parameters. The lower frequency components are selected to ensure that the magnetic flux penetrates deeply. The high frequency components offer superior signal-to-noise ratio. The choice of excitation frequency shown in Table 2 are carefully selected after many attempts. The input voltage and its spectrum are shown in Figs. 6 (a) and 6 (b), respectively.

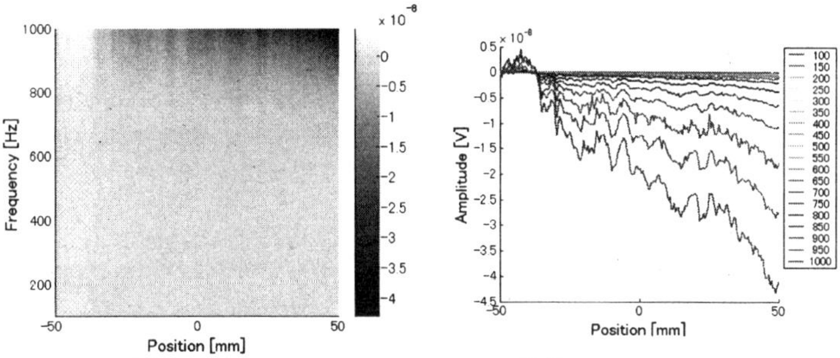

(a) Spectrogram (b) Frequency components

Fig.7 Spectrogram and frequency components (hardening depth : 0.0 mm)

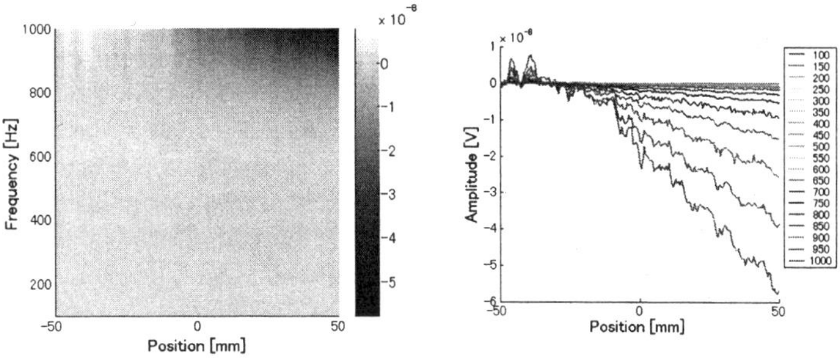

(a) Spectrogram (b) Frequency components

Fig.8 Spectrogram and frequency components (hardening depth : 2.5 mm)

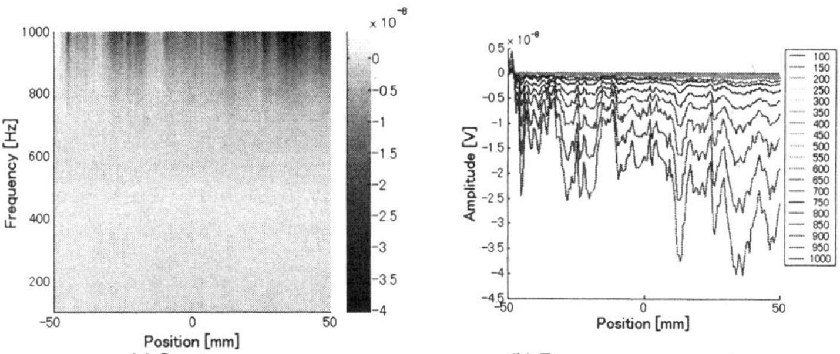

(a) Spectrogram (b) Frequency components

Fig.9 Spectrogram and frequency components (hardening depth : 3.5 mm)

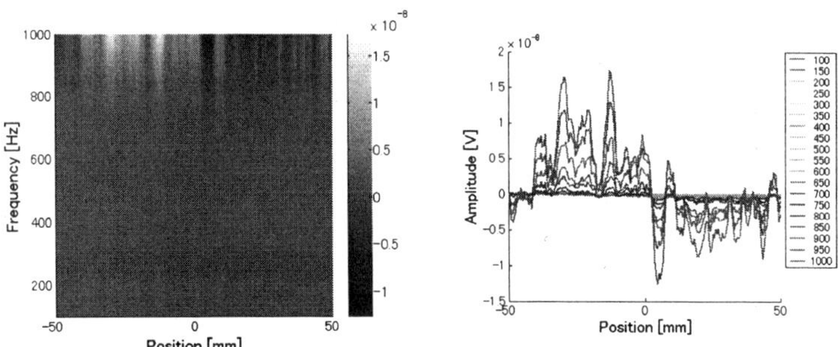

(a) Spectrogram (b) Frequency components

Fig.10 Spectrogram and frequency components (hardening depth : 4.5 mm)

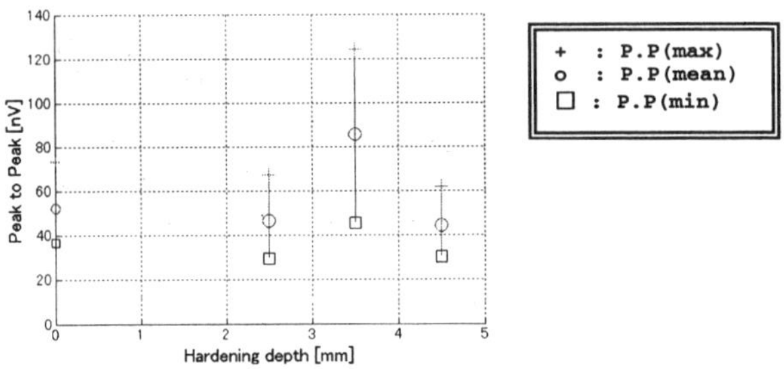

Fig.11 Peak to Peak upon the hardening depth

The spectrograms and each frequency component of the spectrogram are shown in Figs.7, 8, 9, and 10, respectively. As shown in these figures, the signals are different depending on the hardening depth. When the hardening depth is 0.0 mm, the material properties are same from the surface to the inside. Therefore, the signal decreases gradually. Here, the signals starts from the 0 volt for all frequency components because the signals at the starting point are assumed to be 0 level and all signals are referenced with respect to the starting signals. In comparison with these signals, the signals fluctuate considerably after hardening, especially in the case of the 4.5 mm hardening depth. The signals in the cases of the 0 mm and 2.5 mm vary gradually. The surfaces of these samples appear similarly, so the differences of the signals shown in these figures can be described the changes of the material properties due to hardening.

Fig. 11 shows the peak-to-peak voltages of the spectrograms depending on the hardening depth. The peak-to-peak voltages are different in the four arbitrary lines, then the maximum, the minimum and mean values of the four lines are plotted in this figure. From this figure, the peak-to-peak voltages are different depending on the hardening depth. However it is difficult to distinguish the hardening depth with this figure. The value of the 3.5 mm depth is bigger than the others, an issue that needs to be examined.

The detection of the material property's changes after case hardening is very difficult. The MFES method has shown that it can offer some information about the hardening depth.

6. Conclusions

In this paper, the evaluation of the case hardening depth using the multi-frequency excitation spectrogram (MFES) method is presented. The signals of the spectrograms were different depending on the hardening depth. To carry out practical evaluation using the method, the quantitative examination from the spectrogram is necessary. Future work will be focused on establishing a relationship between hardening and the spectrogram signals obtained using the MFES method.

References

[1] B. Zhu, M. J. Johnson and D. C. Jiles, Evaluation of Wear-Induced Material Loss in Case- Hardened Steel Using Magnetic Barkhasen Emission Measurement, *IEEE Trans. on Magn.*,**36** (2000) 3602-3604

[2] T.Chady, M.Enokizono, R. Sikora, T.Todaka and Y. Tsuchida, Evaluation of the New ECT Sensor Using JASEM Round Robin Test Samples in Presence of Fastener Interferences, *Studies in Applied Electromagnetics and Mechanics*, **23** (2002) 166-173

Material Evaluation of Cast Iron Using the Eddy Current Method

Tetsuya UCHIMOTO, Toshiyuki TAKAGI, Makoto KUROSAWA
and Sergiy KONOPLYUK
Institute of Fluid Science, Tohoku University, Katahira 2-1-1, Aoba, Sendai 980-8577, Japan

Toshihiko ABE
Tohoku Center, National Institute of Advanced Industrial Science and Technology, 4-2-1 Nigatake, Miyagino, Sendai 983-8551, Japan

Abstract. This paper reports non-destructive evaluation of mechanical characteristics of cast irons based on eddy current measurement. There are two factors affecting mechanical characteristics of cast irons: graphite shape and matrices. Ultrasonic testing (UT) can evaluate characteristics related to graphite shapes. However, UT can not evaluate characteristics related to matrices. In this paper, matrix structure and hardness of ductile cast irons were estimated with use of eddy current evaluation. The authors confirmed adequate correlation between hardness and eddy current signals and showed that the present method can estimate hardness of ductile cast irons with sufficient accuracy.

1. Introduction

Mechanical properties of cast irons are mainly determined by two factors, morphology of graphite in cast irons and type of matrices. Ultrasonic techniques are most sensitive to graphite morphology and used for evaluation of mechanical properties in gray cast iron and mechanical properties associated with nodularity of graphite in ductile cast iron[1, 2]. However, in ductile cast iron mechanical properties also depend on properties of matrices which can not be assessed by ultrasonic techniques. There are some methods to evaluate matrices through harmonic analysis of eddy current signals[3]. It focuses on changes in shapes of hysteresis loop of cast irons, hence it requires complicated signal processing. In this paper, the authors consider reversible region of initial magnetization curve, and smaller excitation than the previous work is applied, which means eddy current signals reflect only two parameters of conductivity and initial permeability. The advantage of the present method includes direct coupling with physical parameters and simple data acquisition, which accommodates its practical application [4]. This study shows that eddy current (EC) evaluation can be an effective method to assess the properties of matrices. It is expected that complete evaluation of mechanical properties of cast irons will be realized through combination of the present methods for matrices and ultrasonic techniques for graphite morphology.

2. Eddy Current Measurement

EC evaluation of cast irons was carried out with experimental setup consisting of a tandem type probe and an EC instrument. The probe consists of identical two coils arranged in tandem. The each coil has outer and inner diameters of 14 and 8 mm, respectively, height of 4 mm and 200 turns. Upper coil is separated by a distance of 6 mm from the lower coil placed above specimens, and it is not affected by the specimens. The two coils are connected inversely; the output voltage is difference between ones of two coils.

In this study, two series of specimens were prepared. One consists of step block specimens and we call it Series A hereafter. 19 step blocks of different chemical composition, heat treatment were prepared with shell mould castings and a high frequency induction furnace, as shown in Figure 1. Both width and length of each step are 100 mm. Thickness of 5 steps are 5, 10, 20, 30 and 50 mm, respectively. Difference of thickness gives change in cooling rate during casting, which leads different structures and graphite sizes. Preset parameters of 19 specimens are listed in Table 1.

The other series consists of 58 specimens for product inspection of nodular cast irons, and we called it Series B hereafter. Brinell hardness and ultrasonic velocity of all specimens were measured in advance. Brinell hardness of the specimens varies from 140 HB to 270 HB.

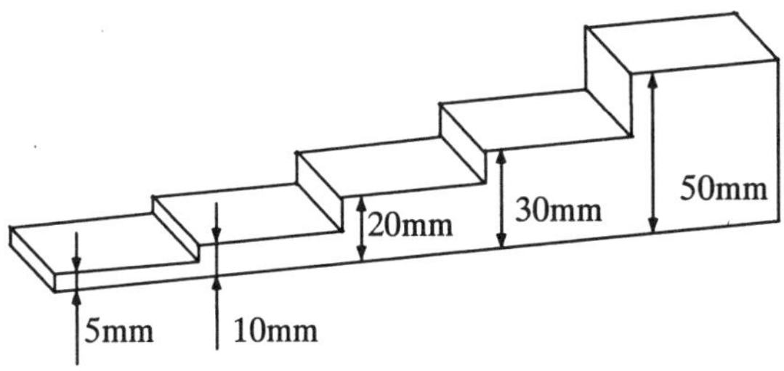

Figure 1: Step block specimens of cast irons.

Table1 Outline and Chemical composition of step-bar specimens.

No.1	Carbon equivalent (3.8%CE)	3.05%C, 2.041%Si, 0.049%P
No.2	Carbon equivalent (4.0%CE)	3.17%C, 2.341%Si, 0.054%P
No.3	Carbon equivalent (4.3%CE)	3.44%C, 2.636%Si, 0.058%P
No.4	Heat treatment (4.3%CE)	900℃→Air-cooling
No.5	Heat treatment (4.3%CE)	900℃→Furnace-cooling
No.6	Heat treatment (4.3%CE)	900℃→775℃
No.7	Heat treatment (4.3%CE)	900℃→750℃
No.8	Heat treatment (4.4%CE)	900℃→725℃
No.9	No Inoculation (4.3%CE)	
No.10	In mold Inoculation (4.3%CE)	
No.11	Residual Mg content	0.00%
No.12	Residual Mg content	0.02%
No.13	Residual Mg content	0.05%
No.14	Addition of Cu	0.5%Cu
No.15	Addition of Cu	1.0%Cu
No.16	Addition of Cu	1.5%Cu
No.17	Addition of Cu, Sn	0.5%Cu, 0.025%Sn
No.18	Addition of Cu, Sn	0.5%Cu, 0.05%Sn
No.19	Addition of Cu, Sn	0.5%Cu, 0.075%Sn

3. Mechanism of Material Evaluation Using EC

Electric resistivity and initial permeability were measured in order to discuss mechanism of the present eddy current method. Some specimens whose Rockwell hardness varies from 150 to 240 were chosen from the Series A. Figure 2 shows correlation between hardness and electric resistivity. The resistivity was measured by the standard four-probe method. Figure 3 shows correlation between hardness and initial permeability. The initial permeability was measured with two concentric solenoid coils and impedance analyser (4294A, Agilent Technologies). Both electric resistivity and initial permeability have linear relationships with hardness. Since eddy current probe voltage reflects both electric resitivity and initial permeability of target materials, it is expected that the probe voltage would be a simple function of hardness.

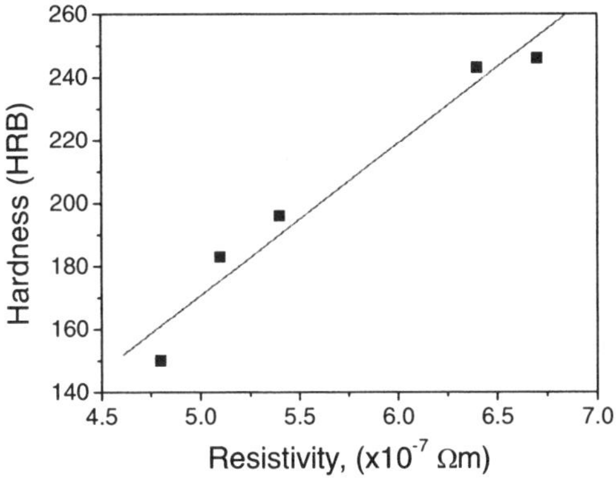

Figure 2: Dependence of hardness on electric resistivity of ductile cast irons.

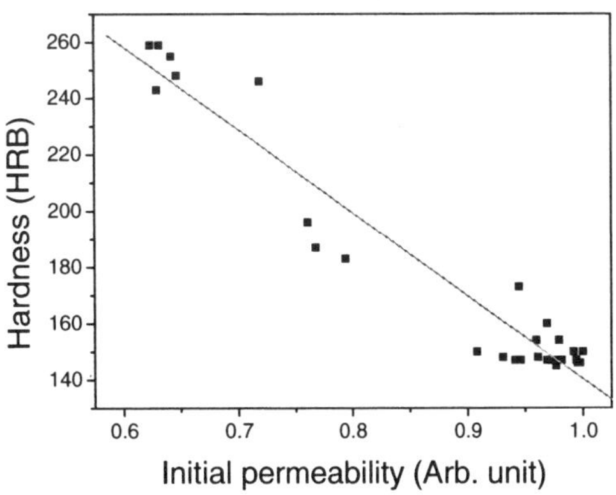

Figure 3: Dependence of hardness on initial permeability of ductile cast irons.

4. Experimental Results

Figure 4 shows correlation between hardness and probe voltage for specimens of the Series A. The test frequency and lift-off were 100 kHz and 0.5 mm, respectively. Here, the probe voltage is the Y component of complex measured signals. The coefficient of correlation is –0.93. For X component of complex measured signals, similar relationship to hardness was obtained except that it is positive correlation. In Figure 4, only one specimen of 354 HB in hardness significantly deviates from the linear fitting curve of experiemtal data shown in Figure 4. The specimen is cut from step with thickness of 5 mm of No. 1 step block. Figure 5 shows optical microscopy photo of the specimen. Surface of the specimen was etched by a solution consisting of 5% nitric acid, 95 % ethanol. The photo indicates that it includes ledeburite structure which is not included in all other specimens. Hence, it is supposed that cementite existed in the specimen affects eddy current signals.

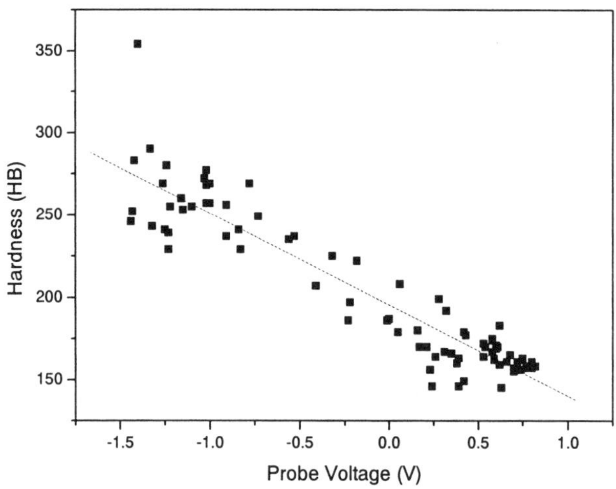

Figure 4: Hadness of 19 step block speciens as a function of eddy current probe voltage.

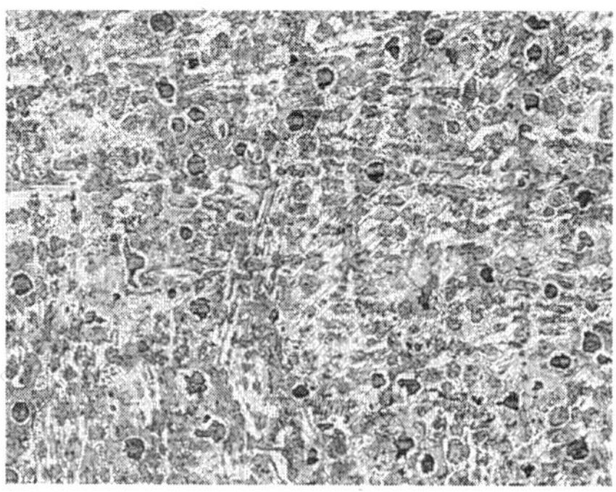

Figure5: Microstructure of specimen whose eddy current signal deviates from the fitting curve of experimental data.

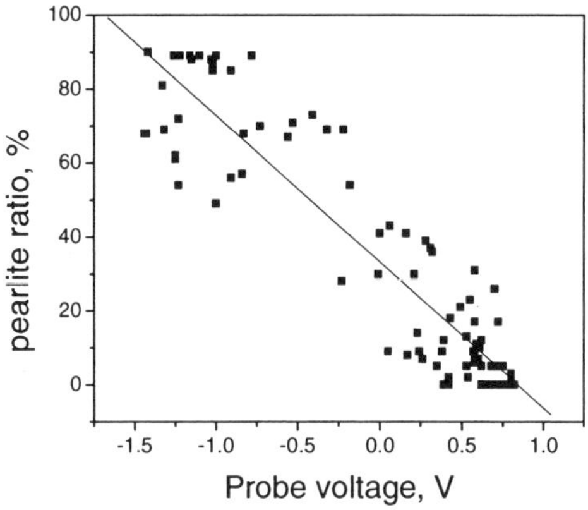

Figure 6: Pearlite ratio of 19 step block specimens as a function of eddy current probe voltage.

Figure 6 shows pearlite ratio of 19 step block specimens as a function of eddy current probe voltage. The pearlite ratio was estimated based on images of microscopy observation, as follows. Image of each field was analyzed by a 1521 square grid. Count of intercepts of ferrite, pearlite and graphite areas with the corners of this grid was performed. Pearlite percentage *P(%)* was calculated by the expression: $P\ (\%) = p/(p+f+g)$ x 100, where p-number of grid intercepts with pearlite fields, *f*-number of grid intercepts with ferrite fields, *g* - number of grid intercepts with graphite ones. Pearlite ratio does not correlate with hardness well, compared with hardness. This is because pearlite structure is alternating layers of ferrite and cementite and has variation depending on the thickness of ferrite and cementite phases. In the sense, hardness is supposed to be an essential parameter to express matrix structure.

Figure 7 shows relationship between ultrasonic velocity and hardness. Ultrasonic velocity of all specimens are more than 5570 m/s, which means graphite of all specimens are well spheroidized [2] and mechanical properties of the specimens will be dominated their matrices. In addition, it is clear that ultrasonic velocity does not depend on hardness, that is, matrix structure. Figure 8 depicts the correlation between hardness and the probe voltage. Measurement conditions and definition of the probe voltage are the same as those of the series A. The coefficient of correlation is –0.91. Among the specimens, four pieces, which are displayed as open squares in Figure 8, are selected and their micro-structure were observed with optical microscopy. Figure 9 shows optical microscopy photos and one can see that specimen 1 and 4 are cast irons with ferrite and pearlite matrices, respectively, and pearilte ratio increases as hardness increases. It can be concluded that eddy current signals reflect the change of matrices and the EC evaluation method developed here can easily distinguish matrix structure of the ductile cast iron.

In Figure 10, data of series A and B are shown in the same plot. Specimens of different casting have the same correation between hardness and eddy current signals, which is quite important in view of practical application of the present method. Figure 11 shows histogram of errors. Here, error is defined as deviation from the linear fitting curve. The present method enable estimation of hardness of ductile cast irons with sufficinent accuracy, taking the consideration of the fact that Brinell hardness measurement usually includes errors of about ± 10 HB.

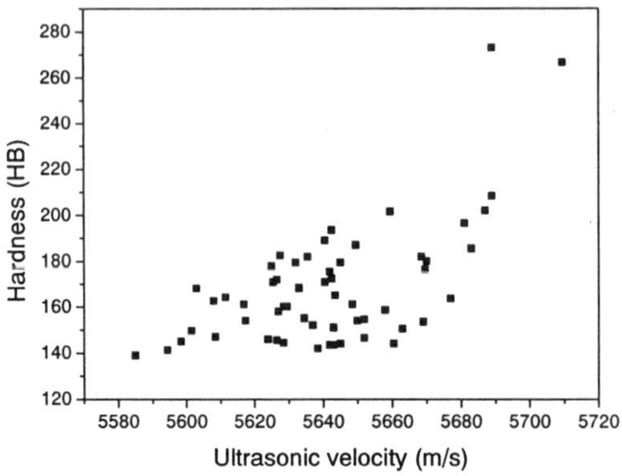

Figure 7: Hardness of Series B specimens as a function of ultrasonic velocity.

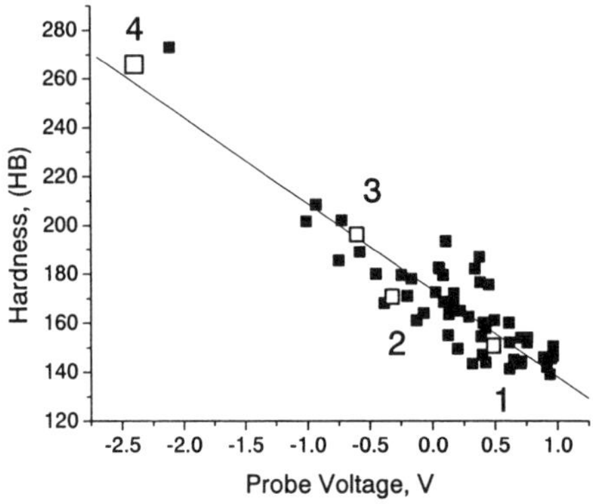

Figure 8: Mechanical properties of Series B specimens as function of EC signals.

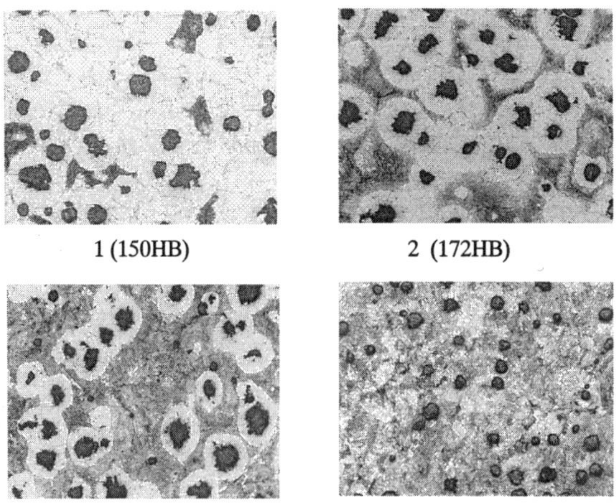

1 (150HB) 2 (172HB)

3 (204HB) 4 (259HB)

Figure 9: Microstructure of 4 specimens. Surfaces of specimens were etched by a solution consisting of 5% nitric acid, 95 % ethanol.

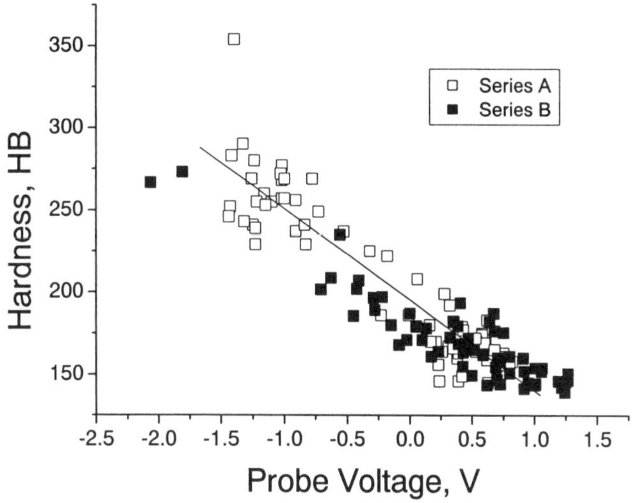

Figure 10: Hardness of Series A and B specimens as a function of eddy current probe voltage.

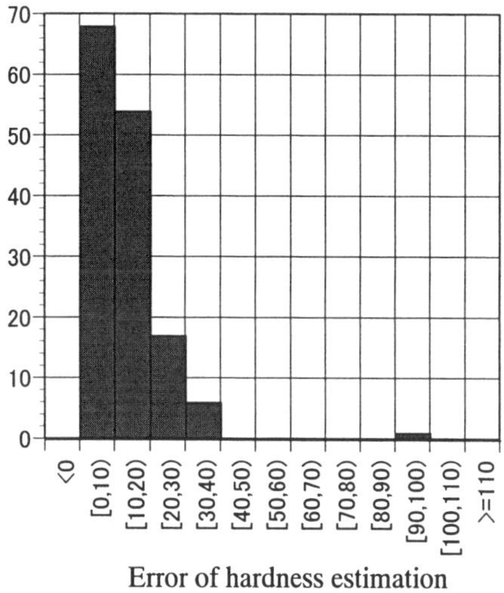

Figure 11: Histogram of estimation errors of hardness. Error is defined as deviation from the linear fitting curve in Figure 10.

5. Summary

In this study, extensive eddy current evaluation of ductile cast irons was performed to establish a novel nondestructive material evaluation method of cast irons. For the purpose, two series of specimens of ductile cast irons were prepared with different casting. The followings can be drawn from eddy current measurements of specimens.

1. The eddy current method can assess hardness of the ductile cast irons independently on their chemical composition, thermal treatment and casting method.
2. The method presented in this paper provides non-destructive evaluation of hardness of ductile cast irons with sufficient accuracy in view of practical applications.

Acknowledgements

The authors would like to thank Research Committee of Non-destructive Evaluation of Cast Iron, Japan Foundry Engineering Society for kindly providing specimens. We are also grateful to Professor T. Noguchi of Hokkaido University and Dr. H. Kage of Kusaka Rare Metal Products Co., Ltd. for fruiteful discussions and T. Satoh of Institute of Fluid Science, Tohoku University for his technical assistance.

References

[1] D.N. Collins, W. Alcheikh, "Ultrasonic non-destructive evaluation of the matrix structure and the graphite shape in cast iron" Journal of Materials Processing Technology, Vol. 55, (1995), pp.85-90.

[2] H. Kage, Y.Tanaka, "Evaluation of Cast Iron Quality by Ultrasonic Testing", Journal of the Japan Foundrymen's Society, Vol.56, No.7 (1984), pp.408-414, in Japanese.

[3] K.L. Feiste, P. Marques, Ch. Reichert, W. Reimche, D. Stegemann, A.J.M. Rebello, E.S. Kruger, "Characterization of nodular cast iron properties by harmonic analysis of eddy current signals", The e-Journal of Nondestructive Testing & Ultrasonics, Vol. 3, No.10 (1998), pp.245.

[4] T.Uchimoto, T.Takagi, S.Konoplyuk, T.Abe, H.Huang and M.Kurosawa, "Eddy Current Evaluation of Cast Irons for Material Characterization", Journal of Magnetism and Magnetic Materials, Vol.258-259 (2003), pp.493-496.

Non-destructive Detection of Creep Void Using the AC Magnetic Method

Mitsuharu SHIWA, Weiying CHENG, Ryouichi KUME, Hiroshi YONEYAMA
Tsurumi R&D Center, JAPEIC,
14-1 Benten-cho, Tsurumi, Yokohama, 230-0044, Japan

Abstract. AC magnetic detection of creep void is conducted for 2.25Cr-1Mo steel base metal (BM) and simulated heat affected zone (HAZ) samples. Parameters from the analogy of hysteresis loop, equivalent remanence, coercity, and hysteresis loss, were used for creep evaluation. Stress-induced damage (SID) parameter is proposed to separate stress-induced damage factor and aging factor. Creep void is observed in all specimens when the SID value is smaller than 0.8.

1. Introduction

Creep damage occurs when structural materials are operated under high temperature and high stress for a long period of operation. Creep void nucleation is formed in the initial stage of creep damage. Void detection is of significant importance for damage evaluation. Magnetic property measurement techniques have been utilized as Non-destructive Evaluation (NDE) method to detect creep damage and evaluate post weld heat treatment (PWHT) temperature [1, 2]. Magnetic hysteresis parameters, such as coercivity, remanence and hysteresis loss were measured and analyzed by using Jiles-Atherton model [3, 4] to evaluate creep damage of Cr-Mo steel pipe base metal [5], seam weld and heat affected zone [6].

In this paper, the AC magnetic testing method is proposed to detect creep void in BM and HAZ specimens of 2.25Cr-1Mo steels. An analogy is drawn between magnetic hysteresis curve and the Lissajous obtained from our measurement. Equivalent remanence, coercity, and hysteresis loss (including eddy current signal) are analyzed. Different magnetic characteristic are shown for as-received (AR), aging and creep samples.

Changes of magnetic properties by creep damage are contained in both thermal aging effect and stress-induced damage effect. It is important to separate the two factors from detected signal in order to evaluate creep damage itself. A creep damage evaluation method using a SID parameter is developed. The method can detect typical creep damage initiation such as creep void in different type of materials.

2. Experimental Procedure

Two types of specimen are used in this experiment, they are respectively base metal (Type A) and simulated HAZ of different PWHT temperatures made by Greeble testing machine (Type B I-IV). Chemical composition of 2.25Cr-1Mo steels, from which the specimens used in this investigation is made of, is shown in Table 1. The Specimens type and applied aging and creeping conditions are presented in Table 2.

The aging temperature was 873K. The geometry of creep damage specimen is shown in Fig. 1. The applied creep times are set to be 1/8, 1/4 and 1/2 of predicted fracture

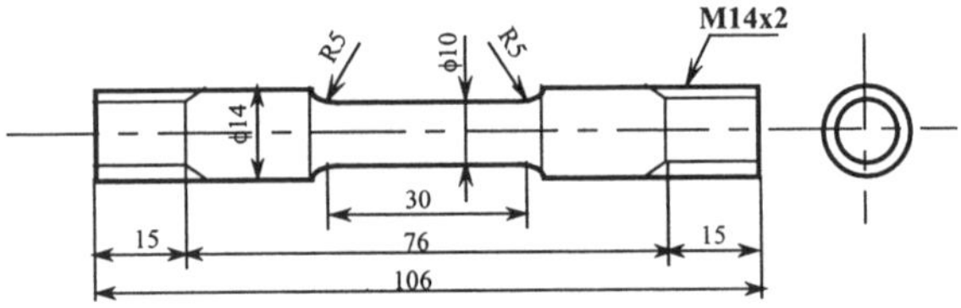

Fig. 1. Geometry of creep damage specimen.

Table 1. Chemical contents of 2.25Cr-1Mo Steel (mass %).

C	Si	Mn	P	S	Cr	Mo
0.09	0.24	0.44	0.015	0.009	2.1	0.95

Table 2. Specimens type and applied aging and creeping conditions.

Specimens type				**Applied condition**		
Type		**PWHT** (K)/ 2H	**Aging or Creep**	**Temp.** (K)	**Stress** (MPa)	**Time** (s)
A	BM	-	Aging	873	-	3600 - 1.1 x 10^8
B-III	HAZ	993				
A	BM	-	Creep		39 & 54	1.8 x 10^7 (1/8PFT) 3.6 x 10^7 (1/4PFT) 7.2 x 10^7 (1/2PFT) 1.44 x 10^8 (PFT)* or Breakage
B-I	HAZ	-				
B-II		953				
B-III		993				
B-IV		1033				

* Predicted fracture time

time (1.44×10^8 s), the creep temperature was 873K and the stresses applied are 39 MPa and 54 MPa.

After being taken out of electric furnace or creep testing machine, the specimens are cut into pieces of 25 mm by 10 mm in size. The macro- and micro- structures of specimens are observed using optical microscope and scanning electron microscope (SEM), respectively.

AC magnetic properties are measured after microstructure observation. Figure 2 shows the block diagram of AC magnetic measurement system[7)]. A differential type probe is used to detect AC magnetic signals. The exciting and detecting coils are coaxially arranged with a ferrite core of 2.5 mm diameter and 20 mm length. The exciting frequency is 10 kHz and the maximum exciting voltage is 8V. Signals are recorded using a 2-channel waveform recorder.

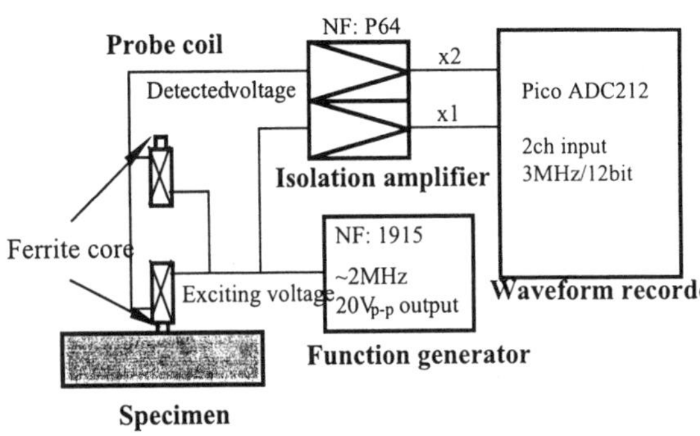

Fig. 2. Block diagram of the AC magnetic measurement system.

3. Experimental Results

3.1 Microstructure observation

Macroscopic observation shows that there is no big change of the grain size in both BM and HAZ specimens during aging. However, during creep, the grain elongates after creep time 7.2×10^7 s of creeping with

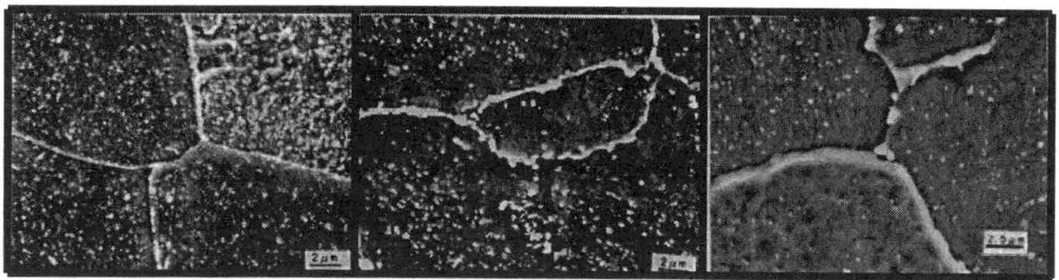

(a) As Received (b) Aging/1.1 10^8s (c) Creep/39MPa/7.2X10^7s

Fig. 3. SEM observation of Base Metal.

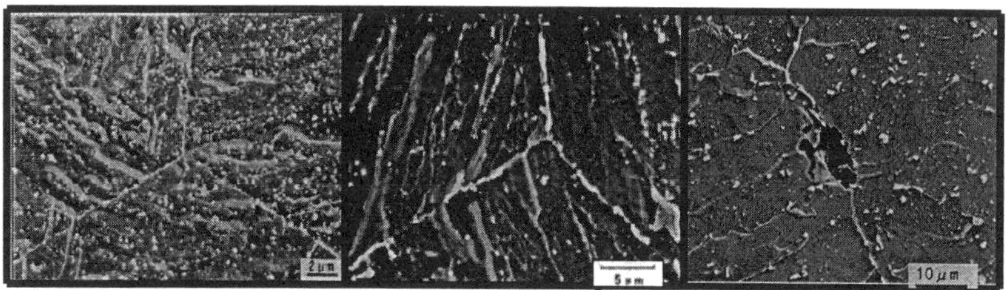

(a) As Received (b) Aging/7.2X10^7s (c) Creep/39MPa/1.0X10^8

Fig. 4. SEM observation of HAZ.

applied stress of 53MPa. When the applied stress is 39MPa, the grain elongation can be observed just before the breakage of both BM and HAZ specimens.

The microstructure is observed using SEM. SEM observation of BM metal is presented in Fig. 3, where Figs. 3(a), (b) and (c) are corresponding to specimens of AR, 1.1×10^8 s of aging, and 7.2×10^8 s of creeping under 39MPa of stress. SEM observation of HAZ is presented in Fig. 4, where Fig. 4(a), (b) and (c) corresponding to specimens of AR, 7.2×10^7 s of aging, and 1.0×10^8 s of creeping under 39MPa stress. Carbide precipitations are observed in the sub grain in both aging and creep samples of BM. The volume of carbide precipitation increases with aging and/or creep time. While in case of HAZ, large scale of carbide precipitations is observed along the grain boundary in long-term aging (Fig. 4 (b)). Figure 4(c) shows the coalescence of creep voids in creep of HAZ. Smaller creep void is observed in BM (Fig. 3 (c)).

3.2 Basic properties of Lissajous

Taking the exciting voltage as the horizontal input and detected voltage as vertical input forms a Lissajous. An analogy is drawn between the magnetic hysteresis loop and the Lissajous from this experiment. In this paper, the concepts of coercivity, remanence, hysteresis loss, and permeability in magnetic hysteresis property were used in AC voltage analysis as an equivalent value of DC magnetization. The equivalent values include an effect of eddy current. The equivalent remanence and coercivity are defined as the zero-cross value of exciting and detected voltages, and the equivalent hysteresis loss (including eddy current) is defined as the area formed by the Lissajous by equation

$$HL = \sum_{V_E^{\max}}^{V_E^{\min}} \frac{1}{Z} V_D dV_E - \sum_{V_E^{\min}}^{V_E^{\max}} \frac{1}{Z} V_D dV_E \,, \qquad (1)$$

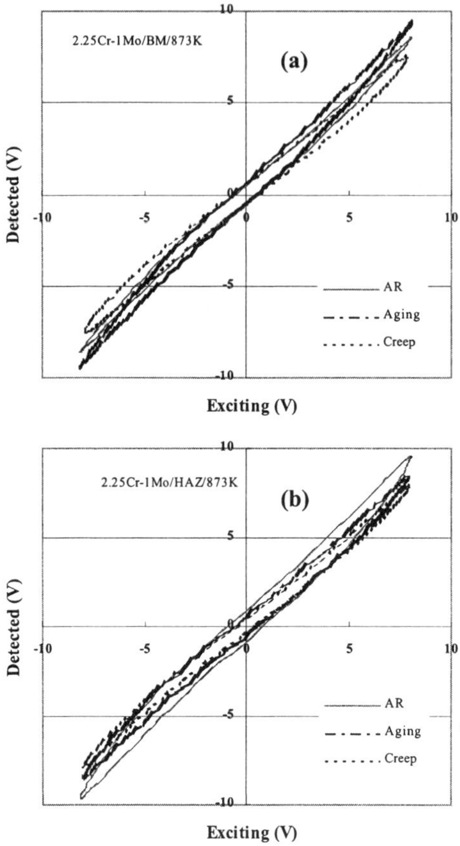

Fig. 5. Lissajous of typical AR, aging and creep samples, BM(a), HAZ(b).

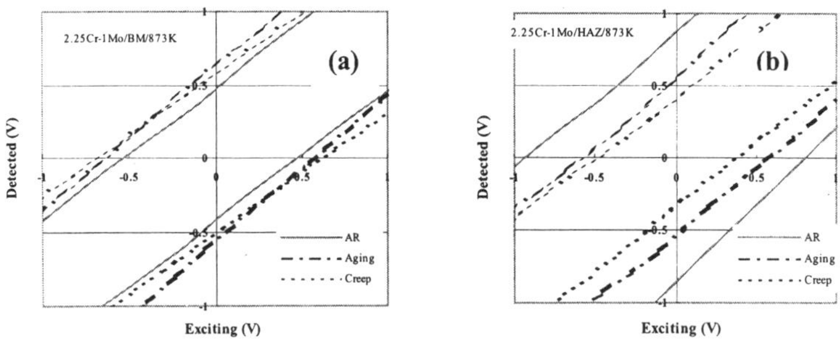

Fig. 6. The exciting and detected voltages near zero-cross region corresponding to Figs. 5 (a) and (b).

where V_E and V_D are respectively exciting and detected voltages, and Z is impedance of the probe.

AC magnetic measurement is carried out to AR, aging and creep samples of both BM and HAZ material that are used in SEM observation in Section 3.1. Lissajous are obtained from the measurement. The Lissajous clearly show that, for BM specimens the

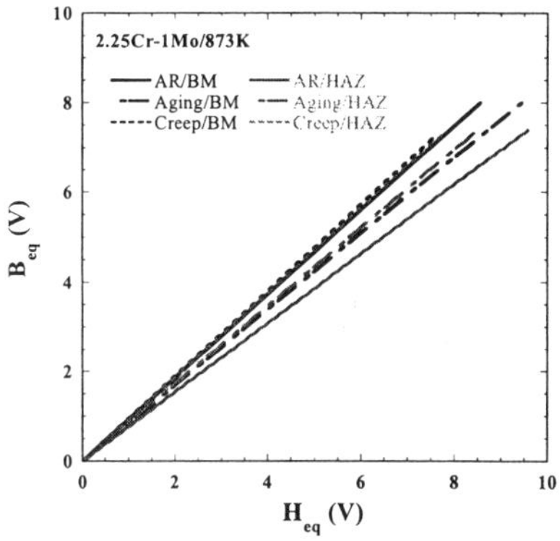

Fig. 7. Equivalent linear B-H curves of BM and HAZ samples.

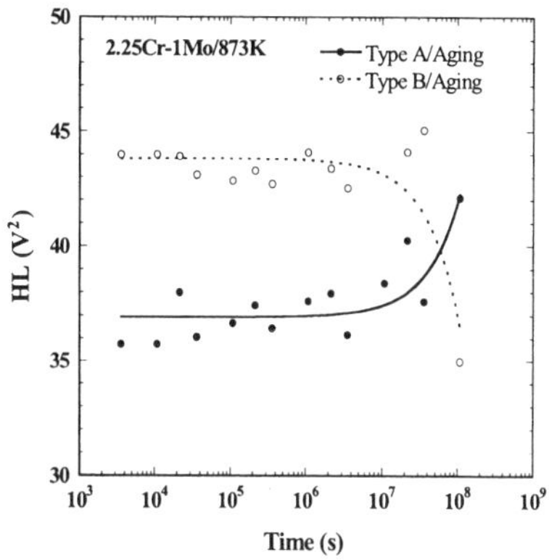

Fig. 8. HL of BM and HAZ samples during aging.

detected voltage value of aging sample is bigger than that of AR sample and bigger than that of creep sample. However, for the HAZ specimen, the detected voltage decrease for aging and creep samples compared to AR. Figures 6 (a) and (b) show zero-cross region of exciting and detected voltage corresponding to Figs. 5 (a) and (b). It is observed that there is no big difference of residual detected voltage $V_D(V_E=0)$ and residual exciting voltage $V_E(V_D=0)$ among the BM samples. However, a relatively bigger difference is found from HAZ samples. Both the exciting and detected voltages decreased for aging and creep compared to AR.

Figure 7 plots the equivalent linear B-H curves based on maximum detected voltage and maximum exciting voltage in a current-exciting detection on BM and HAZ samples. In the current-exciting condition, V_D and V_E correspond to H_{ep} and B_{eq} respectively. The tangential of the curves stand for equivalent permeabilities.

The equivalent coercivity H_0^{AC}, remanence B_0^{AC}, maximum exciting H_{max}^{AC}, maximum magnetic flux density B_{max}^{AC} and magnetic permeability μ_r^{AC} of BM and HAZ samples are normalized by taking the ratio of each creep and aging specimen's voltage to that of AR's. The normalized values are listed in Table 3. It shows that the H_0 and B_0 of BM increase after aging and creep; on the contrary, the H_0^{AC} and B_0^{AC} of HAZ decreased after aging or creep. The μ_r^{AC} of BMs was almost the same as that of AR. However, the μ_r^{AC} of HAZ increases compare to that of the AR.

3.3 Hysteresis Loss

Figure 8 shows HL of BM and HAZ samples during aging. The HL of BM and HAZ specimens change in opposite direction, that is, HL of BM increases and HL of HAZ decreases with aging time.

Table 3. Equivalent typical magnetic properties of BM and HAZ samples normalized by taking the ratio with their ARs voltages.

	Base Metal		HAZ	
	Aging	Creep	Aging	Creep
H_0^{AC}	1.34	1.23	0.63	0.46
B_0^{AC}	1.37	1.29	0.69	0.53
H_{max}^{AC}	1.10	0.88	0.89	0.82
B_{max}^{AC}	1.00	0.90	1.00	1.00
μ_r^{AC}	0.91	1.03	1.13	1.22

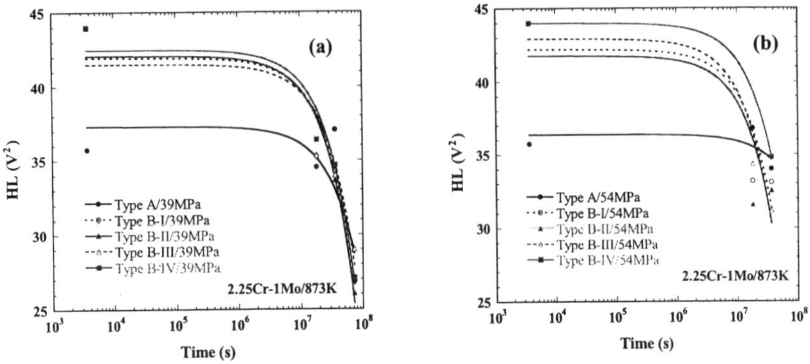

Fig. 9. HL of BM and HAZ during creep under 39 MPa (a) and 54 MPa(b).

Figure 9 shows HL of BM and HAZ during creep under stress 39 MPa (a) and 54MPa (b). The HLs of all creep-damaged specimens decrease with the creeping time.

4. Discussion

4.1 Regression curves of Hysteresis Loss

Regression curves based on the measurement results in Figs. 6 and 7 are formed from the following equation,

$$HL = a + bt \qquad (2)$$

where t is time of creep or aging, a and b are constant.

Table 5 lists the parameters a, b of Eq. (2) and the correlation coefficient R between measurement and regressed data of aging or creep specimens. Relatively high value of R is obtained. The high correlation means that the regression curves can be used as the master curves for aging and/or creep damage evaluation.

4.2 Creep damage Evaluation parameter of SID

Figures 8 and 9 also show that the measurement value depends on the initial value of

Table 4. Regression curve's parameters and the correlation coefficient R of aging and creep.

Type		a	b (10^{-7})	R
Aging	A	36.929	0.4933	0.81
	B-III	43.819	-0.6773	0.83
Creep/39MPa	A	37.294	-1.1930	0.80
	B-I	41.930	-2.1490	0.96
	B-II	42.070	-2.3165	0.97
	B-III	41.513	-1.8869	0.93
	B-IV	42.470	-2.2029	0.98
Creep/54MPa	A	36.374	-0.4770	0.61
	B-I	42.200	-3.0204	0.87
	B-II	41.764	-3.1786	0.83
	B-III	42.917	-3.5512	0.96
	B-IV	43.988	-2.5430	1

materials and trend of change. Changes of properties by creep damage are contained in both thermal aging effect and stress-induced damage effect. In order to evaluate creep damage itself, it is necessary to separate the two factors from detected signals.

The measured creep signal $M^C(L,\sigma)$ can be characterized by

$$M^C(L,\sigma) = \beta(A(L) \otimes D(\sigma)) \qquad (3)$$

where L is Larson-Miller parameter for aging given by $L = T(\alpha + \log t)$ in which T is temperature, t is time, α is a constant of material, σ is applied stress, $A(L)$ is aging factor, $D(\sigma)$ is stress-induced damage factor and β is a transfer function, $\otimes$ stands for convolution mathematically. M can be any measured signals, such as equivalent remanence, coercity, or hysteresis loss. Eq. (3) means that creep damage is a convolution of aging factor $A(L)$ and stress-induced damage factor $D(\sigma)$. On the contrary, the stress-induced damage factor can be deduced from the deconvolution of $M^c(L,\sigma)/\beta$ and $A(L)^{-1}$. The aging factor $A(L)$ can be obtained from the transformation of the measured signal of aging, $M^A(L) = \beta A(L)$ as $A(L) = M^A(L)/\beta$.

Consequently, provided the measured creep damage signal $M^c(L,\sigma)$, the measured aging signal $M^A(L)$ and the transfer function β are known, the stress-induced damage factor $D(\sigma)$ can be obtained from the deconvolution of $M^c(L,\sigma)/\beta$ and $(M^A(L)/\beta)^{-1}$, and the creep damage can be evaluated from factor $D(\sigma)$.

The stress-induced damage ratio of creep $D_R(\sigma)$ is defined by

$$D_R(\sigma) = \frac{D^C(L,\sigma)}{D^A(L)} \qquad (4)$$

where the characteristic damage value of creep and characteristic degradation value of aging is defined by $D^C(L,\sigma) = \dfrac{M^C(L,\sigma)}{M_i}$ and $D^A(L) = \dfrac{M^A(L)}{M_i}$ respectively, where M_i is detected signal for material in initial condition.

Finally, we have the stress-induced damage ratio of creep $D_R(\sigma)$ as

$$D_R(\sigma) = \frac{\dfrac{M^C(L,\sigma)}{M_i}}{\dfrac{M^A(L)}{M_i}} = \frac{M^C(L,\sigma)}{M^A(L)} \qquad (5)$$

$D_R(\sigma)$ is also named as parameter SID (stress-induced damage) ratio of creep. It can be found that both the effect of initial condition of material and aging degradation factor are rejected in Eq. (5).

4.3 Creep Void evaluation by SID

It's important to detect creep void nucleation in 2.25Cr-1Mo steel. This is the initiation of creep damage of the material. $D^N{}_R(\sigma)$ is a measure of damage value at the point when creep void nucleation starts forming. Initial creep damage occurs when $D^N_R(\sigma)$ is smaller than $D_R(\sigma)$,

$D^N_R(\sigma) \le D_R(\sigma)$. $D^N_R(\sigma)$ have to be decided from microstructure observation.

The regression of $M^C(L,\sigma)$ and $M^A(L)$ is also carried out according to Eq. (2) using the parameters in Table 5. Figure 10 shows creep void detection line of both BM and HAZ specimens. The SID value is calculated by Eq. (5). Creep voids are observed in all creep damage specimens when SID is less than 0.8.

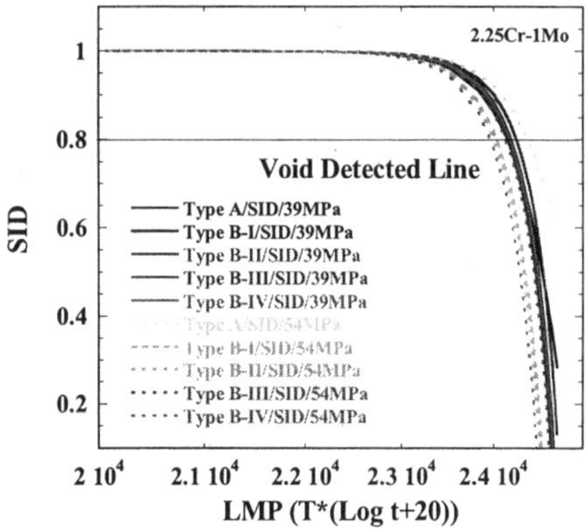

Fig. 10. Creep void detection line by SID value,LMP is Larson-Miller Parameter, T is temperature, t is time.

5. Conclusions

Nondestructive detection of creep void is developed for low alloy steel by using AC magnetic method. Two types of 2.25Cr-1Mo steel specimens, base metal and simulated heat affected zone under aging and creep damage, were prepared for the tests.

It was observed that the equivalent hysteresis loss HL of both BM and HAZ specimens changed in opposite direction, that is, HL of BM increases and HL of HAZ decreases with aging time. On the other hand, the HLs of both BM and HAZ decrease with creeping time. The HL of creep samples is affected by both aging and stress-induced damage. In order to evaluate creep damage, stress-induced damage (SID) ratio parameter is proposed to remove aging factor of materials from creep damage. Creep void is observed from creep damage samples of SID value under 0.8.

References

[1] M. Shiwa, *et al*,"Evaluation of Creep Damage and SR Temperature of Low Alloy Steel Using AC Magnetic Method", 2nd International Conference on NDE in relation to Structural Integrity for Nuclear and Pressurized Components, New Orleans, (2000).

[2] W. Cheng, M. Shiwa, et al, "Finite Element Simulation of Magneto Inductive Evaluation of PWHT Temperature of Cr-Mo Welded Joints," ISEM 2003, (2003), Paris.

[3] G. Chen, M.Shiwa, and Yoneyama, H., "Theoretical Investigation of Change of Magnetic Property Due to Creep Damage Using Jiles-Atherton Magnetization Model", *International Journal of Applied Electromagnetics and Mechanics*, Vol.13 (1), (2002), IOI Press, to be published.

[4] D. Jiles and D. Atherton. *J. of Magnetism and Magnetic Materials*, 61, (1986), 48-60.

[5] Z. Chen, M. Govindaraju, *et al*, *IEEE Transaction on Magnetics*, 30, 6(1994), 4596-4598.

[6] M. Sablik, D.Jiles and M.Govindaraju, *IEEE Transaction on Magnetic,* **34**, 1998, 2156-2158

[7] M. Shiwa, *et al*,"Evaluation of PWHT temperature for Cr-Mo steel weld joints by AC magnetic harmonic analysis", The 8th International Workshop on Electromagnetic NDE, Saarbrucken, (2002).

Electromagnetic Nondestructive Evaluation (VIII)
T. Sollier et al. (Eds.)
IOS Press, 2004

Magnetic Adaptive Testing of Thermally Treated Construction Steel

O. STUPAKOV[1,2], B. SKRBEK[3] and I. TOMÁŠ[1]
[1]*Institute of Physics, ASCR, Na Slovance 2, 18221 Praha 8*
[2]*Faculty of Mathematics and Physics, Charles University, Ke Karlovu 3, 120 16 Praha 2*
[3]*Technical University of Liberec, Hálkova 6, 46117 Liberec 1*
Czech Republic

Abstract. Results of investigation of magnetic response to structural alterations, caused by quenching and by consequent tempering, in the case of two series of industrially widely used low-alloyed construction steels are presented. The Magnetic Adaptive Testing method is introduced and tested as an alternative technique for non-destructive evaluation of ferromagnetic construction materials. Advantages and drawbacks of application of the method are discussed.

1. Introduction

The issue of nondestructive examination of compositional and structural variations of construction ferromagnetic steels is one of the very topical problems in metal industry. General simplicity and economical feasibility of magnetic testing techniques attracts broad attention to inspection methods based on coupling of the magnetization process of such materials to the details of their structural variation.

However, the measurement of the traditional parameters of the saturation magnetization process (maximal permeability, remanent magnetic induction, coercive force, etc., frequently named “magnetic hysteresis testing”) [1-2] requires as a rule a rather high applied field and these descriptors are often not very reliable in their response to subtle structural alterations of ferromagnetic materials. That is why the research and development of new techniques, able to be *optimized* to particular structural degenerations, *sensitive equally to or better than* the classical methods, are the necessities of modern days, see e.g. [3-4].

This paper belongs to a starting series of studies carried out with the main purpose of introduction of the Magnetic Adaptive Testing (MAT) [5] as an experimental method for non-destructive evaluation of ferromagnetic construction materials (mainly steel and cast iron). The applicability and sensitivity of the MAT method will be investigated in a number of different industrially relevant situations where ferromagnetic construction materials undergo smaller or larger structural changes. The structural variations of interest affect the quality of the material, which may be important to monitor either for documentation of the quality itself (cast iron quality, surface hardened steel, deformation, etc.) or, in limiting cases, for prevention and early warning when the loss of quality can result in safety losses, risk for damage of machinery and danger for human beings.

In the paper, the applicability of MAT to the assessment of structural variations of industrially broadly used construction steels due to quenching and sequential tempering is tested. Its sensitivity and field requirements are compared with the traditional magnetic parameters as well as verified and confirmed by metallographical investigation.

2. Principles of the Magnetic Adaptive Testing method

It is well known that the process of magnetization of ferromagnetic objects is strongly influenced by the composition and structural uniformity of the material, by the presence of large-scale or local mechanical tensions, by the occurrence of precipitates, inclusions, voids, different phases, by the size and shape of grains, etc. In fact, the magnetization process can be influenced by everything which has an effect on the freedom of motion of magnetic domain walls and on the constraint of rotation and size of the magnetization vectors within the domains. The study of this interaction process of the domain walls with microstructure parameters, respectively with the lattice defects, is the traditional way to assess microstructure properties variations of ferromagnetic materials by means of their magnetic response by the determination the *major* hysteresis loop (saturation-to-saturation) and use the most suitable of its parameters for the characterization (for reviews, see [1-2]).

The currently presented Magnetic Adaptive Testing method employs a similar approach but with a significant distinction: it measures a *family of minor* loops and makes use of a simple software to pick up *the* minor loop from the whole family and *the* descriptor of the selected minor loop, which is the most sensitive and the most reliable for the used sample material and for the followed structural change. As a result, the MAT descriptors of the structural variations of ferromagnetic materials are generally measured at minor loops the amplitude of which is *far below saturation*. Since the descriptors are chosen to be optimal with respect to the material, with respect to the inspected sample alteration and with respect to the specific inspection task requested (i.e. the method is *adaptive* towards the *material*, the structural *variation* and the particular *task*), and they are frequently *more sensitive* than the traditional parameters.

A typical measurement for the MAT begins with demagnetization of the sample in order to achieve a reliable starting magnetic state. Then the sample is magnetized by an applied magnetic field, typically by a number of "triangular" sweeps of the field with their amplitudes growing by step by step from zero up to a given maximum. The magnetic response of the sample is recorded, resulting in a large database of minor loops data, essentially equivalent to the datasets handled in the Preisach model of the ferromagnetic hysteresis [3-4]. As the collected dataset contains complex information about the magnetic hysteresis of the sample [6], it should reflect even subtle variations of the magnetization process due to the features and feature changes of the material microstructure.

At the beginning, a reference series of samples - which are different from one another by a "degeneration parameter, T" corresponding to the structural changes of interest - is measured. The optimum descriptor, D, (and/or descriptors) with the most advantageous dependence, $D(T)$, is (are) determined from this series of measurement. The descriptor is specified by its "choice at the field-point". In other words, the MAT method picks up from the series of the recorded datasets an element measured for all the samples of the series on a minor loop with the applied field *amplitude*, $h_B=h_{B0}$, at the applied field *current value*, $h_A=h_{A0}$, which was found to provide *the most sensitive and the most descriptive* "degeneration function" $D(h_{A0},h_{B0},T)$ with respect to the investigated variation of the degeneration parameter, T, of the studied material.

The actual content of the experimentally measured dataset depends on the performance of the magnetic measurement. Using the inductive measurement as described in the next section, the basic experimental data are the permeability minor loops, $\mu(h_A,h_B)$, and the degeneration function of the descriptor chosen at the field point (h_{A0},h_{B0}) is $\mu(h_{A0},h_{B0},T)$. By a following numerical differentiation or integration of the μ-dataset with respect to h_A or h_B, complementary datasets can be obtained. Such complementary datasets contain essentially a volume of information equivalent to the basic one. However, their presentation can be more advantageous in particular cases depending of the inspection task.

3. Measurement

The datasets of the minor loops concerning the MAT procedure of material evaluation can be measured for instance by the electromagnetic induction method in a simple permeameter. The magnetic driving field is produced in a driving coil by a current, which is waveform controlled by a steering computer and amplified. The voltage induced in the pick-up coil is recorded, digitized, and processed in the same computer. During the experiment, a dataset file and a text file containing information about the parameters of the measurement are stored.

The block scheme of the permeameter, together with the triangular waveform of the applied magnetic field, used for the presented measurement, are shown in fig.1. A typical experimental output is shown in fig.2.

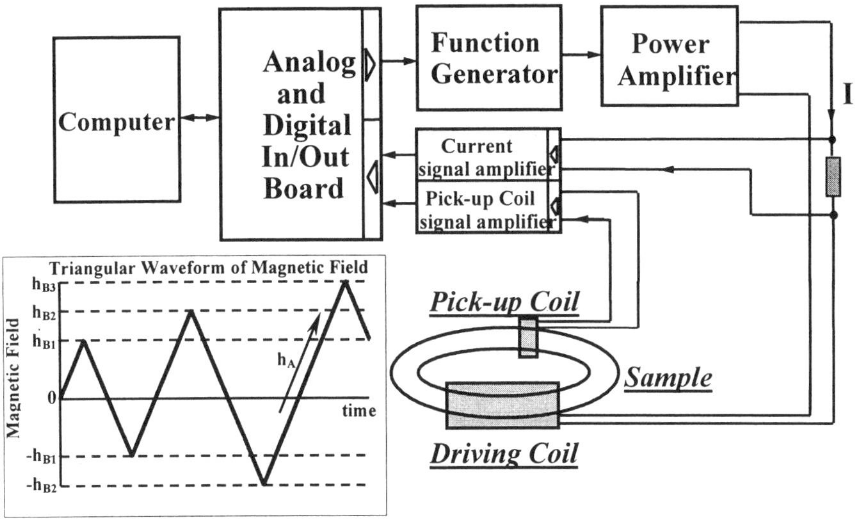

Fig.1 Principal block-scheme of the permeameter setup with the triangular waveform of the applied magnetic field h_A

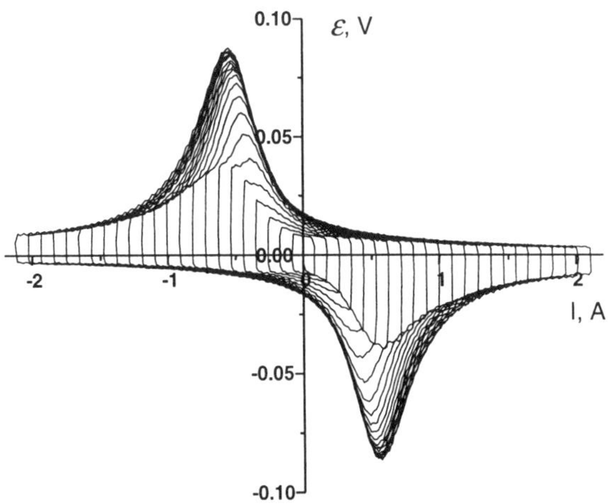

Fig.2 Typical experimental dependence of the pick-up coil voltage (proportional to the differential permeability, μ) on the driving coil current (proportional to the applied magnetic field, h_A)

The inductive measurement method combines in the permeameter with the triangular shape of the driving current. This means that - for the closed magnetic circuits of the used samples - the output signal, ε, is proportional to differential permeability $\mu\mu_0 = dB/dh_A$ of the investigated material:

$$\varepsilon = -\frac{d\Phi}{dt} = -kS\frac{dB}{dt} = -kS\frac{dB}{dh_A}\frac{dh_A}{dt} = -\left(kS\mu_0 n\frac{dI}{dt}\right)\mu \qquad (1)$$

where Φ is the magnetic flux, B is the magnetic induction, k and S are the number of turns and the cross-section area of the pick-up coil, n is the winding density of the driving coil, dI/dt is the *constant* driving current slope with time t, μ is the relative differential permeability, and μ_0 is permeability of vacuum.

4. Application of MAT to tempered steel

Two series of thermally processed samples were investigated. In both cases low alloyed, industrially widely used construction steels were treated, their chemical composition is shown in Table 1.

Table 1 Chemical composition of investigated steels

Ser.	CSN	EN	% C	% Mn	% Si	% Cr	% Ni	% Cu	% P	% S
A	414260	54SiCr6	0.50-0.60	0.50-0.80	1.30-1.60	0.50-0.70	max 0.50	max 0.30	0.035	0.035
B	412050	C45	0.42-0.50	0.50-0.80	0.17-0.37	max 0.25	max.0.30	max 0.30	0.04	0.04

All samples were machined into the shape of flat rectangular bars shown in Fig.3, which made it possible to measure the magnetic response of the material in the *closed magnetic circuit* around the "window" in the middle of the sample. At the same time the extended pads of the samples allow to clamp them into an electromechanical loading machine (Instron) for investigation of the influence of their mechanical properties on the magnetization process.

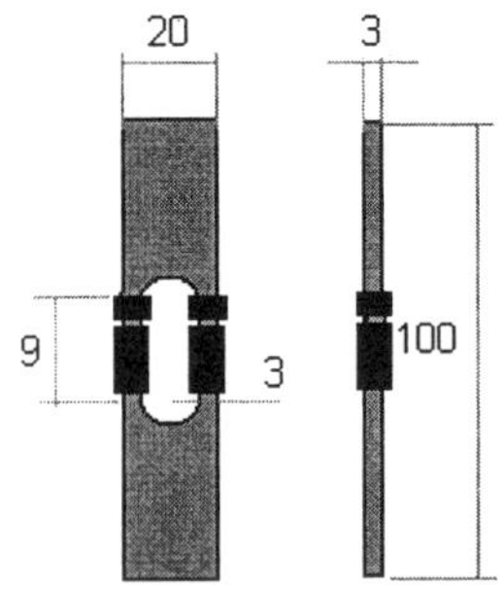

Fig.3 Shape of the samples with the magnetically closed window in the middle and with the schematic positions of the driving coils and the pick-up coils at each leg of the window-frame.

The structural variations in each Series of the samples were produced by thermal processing (quenching in oil from 840°C down to 20°C, followed by consequent tempering for 30 minutes at temperatures T=300, 420, 500, 580, 660°C). The tempering temperature, T, was chosen to be the MAT "degradation parameter". This process varied the steel structure from needle-like brittle highly-distorted martensite structure (quenched sample) over mixtures of partially relaxed martensite and sorbite to pure sorbite in all samples Series A, and in most samples of Series B (see Discussion).

The thermally processed samples were then given a pair of magnetizing coils and a pair of pick-up coils positioned approximately according to Fig.3 (actually the pick-up coils were wrapped straight around the sample legs and the driving coils were wound on the top), and they were magnetically measured by the induction method: The two driving coils connected in series magnetize the samples in the closed circuit around the window. The triangular-shaped slow oscillation of the driving current with constant slope (1A/sec for the current or 2kA/m/sec for the driving field), and with step-by-step increasing amplitude, magnetizes the samples up to their saturation, which was attained at the maximal applied field values of about 7kA/m for the quenched samples. The used slope value corresponds to the quasi-static magnetization for Series A.

The main results of the investigation are illustrated on Series A and presented in Figs. 4-6. The behavior of MAT descriptors (differential permeability $\mu(h_{A0},h_{B0},T)$, Fig.4, and magnetic induction $B(h_{A0},h_{B0},T)$, Fig.5) are compared with the correspondent traditional parameters (maximal differential permeability and remanent induction of saturated loop, respectively) and with the monotonically decaying hardness in Fig.4. It is worth noting that the presented MAT-descriptors are certainly not less sensitive than the correspondent traditional ones and that the working field-points of the MAT-descriptors is about four time less than the saturation field. The option of shape optimization of the degeneration functions is presented in Fig.6 for the Preisach-like $d\mu/dh_B$ MAT descriptor.

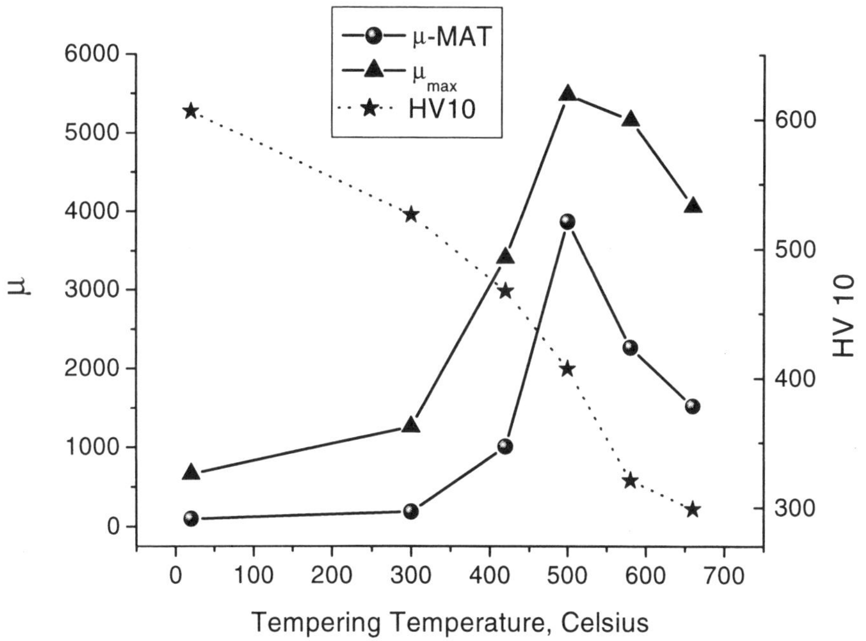

Fig.4. Dependences of the μ-MAT descriptor (chosen at the field point h_A=1130A/m, h_B=1750A/m), the maximal permeability, μ_{max}, and Vickers hardness (10kg) on the tempering temperature, T, for Series A

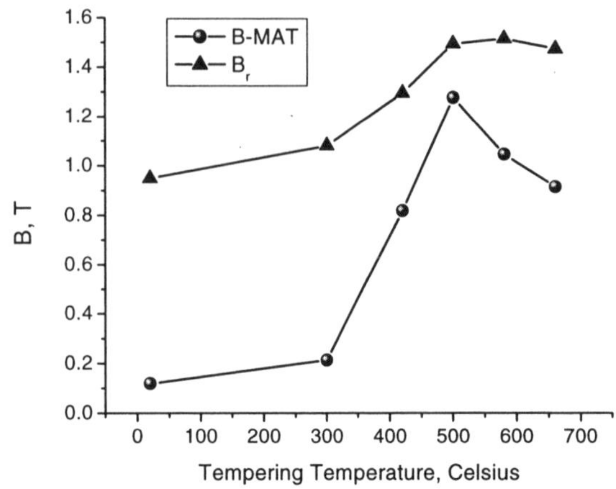

Fig.5 The remanent magnetic induction B_r (traditional parameter) with the corresponding MAT optimum descriptor (at the field point h_A=1100A/m, h_B=1700A/m)

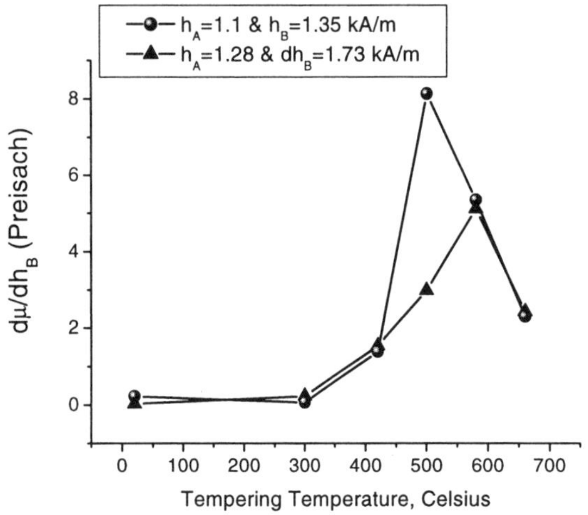

Fig.6 Two different options of the shape of the Preisach-like $d\mu/dh_B$-MAT degeneration function (at field points h_A=1100A/m, h_B=1350A/m, and h_A=1280A/m, h_B=1730A/m)

5. Discussion

All the results presented in the plots correspond to Series A, which provided more reliable data (scatter of "identical" samples is smaller than approx. ±5%) than Series B. The outcome of Series B was qualitatively very similar to Series A but the scatter was (at some samples) substantially larger (up to about ±15%). This poor behavior of the investigated samples of Series B originates very probably in the substantially larger *critical* cooling speed of the material B than A. The applied quenching in oil (used for all the samples in order to avoid unwanted geometrical distortion and/or cracking of tiny frames of their "windows") apparently did not reach the proper cooling speed at all samples B. This resulted in irregular appearance of bottom bainite in some of B-samples. This suspicion was confirmed by metalographic inspection.

Based on the similar magnetic response of the investigated two kinds of low-alloyed steel samples, and on the well-known *principally similar structural behavior of the whole family of low-alloyed construction steels* (which stands for the by far largest volume of the industrially used steel), the observed magnetic behavior (see Fig.4 and 5) is probably typical for the process of "martensite toughening" of most of the low-alloyed steel kinds: As the tempering temperature is increased, the martensite toughening first improves the magnetic response of the tempered steel, then a maximum of the response is reached (this occurred around T=500^{0}C in both steels A and B), and the magnetic response drops down again at still higher T. Based on the knowledge of what happens with the material structure during the martensite toughening process, it can be argued, that the increase of the tempering temperature improves mobility of the domain walls by decreasing the number and the pinning strength of the local micro-tensions (present due to the frozen-in carbon atoms within the martensite cells) and of the local tensions among the needle-like grains. However, at the same time, the carbon atoms released by the tempering from the martensite cells start to produce another system of the domain wall traps, i.e. the carbide precipitates. As a result the magnetic response (reflecting freedom of the magnetic domain walls) reaches its maximum at the tempering temperature around T=500^{0}C, when the combination of the local-tension-pinning plus the carbide-precipitates-pinning reaches its total minimum.

There is an open question if this "magnetic optimum tempering temperature (≈500^{0}C)" corresponds to an important mechanical-structural property of the steel material. As far as we know, the tempering temperature T=500^{0}C does not have any specific importance from the point of view of a general industrial application. The material is the hardest just after quenching and any tempering decreases its hardness and increases ductility. The material becomes softer with the tempering temperature increase and for specific applications specific tempering is used.

One of the ideas accompanying the present exploration was our curiosity about available shapes of the obtainable degeneration functions of the MAT descriptors, $D(h_{A0},h_{B0},T)$. The curves in Figs.4-6 show the most typical of them and discover that degeneration functions of any of the investigated descriptors are not monotonous and contain their maximum mentioned above. Figs.4-5 present the optimum MAT descriptors μ and B compared to the correspondent traditional parameters (but at substantially lower levels of magnetization of the samples, though), which do not differ principally from each other. The MAT descriptor $d\mu/dh_B$ shows, however, an interesting potentiality of the MAT method, namely the extended choice of degeneration curves (of their shape and/or sensitivity) available by a task-oriented preference of the descriptor's working field-point. The curve $d\mu/dh_B$*(1100A/m, 1350A/m)* corresponds to the typical shape similar to Figs.4-5. However, the maximum of the curve $d\mu/dh_B$*(1280A/m, 1730A/m)* in Fig.6 is shifted towards so high values of T that the curve becomes monotonous in the main part of the scope of interest. This later curve broadens the monotonous range up to T=580^{0}C and approaches the attempted task - of employing the MAT measurement of the martensite toughening process as a tool for non-destructive quality testing of tempered steel - closer than any of the traditional magnetic parameters. The presently applied testing technique of the martensite-toughening is based on a cumbersome mechanical detection of the material hardness, see the HV10-curve in Fig.4. The classical coercive force seems promising from this point of view, as it is a more-or-less monotonous not-increasing function of T. However, sensitivity of this curve at the interesting range of T is very poor.

The MAT method employs mainly relative measurements and compares values obtained on the investigated sample with those obtained on a reference one. This requires to keep the measurement conditions always as close to each other as possible. One of the important features of the measurement is the constant value of the triangular current slope from one compared measurement to another and also within a single measurement from one "triangle" to another. In the case of magnetically closed samples, this precaution makes the measured signal proportional to differential permeability, see (1), and helps in understanding of physical meaning of the considered descriptors. However, their physical meaning should not be identified with that of the corresponding traditional magnetic parameters (if they exist) and should be taken with some reserve. The descriptors are generally values of the minor loops magnitude (and/or slope and/or curvature) at the selected field-points, which are helpful for the shape and sensitivity of their degeneration functions.

The value of the slope of the triangular current represents also speed of the magnetization process in the sample with all relevant consequences. It influences on level of the induced signal and time-stretch of the measurement, but also level of the created eddy currents and shape of the output signal. For the sake of simplicity and of better understanding of what magnetic processes take place during the measurements, the slope is now kept as low as possible in order to stay close to the quasistatic behavior of the material. Application of higher slopes can be found beneficial in special cases, however, and we assume to use them later.

6. Conclusions

It can be concluded that the MAT method was successfully applied to monitoring of magnetic response of thermally processed low-alloyed construction steel. The optimum MAT descriptors shown in Figs. 4-5 were found at least *not less sensitive* and they were determined at substantially *lower levels of the samples' magnetization* than the corresponding traditional parameters.

The magnetic response of the thermally processed construction steels is characterized by a typical peak at about T=500^{0}C and a qualitative explanation of such a conduct was suggested. This non-monotonous behavior of the magnetic parameters (both traditional and MAT) was found not quite satisfactory for the originally intended industrial application (magnetic testing of quality of steel tempering), which would prefer a monotonous dependence on T up to T=660^{0}C.

However, the $d\mu/dh_B$ MAT descriptor shown in Fig.6 offered the monotonous span of the curve up to T=580^{0}C. Even though this is not satisfactory for the desired application yet, it illustrates the potential power of adaptation of the shape of the MAT degeneration curves as close to the intended task as possible.

Acknowledgement

The authors appreciate financial support of the project 101/02/0236 by the Grant Agency of the Czech Republic and of the project K1010104.

References

[1] D.C.Jiles, Review of magnetic methods for nondestructive evaluation. NDT International, **21** (1988) 311
[2] M.K.Devine, The magnetic detection of material properties. J.Min.Met.Mat. (JOM), (1992) 24
[3] V.Basso, G.Bertotti, A.Infortuna, M.Pasquale, Preisach model study of the connection between magnetic and microstructural properties of soft magnetic materials, IEEE Trans.Magnet., **31** (1995) 4000
[4] O.V.Perevertov, Application of the Preisach model to the magnetization process in steels. J. Phys. D: Appl. Phys. **35** (2002) 2467
[5] I.Tomáš, Non-Destructive Magnetic Adaptive Testing of Ferromagnetic Materials, submitted for publication in the J. Mag. Mag. Mat.
[6] I.D. Mayergoyz, Mathematical models of hysteresis, Springer-Verlag, New York, 1991

Electromagnetic Nondestructive Evaluation (VIII)
T. Sollier et al. (Eds.)
IOS Press, 2004

Low Frequency Excitation of MFES

S. Nagata[1], M. Enokizono[2], T. Chady[3] and R. Sikora[3]
[1]*Miyazaki University, 1-1 Gakuenkibanadai-nishi, Miyazaki, 889-2192, JAPAN*
[2]*Oita University, 700 Dannoharu, Oita, 870-1192, JAPAN*
[3]*Szczecin Technical University, ul. Sikorskiego 37, 80-310, Szczecin, POLAND*

Abstract. This paper discusses about the low frequency excitation using a multi-frequency excitation and spectrogram (MFES) method. The relation between frequency characteristics of Hall sensor and excitation are discussed. In order to improve the frequency characteristics of Hall sensor, V-I converting amplifier for low frequency excitation is proposed. From the measurement results, introducing the Hall sensor array leads high resolution to detect small cracks.

Introduction

A demand for NDT technique recently increases. Especially, developments for detecting and estimating a very small crack such as the stress corrosion crack (SCC) seem to become more and more important. In order to detect a very small crack, comparatively small sensor is needed.

MFES method had given a lot of good results for precise crack characterization [1][2]. Those results were, so far, picked by the search coils wounded at the sensor yokes. We had already proposed a combination of the MFES and very small Hall element sensor for lower frequency excitation [3]. It is because that the search coil would not be so sensitive when driven by lower frequency (under 100 Hz) waveform. When the magnetic specimens such as iron or steel plates are under testing, lower frequency excitation is necessary due to the serious skin effect. The combination of the MFES method and Hall element sensor is considered to be also suitable for detecting very small crack and estimating the crack shape, therefore, we study the validity of this combination.

On the other hand, MFES requires stable and wide frequency property for the sensor. More than ten sinusoidal waves are superimposed for an excitation waveform in MFES system. It means that the highest component frequency should be more than 10 times of the lowest one.

However, Hall element sensor does not show good characteristics under high frequency excitation. It shows a good sensitivity under 100 Hz rather than search coil, however, shows the remarkable decay according to the frequency increase.

In this paper, we present the V-I converting amplifier that improves the frequency characteristics of Hall element sensor, and discuss about the relation between the positions of Hall element sensors and the spectrograms.

1. MFES system

Figure 1 shows the constructions of the developed block scheme of whole MFES system. The waveform data are stored in EPROM. In this study, the system clock frequency provided by the A/D converter is fixed to 62.5kHz. This clock frequency determines the lowest frequency component of MFES as 61.0Hz (62.5kHz/1024 samples). 19 components from 61Hz through 610Hz every 30.5Hz are selected and superimposed into the excitation waveform.

Figure 2 illustrates the construction of sensor block. Under the sensor yoke leg, Asahi-Kasei's GaAS Hall elements HG-106C[4] are set. The package size of an HG-106C is about

1.5 mm x 1.5 mm (not includes sensor leads), therefore they can be set almost everywhere near the sensor yoke. However, the heights are 0.5 mm, the lift off should be larger when the Hall sensors are set between the sensor yoke and the specimen. In our case, the lift off is about 1.3 mm.

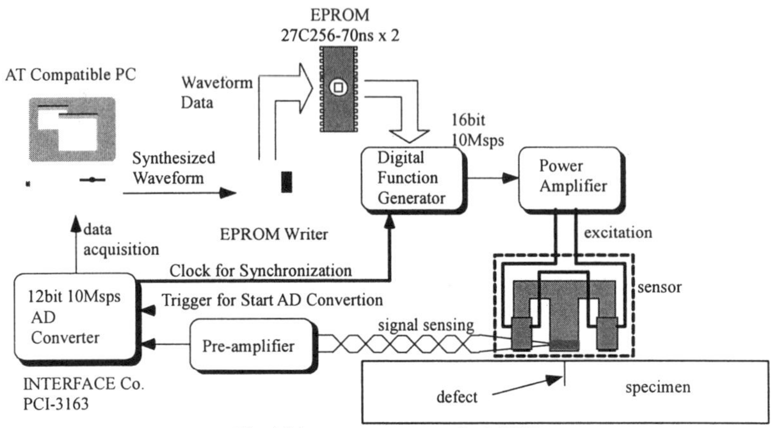

Fig. 1 Measurement system for MFES.

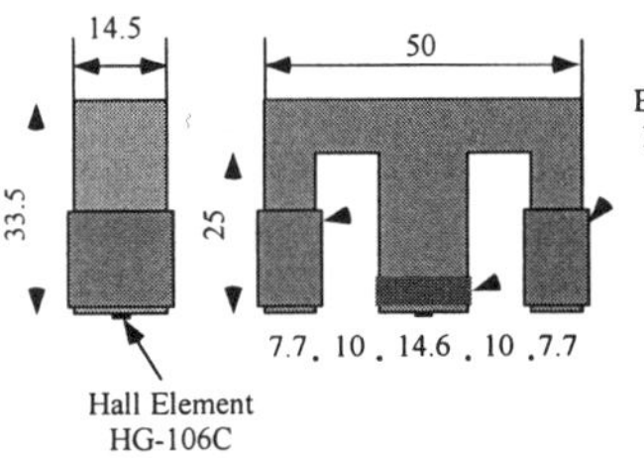

Fig. 2 Constructions of sensor block

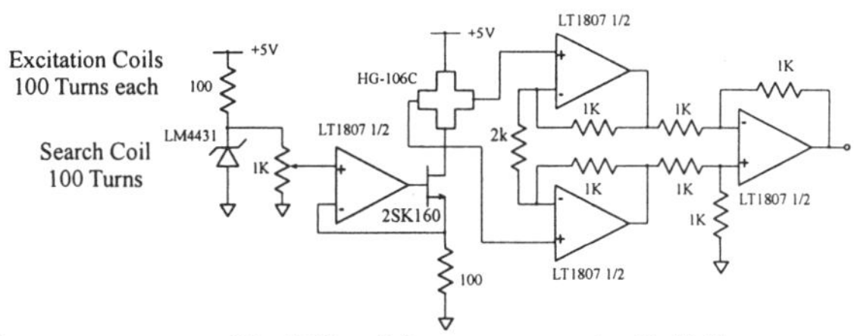

Fig. 3 Circuit for measurement with Hall sensor

Fig. 3 shows the electronic circuit for each Hall sensor. A Hall element is driven by constant current (5.0mA) due to the improvement of the thermal characteristics of GaAs Hall element.

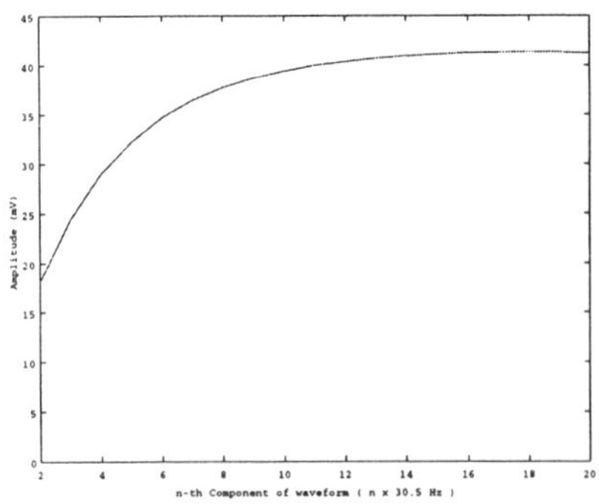

Fig. 4 Frequency dependencies of induced voltage of the search coil. The exciting winding is driven by the voltage source.

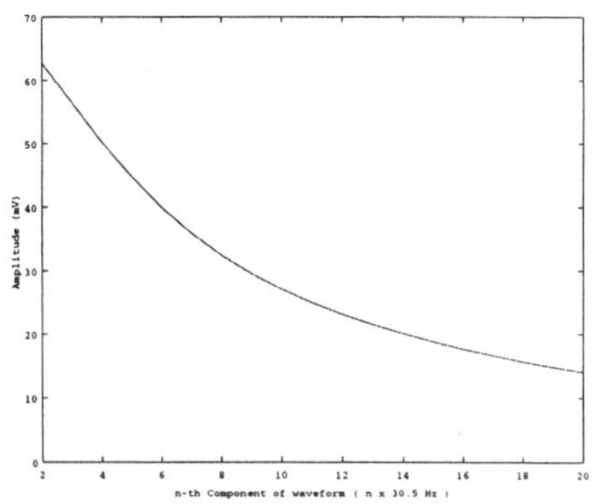

Fig. 5 Frequency dependencies of the Hall sensor output voltage. The output voltage is in proportion to the magnetic flux density B. The exciting winding is driven by the voltage source.

When the specimen under testing is made of the magnetic material such as iron or steel, it is well known that a serious skin effect occurs. Even when exciting frequency is only 50 Hz, the skin depth might be two or three (mm). It means that either flux flow or eddy current flow is not perturbed by the opposite side cracks. In order to penetrate the flux and the eddy current, the exciting frequency should be decreased.

Figure 4 and Figure 5 show the frequency characteristics of two kinds of sensor, the search coil and Hall sensor. These characteristics are measured when the sensor block is constructed as shown in Fig. 2, and is driven by the MFES waveform including 19 components mentioned above. The horizontal axes are the component number.

When low frequency excitation less than 100(Hz), the search coil is not suitable for sensing device. Figure 4 shows the frequency dependencies of the search coil when the MFES waveform is applied. The frequency characteristics of the search coil are almost constant above 500Hz up to about 100kHz. Within this frequency range, the search coil is suitable for MFES. In low frequency (less than 100Hz) excitation, however, the sensitivity of the search coil decreases as shown in Fig.4.

On the other hand, the frequency dependencies when the Hall sensor is used are shown in Fig. 5. The Hall sensor is sensitive in lower frequency range, otherwise, in higher frequency range the characteristics decay. The desirable frequency dependencies for MFES method are almost constant ones within every frequency range.

However, we think that the frequency characteristics shown in Fig. 5 depend on the excitation source.

When using search coils, the excitation by the voltage source is reasonable. The search coil picks up the induced voltage E written in equation (1),

$$E = -n\frac{\partial \mathbf{B}}{\partial t}S \propto V\,, \tag{1}$$

where, n is the number turns of search coil winding, $\mathbf{B}$ the magnetic flux density, S the interleakage cross section, and the exciting voltage V. Induced voltage E should be proportion to the exciting voltage V. Usually, the exciting winding is driven by the voltage value, which is supplied by the power amplifier.

On the other hand, when using the Hall sensor, exciting winding should be driven by the current source. The Hall sensor picks up the magnetic flux density $\mathbf{B}$. Assumed that the sensor yoke has a linear magnetic property, $\mathbf{B}$ would be in proportion to the exciting current I. Exciting current I is represented by the following simple Ohm's law, and it becomes lower and lower by a angular frequency ω increase when exciting winding is driven by constant voltage.

$$|\mathbf{B}| \propto I = \frac{V}{Z} = \frac{V}{R_L + j\omega L}\,, \tag{2}$$

where, Z is the impedance, R_L the ohmic resistance, and L the inductance of exciting winding, respectively. It is the main reason that the frequency characteristics shown in Fig. 5 decrease as the frequency increase.

2. V-I converting Amplifier and Hall sensor

In order to improve the excitation suitable for Hall sensor, V-I converting amplifier (shown in Fig. 6) is employed for excitation power amplifier. The input of this amplifier is voltage V_{in}. The amplifier controls the current value so that the current I_{out} flows into the load as proportional to input voltage V_{in}. In this circuit, the conversion ratio is 0.1 (A/V). This conversion ratio can be easily changed by replacing the resistor value R (10 (Ω)).

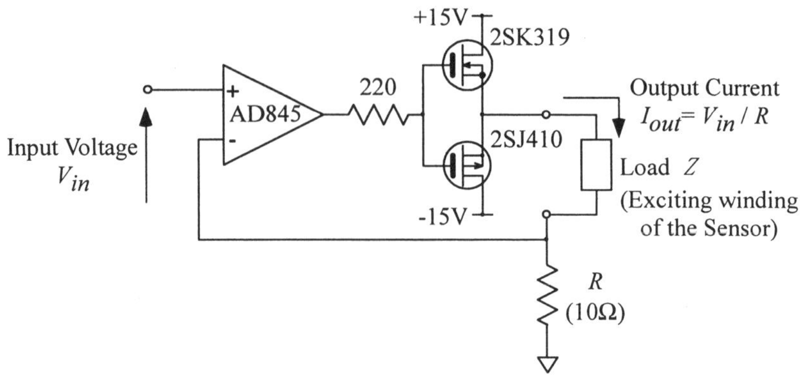

Fig. 6 V-I converting amplifier (0.1A/V).

Figure 7 shows the improved frequency dependencies of the Hall sensor by using the V-I converting amplifier instead of usual power amplifier. The frequency dependencies are adjusted at higher frequency range. Although amplitudes of frequency components become smaller than the ones shown in Fig. 5, it can be solved easily by increasing the gain of the pre-amplifier.

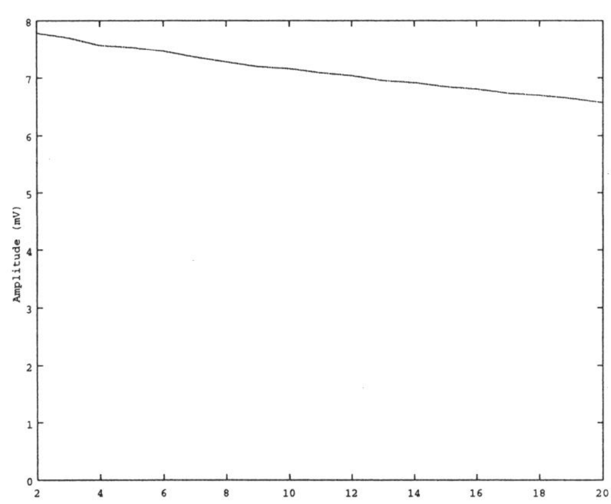

Fig. 7 Frequency dependencies of Hall element output are improved by using V-I converting amplifier. (exciting winding is driven by the current source)

3. Hall sensor array

The effective positions of the Hall element sensors should be clarified. Figure 8 shows the Hall sensor array, which contains 12 Hall element sensors. The Hall element sensor HG-106A is so small that it can be set freely everywhere of sensor block. In order to arrange the surface mount package of HG-106A, we introduce a through hole print circuit board (70 mm x 70 mm, 0.2mm depth).

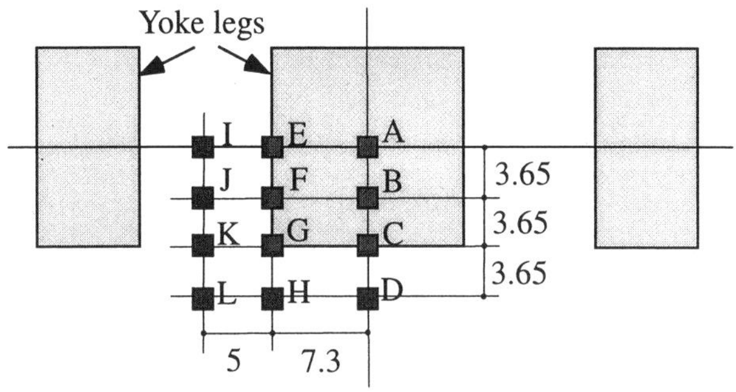

Fig. 8 Arrangement of Hall sensor array (unit: mm).

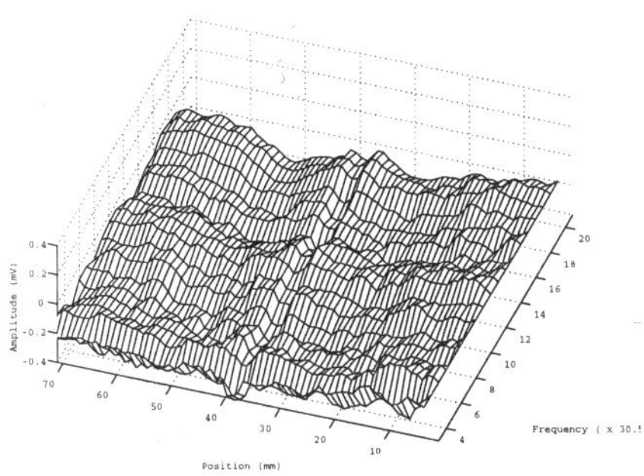

(a) Sensor A (crack at 36mm position)

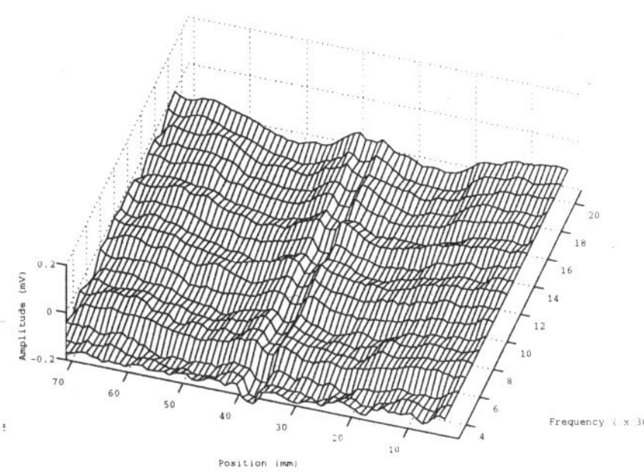

(b) Sensor B (crack at 36mm position)

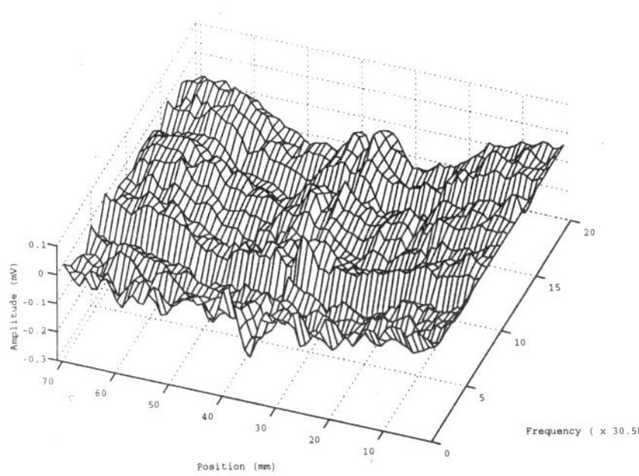

(c) Sensor C (crack at 36mm position)

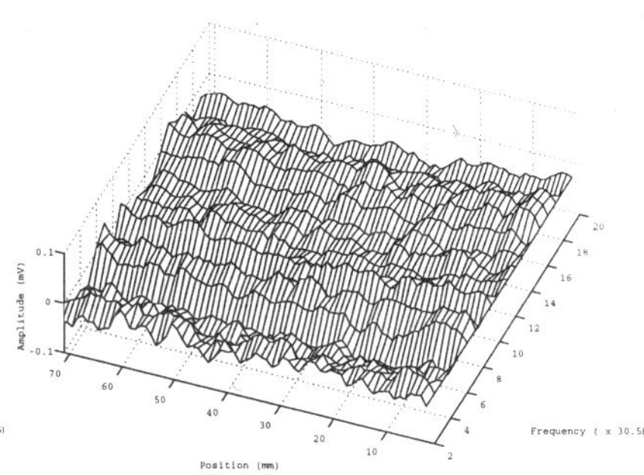

(d) Sensor D (crack at 36mm position)

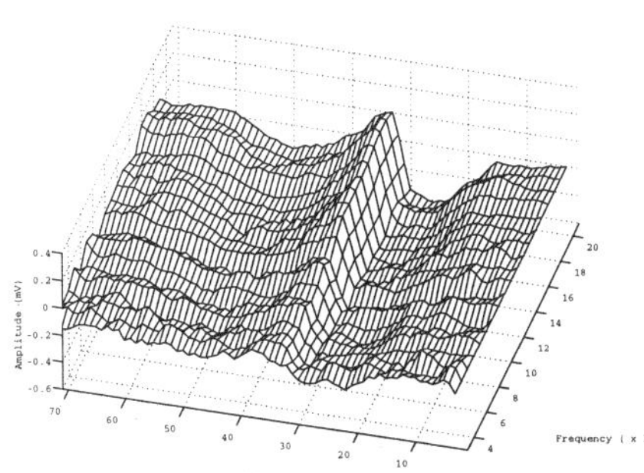

(e) Sensor E (crack at 28.7mm position)

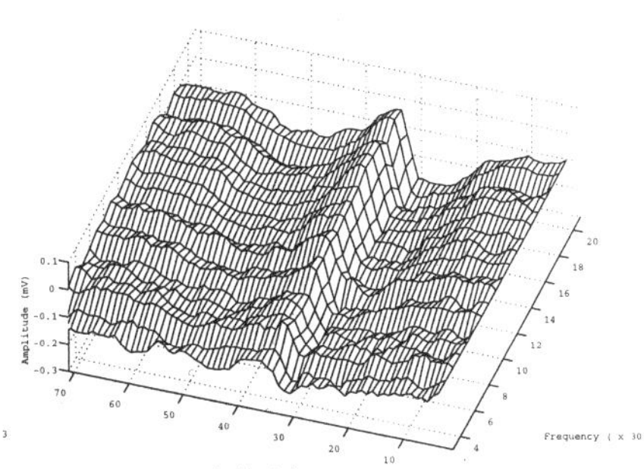

(f) Sensor F (crack at 28.7mm position)

Fig. 9 Spectrograms picked by the sensors from A through L

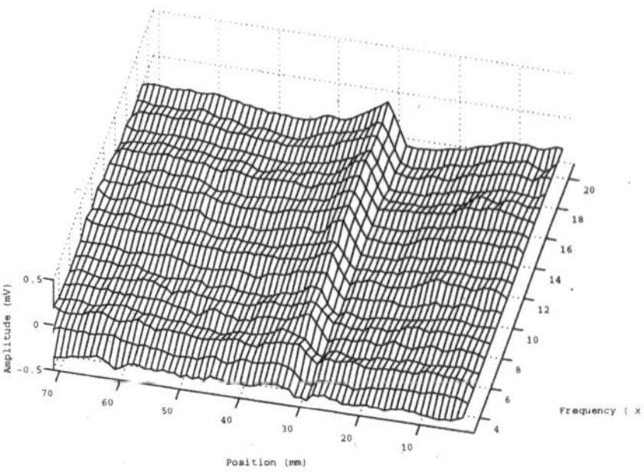

(g) Sensor G (crack at 28.7 mm position)

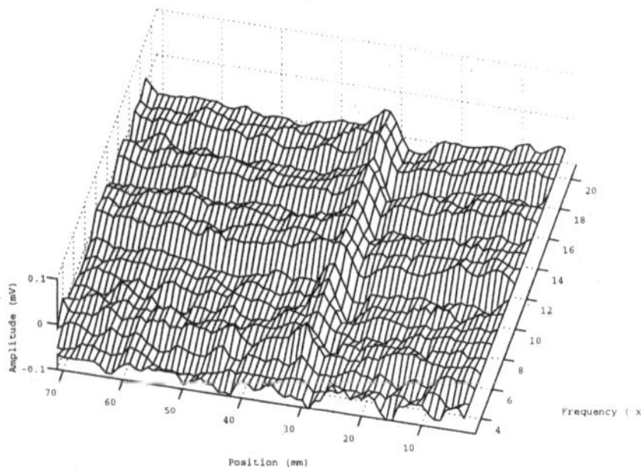

(h) Sensor H (crack at 28.7mm position)

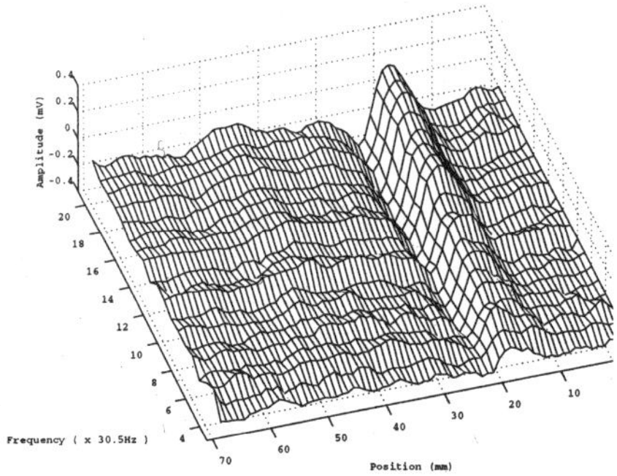

(i) Sensor I (crack at 23.7mm position)

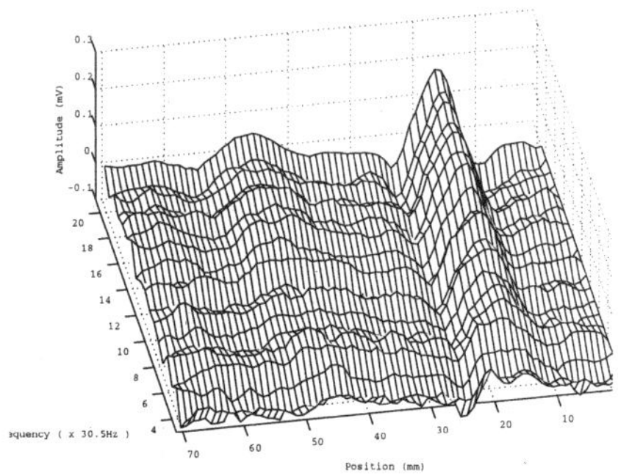

(j) Sensor J (crack at 23.7mm position)

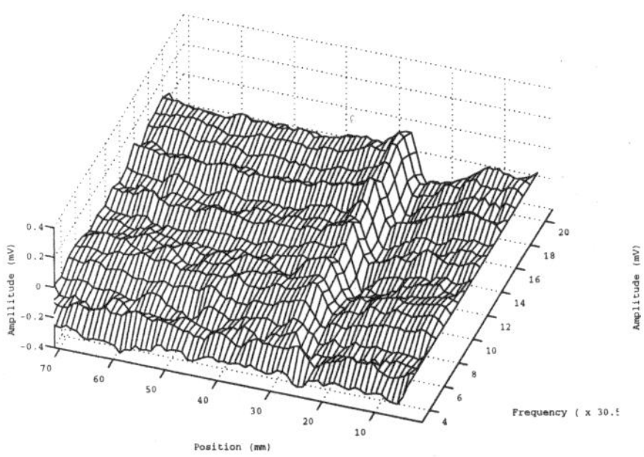

(k) Sensor K (crack at 23.7mm position)

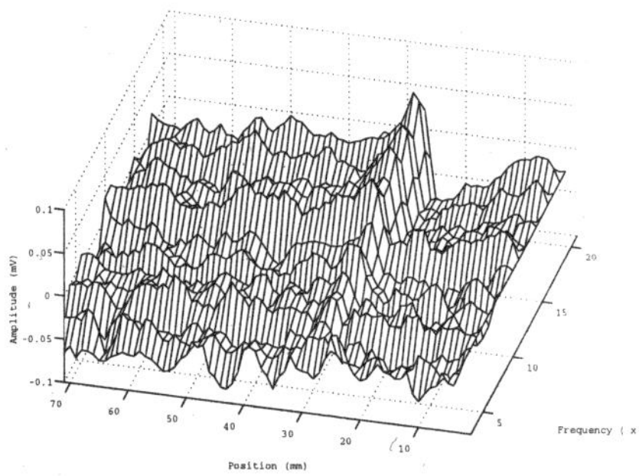

(l) Sensor L (crack at 23.7mm position)

Fig. 9 Spectrograms picked by the sensors from A through L.

4. Results and Discussions

The specimen under testing is a steel plate SS-400, 200mm length x 100 mm width x 5 mm depth, with 0.6mm width x 2 mm depth (40%) surface side crack.

Figures 9 (a)-(l) show the distributions of MFES spectrogram of corresponding Hall sensor A to I.

Figures 9 (a)-(c) show remarkable sharp v-shape signal near the position 36mm (crack position), which indicates the presence of the crack. Also the signal appears within whole frequency range. These signals are not so large relative to the background noise level, but they indicate exact crack position. Furthermore, with their sharp v-shape signal, it can be considered that the width of the crack is easily identified. It may have keys that may be used to identify the crack sizes and shapes. In Figure 9 (d), no signal is shown. The sensor position D is not effective.

Figures 9 (e)-(h) shows the signal picked up near the yoke leg edge. Yoke leg edge is the position which is supposed to sensitive because the magnetic flux may concentrate. Signals indicate the presence of the crack appear from (e)-(h), however, do not seem remarkable sharp signal like shown in (a)-(c). The signals of defect once fall down and arise up according to the sensor moves near the crack. Especially, Figs. 9 (e)-(g) show that the signal values rapidly change very near the crack position. These sensor positions are useful to know the presence of the defect from a signal-to-noise ratio point of view. Figure 9 (h) also shows the small but characteristic signals.

Figure 9 (i)-(l) show the signal distributions when the Hall sensors are set between two yoke legs. These positions are supposed to be more difficult to detect the crack, however, signals indicate the presents of the crack.

5. Conclusions

The combination of MFES and Hall sensor was discussed. The frequency dependencies of Hall element senor have been improved by using the present V-I converting amplifier instead of usual power amplifier. Hall sensor array has been employed and it has enabled MFES to detect the crack within high resolution and has revealed the position dependencies of crack signals.

References

[1] T. Chady, M. Enokizono, "Eddy Current Testing of Stainless Plates by Using Matrix Sensor," *Proceedings of the 7th Magnetodynamics Conference*, pp. 107, April 1998.

[2] T. Chady, M. Enokizono, R. Sikora, "Crack Detection and Recognition Using Eddy Current Differential Probe," *IEEE Transactions on Magnetics*, vol. 35, No. 3, pp. 1849-1852, May 1997

[3] S. Nagata, T. Chady, M. Enokizono, R. Sikora, "Development of Practical System of Multi Frequency Excitation and Spectrogram Method", proceedings of eNDE 2002.

[4] http://www.asahi-kasei.co.jp/ake/en/ms/index.htm

[5] S. Nagata, T. Chady, M. Shidouji, M. Enokizono, "Development of practical MFES system for concrete material", Electromagnetic Nondestructive Evaluation (VI), IOS-Press, pp.104-107, 2002.

Electromagnetic Nondestructive Evaluation (VIII)
T. Sollier et al. (Eds.)
IOS Press, 2004

A New Inspection Methodology for Silicon Steel Sheet Quality Evaluation Based on Image Helmholtz Equation

*Hisashi ENDO, **Iliana MARINOVA, *Toshiyuki TAKAGI,
***Seiji HAYANO, and ***Yoshifuru SAITO
**Institute of Fluid Science, Tohoku University,*
2-1-1 Katahira, Aoba-ku, Sendai 980-8577, JAPAN
***Department of Electrical Apparatus, Technical University of Sofia, Sofia 1756, Bulgaria*
****Graduate School of Engineering, Hosei University,*
3-7-2 Kajino, Koganei, Tokyo 184-8584, JAPAN

Abstract. We have previously proposed an image processing methodology based on image Helmholtz equation to examine the quality of silicon steel sheet which is widely used in most of the magnetic apparatus. This new image processing methodology makes it possible to extract the magnetization characteristics by microscope-based observation, evaluating the local and global magnetization curves and iron loss generating parts as the state transition matrix of the Helmholtz equations. We have succeeded in visualizing and quantifying the domain behaviors depending on the physical conditions in order to replace the human oriented inspection.

1. Introduction

In silicon steel sheets widely used in most of the magnetic apparatus, e.g., rotating machines, transformers etc., defects and closure domains increase iron loss so that microscopical quality evaluation is of paramount importance process to produce high quality silicon steel sheets [1]. Heretofore such a task could only be accomplished by skilful experts' eyes as well as elaborate analysis/simulators. To overcome and replace this human oriented inspection, the present study aims to provide a methodology for visualizing and quantifying the domain motion dynamics of silocon steel sheets.

Previously, we have proposed the image Helmholtz equation approach as a dynamic image analysis methodology [2]. Assuming a digital image to be a scalar field leads to the image governing equations. The image Helmholtz equation is a Helmholtz type of partial differential equations expressing the digital dynamic images. The solution of image Helmholtz equation generates the dynamic images. Its state transition matrix equivalently represents the characteristic values of physical dynamic system visualized by finite number of images as an animation. Thus, visualized systems can be evaluated quantitively with the same mannar as the classical control theory.

In the magnetic domain image analysis, assuming the averaged contrast of domain image as an entire flux density B leads that the characteristics of domain motion deduces from the state transition matrix of the image Helmholtz equation. In case of silicon steel sheets, we must take into account the domain structure in depth of the specimen to clarify the lancet domain generation. In the present paper, a series of SEM (scanning electron microscope) domain images is employed. The imaginary parts of the state transition matrix visualize iron loss generating parts. The solutions of image Helmholtz equation reproduce the magnetic domain SEM images at arbitrary magnetized conditions. Moreover, the solutions give the global and local magnetization curves reflecting on the physical condition of target. As a result, comparison between computed and experimented magnetization curves verifies our methodology.

2. Image Helmholtz Equation

2.1 Formulation

The Helmholtz type equations govern many of the physical dynamic systems. Assuming a frame to be described in terms of a scalar field U, any dynamic image can be presented as a solution of the Helmholtz equation [2]:

$$\nabla^2 U + \varepsilon \frac{\partial}{\partial \alpha} U = -\sigma, \quad (1)$$

where ε and α denote the moving speed parameter and transition variable, respectively. The first and second terms on the left side of (1) represent the spatial expanse and transition of a frame to the variable α, respectively. An image source density σ can be given by the Laplacian of final image U_{Final}, meaning that the static image is obtained by

$$\nabla^2 U_{Final} = -\sigma \, . \quad (2)$$

Note that the parameter ε in (1) is unknown. However, this means that it is possible to extract the characteristic of physical properties from the visualized dynamic image if it is possible to determine ε from given frames of the dynamic image. Calculation of the parameter ε is the key to visualize the characteristics of dynamic system.

2.2 General Solution

Discretization of (1) using pixel arrangement as nodal points derives the system of equations. The modal analysis of discretized (1) gives the general solution:

$$U(\alpha) = \exp(-\Lambda\alpha)(U_{Start} - U_{Final}) + U_{Final}, \quad (3)$$

where U_{Start} and U_{Final} represent the initial and final frames of the dynamic image, respectively [2]. Moreover, $\exp(-\Lambda\alpha)$ denotes the state transition matrix. The values ε and σ in (1) are reduced into the matrix Λ and final image U_{Final}, respectively. Because of the parameter ε in (1), the state transition matrix is unknown. This means that we have to determine the state transition matrix from given frames of dynamic image.

2.3 State Transition Matrix

If the solution $U(\alpha)$ is assumed to be one of the frames of dynamic image, it is possible to determine the elements in matrix Λ by modifying (3):

$$\Lambda = -\frac{1}{\alpha} \ln\left(\frac{U(\alpha) - U_{Final}}{U_{Start} - U_{Final}}\right). \quad (4)$$

Since the matrix Λ is a diagonal matrix, then (4) is carried out at each corresponding pixel value of three distinct images.

When we solve the differential equations, all medium parameters are given. However, in this image analysis methodology, determination of the matrix Λ by means of (4) is capable of extracting the dominant characteristics of visualized physical dynamics. This is a distinguished feature of our image analysis approach. As a result, substituting the matrix Λ into (2) yields dynamic image generation at the arbitrary α.

3. Quality Evaluation of Silicon Steel Sheets

3.1. SEM Domain Images and Image Helmholtz Equations

Figure 1 shows the SEM domain images of a silicon steel sheet under the distinct magnetized states [3]. The specimen is the ORIENTCORE·HI-B (Nippon Steel Corporation product) without surface coating and its thickness is 0.23 mm. The backscattered electron observation (Type-II) is carried out at 160 kV acceleration voltage. At this condition, the domain patterns about 10μm depth from the surface of specimen could be visualized as shown in Fig.1 [4]. The external field is applied to rolling direction with sloping excitation. Table 1 lists the conditions of domain image measurement used in this paper.

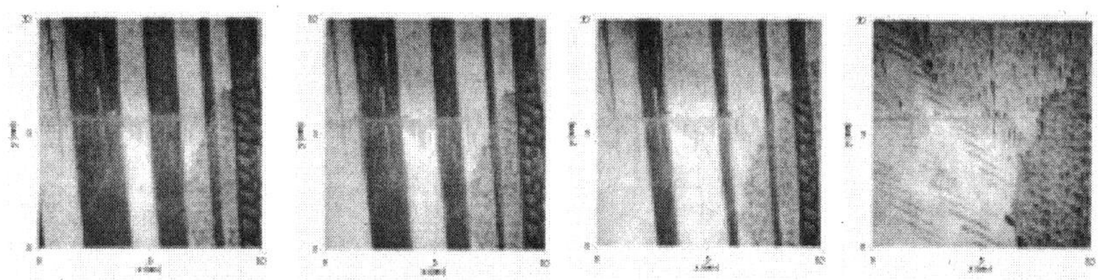

(a)H=0.0 A/m, B=0.0 T (b)H=2.9 A/m, B=0.1 T (c)H=9.3 A/m, B=1.6 T (d)H=115.4 A/m, B=1.88 T

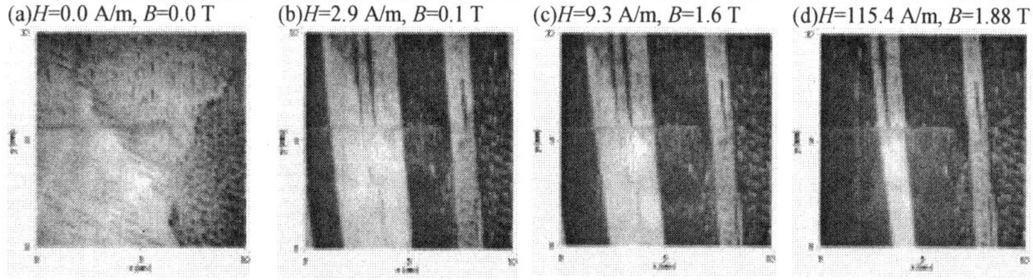

(e)H=236.3 A/m, B=1.9 T (f) H=0.0 A/m, B=1.73 T (g)H=-4.7 A/m, B=-1.22 T (h)H=-9.0 A/m, B=-1.56 T

Fig. 1 Magnetic domain SEM images of the ORIENTCORE HI-B produced by Nippon Steel Co.(100x100 pixels, 0.1 mm/pixel).

Table 1 Measured Domain Images.
H : External magnetic field intensity, B : Flux density

Image No.	H(A/m)	B(T)	Image No.	H(A/m)	B(T)
1	0.00	0.00	13	214.13	1.93
2	2.85	0.10	14	160.37	1.92
3	9.26	1.63	15	98.68	1.91
4	24.16	1.73	16	54.66	1.84
5	30.23	1.78	17	28.53	1.83
6	54.59	1.84	18	3.73	1.77
7	84.92	1.86	19	0.00	1.73
8	115.39	1.88	20	-4.60	1.73
9	160.69	1.90	21	-5.95	-0.06
10	236.32	1.92	22	-7.45	-1.43
11	324.31	1.95	23	-9.07	-1.56
12	269.64	1.95	24	-11.50	-1.62

Supposing a domain image shown in Fig.1 to be composed of a scalar field U as flux density, the dynamics of magnetic domain can be represented by the image Helmholtz equation. In magnetizing state, the domain motion is caused by applied magnetic field H, so that the transition variable α in (3) is replaced by the magnetic field intensity H. Namely, (1) and (3) are respectively rewritten by

$$\nabla^2 U + \varepsilon \frac{\partial}{\partial H} U = -\sigma, \tag{5}$$

$$U(H) = \exp(-\Lambda H)(U_{Start} - U_{Final}) + U_{Final}. \tag{6}$$

The matrix Λ is determined by given domain images. The elements in the i-th matrix Λ_i are determined from a series of distinct SEM images in much the same way as (4), i.e.,

$$\Lambda_i = -\frac{1}{H_{i+1} - H_i} \ln\left(\frac{U_{i+1} - U_{i+2}}{U_i - U_{i+2}}\right), \; i = 1,2,\ldots, p-2, \tag{7}$$

where the subscript i refers to a domain image number; p denotes the number of used domain images; U_i and U_{i+2} correspond to the initial and final domain images, respectively. 24 domain images listed in Table 1 were used, therefore, 22 matrices Λ were computed. As is well known in control theory, the elements in matrix Λ correspond to the modes of dynamic system. Extracted matrix Λ by means of (7) equivalently represents the modes of magnetization. In this case, the real and imaginary parts of the matrix Λ can be represented in-phase and 90-degree phase different components to the applied field H. Therefore, considering Λ derived from the SEM domain images leads to visualization of iron loss generating parts.

3.2. Visualization of Domain Dynamics

Figure 2 shows the real and imaginary parts of the elements in the matrix Λ. The elements are displayed in accordance with these of the domain pattern in Fig.1. In the weak field intensity, the real parts represent magnetic boundary displacements and magnetic domain movements [Figs. 2(a) and (b)]. On the other hand, there are non-zero values in the imaginary part of Fig. 2(b), suggesting the phase delay component to the applied field H. This visualizes the force against the applied field at the grain boundary. In the high field intensity, Figs. 2(c) and (d), the magnetization process is mainly carried out by rotation of magnetization and the iron loss is caused by the lancet domains shown as small-triangular shape domains in Figs.1(d) and (e). Thus, based on the matrix Λ, it has been shown that the magnetic domain motion dynamics, i.e., behaviors of magnetic domain and iron loss generating processes, can be visualized.

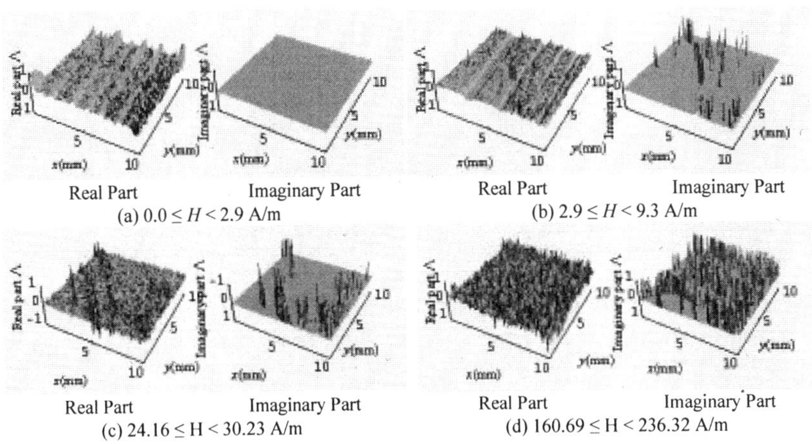

Fig. 2 State transition matrices derived from SEM domain images by means of (7).

3.3. Global Magnetization Curve Generation

Extracted matrices Λ shown in Fig.2 are characteristic value matrices to the applied field H. Equations (6) and (7) generate the domain images at arbitrary excitation. Furthermore, since the contrast of domain images represents the magnetized degree of magnetization, then computing averaged contrast of an entire domain image gives an entire flux density. Fig.3 shows each frame of the animation images generated by means of (6) and the magnetization curves. Even though the domain images represent a limited area of the specimen, the experimental result shown in Fig.4 supports our methodology.

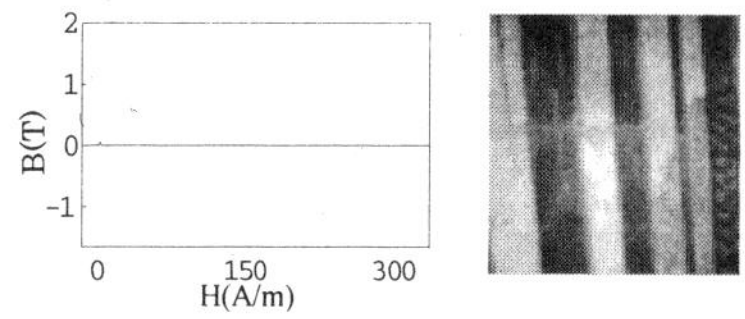

(a) H =1.43 A/m, B=0.05 T

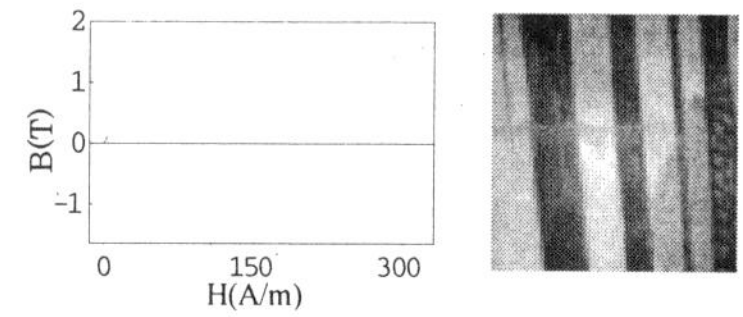

(b) H =3.21 A/m, B =0.1 T

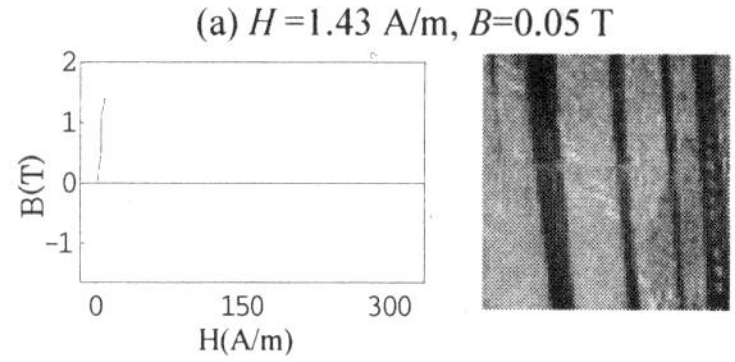

(c) H =7.06 A/m, B =1.39 T

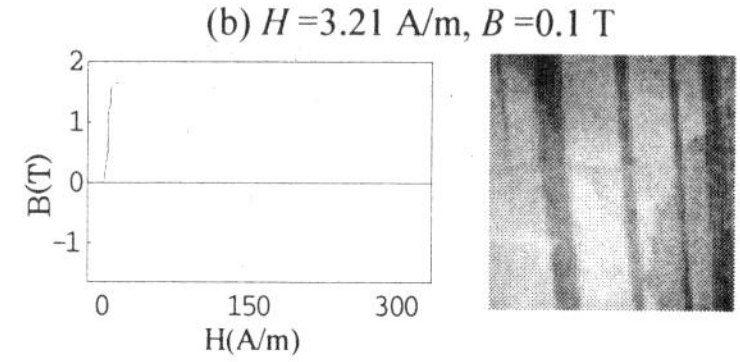

(d) H =13.45 A/m, B =1.65 T

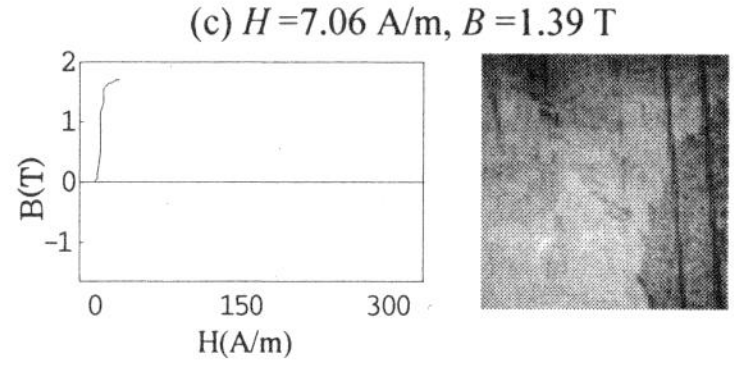

(e) H =20.90 A/m, B =1.71 T

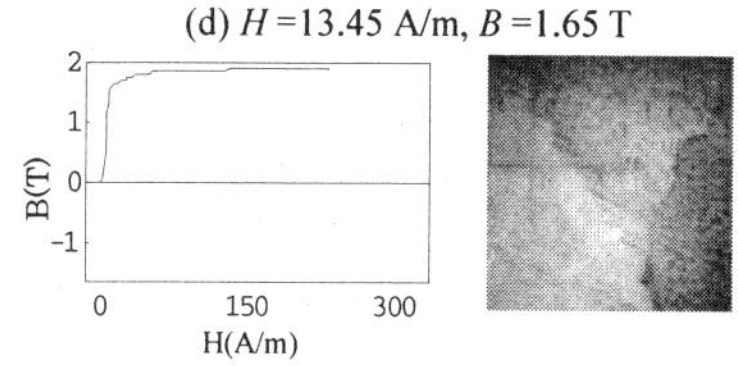

(f) H =228.47 A/m, B =1.92 T

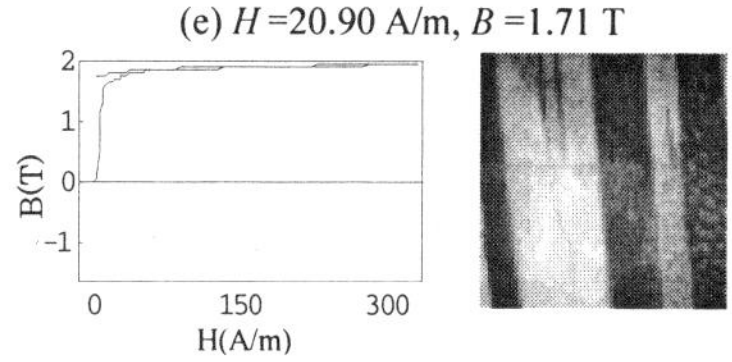

(g) H =2.32 A/m, B =1.76 T

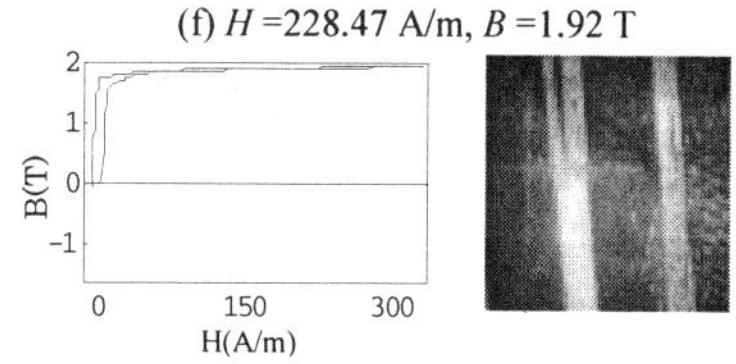

(h) H=-6.39 A/m, B =-0.06 T

Fig. 3 Reconstructed domain patterns and magnetization curves by means of (6).

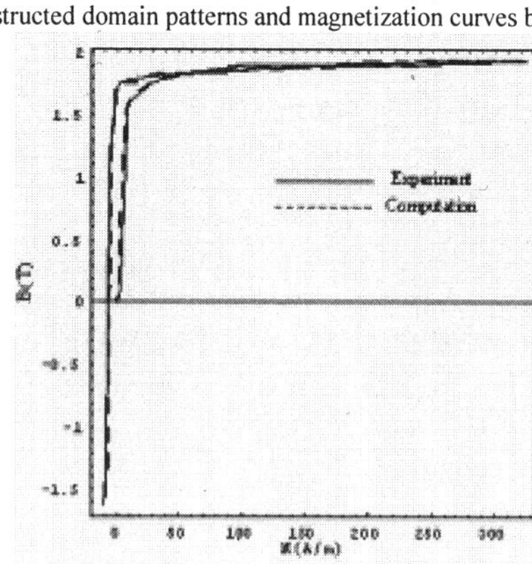

Fig. 4 Comparison between computed and experimented magnetization curves.

3.4. Local Magnetization Curve Generations

Focusing on the particular points on the domain images, it is possible to generate the local magnetization curves as shown in Fig.5. Fig.5(b) shows the selected parts for drawing local magnetization curves. The local magnetization curve in this methodology is based on that of entire specimen described in *Section 3.3*. Our methodology makes it possible to estimate the magnetization processes reflecting on the physical condition of the specimen. Observing the magnetization curves in Fig.5 suggests distinguished features that all of the curves exhibits the discontinuity when transiting from the irreversible magnetic domain movement to reversible rotational magnetization regions. This means that our image Helmholtz methodology is capable of extracting the closure as well as strained domains instead of the skillful experts' eyes.

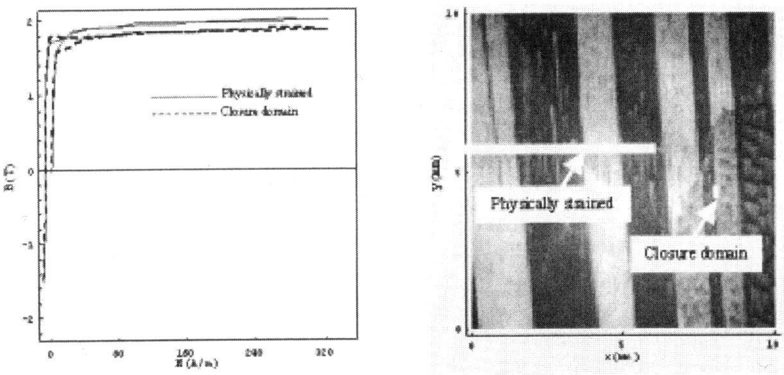

(a) Local magnetization curves

(b) Selected parts for local magnetization curve generation

Fig. 5 Local magnetization curve generations.

4. Conclusion

As shown above, this paper has proposed the novel nondestructive silicon steel evaluation methodology, which is able to extract the lancet as well as strained domains instead of the skillful human eyes. When dynamic image of the silicon steel sheet is obtained, then it is possible to extract the physical parameters of magnetization by means of the state transition matrices. It is obvious that this new nondestructive silicon steel quality evaluation can be implemented in a fully computerized system.

Acknowlededements

The authors are grateful to Dr. C. Kaido, Dr. M. Fujikura and Dr. H. Mogi, Nippon Steel Corporation, for providing the measurement devices and their fruitful discussion.

References

[1] H. Endo, S. Hayano, Y. Saito, M. Fujikura, and C. Kaido, Magnetization Curve Plotting from the Magnetic Domain Images, *IEEE Trans. Magn.*, **37**(4) (2001), 2727- 2730.

[2] H. Endo, S. Hayano, Y. Saito and T. L. Kunii, A Method of Image Processing and its Application to Magnetodynamic Fields, *Trans. IEE of Japan.*, **120-A**(10) (in Japanese) (2000), 913-918.

[3] T. Nozawa, T. Yamamoto, Y. Matsuo, and Y. Ohya, Effects of Scratching on Losses in 3-percent Si-Fe Single Crystals with Orientation Near (110)[001], *IEEE Trans. Magn.*, **15**(2) (1979), 972-981.

[4] K. Tsuno and T. Yamamoto, Observed Depth of Magnetic Domains and Their Walls in Scanning Electron Microscopy, *Phys. Stat. Sol. (a)*, **37** (1976), 437-448.

Advances in Electromagnetic Nondestructive Techniques

Electromagnetic Nondestructive Evaluation (VIII)
T. Sollier et al. (Eds.)
IOS Press, 2004

Enhancement of magneto-optic images

Pradeep Ramuhalli, Fan Yuan, Unsang Park, Jason Slade, Lalita Udpa
NDEL, Dept. of Electrical and Computer Engineering,
Michigan State University, East Lansing, MI 48824 (USA)

Abstract. Magneto-optic (MO) imaging produces images of magnetic flux leakage from surface and sub-surface defects. The presence of domain boundaries in the MO sensor produces a characteristic artefact in the images that renders detection of cracks difficult. This paper presents two different approaches for the removal of domain boundaries in MO images. The first approach uses static methods that rely on various 2D filtering and image processing algorithms to filter out the noise. The second approach is a dynamic method, and uses successive frames from the MO imager to segment out signals from objects of interest in the data. Results of applying the two approaches to several experimentally acquired MO images are presented and compared.

1 Introduction

Magneto-optic imaging (MOI) is a relatively new technology that produces images of magnetic flux leakage from surface and subsurface defects [1]. The magneto-optic/eddy current nondestructive testing instrument relies on the principles of Faraday induction and magneto-optic rotation [1] - [4]. In magneto-optic imaging, an alternating current carrying foil serves as the excitation source and induces eddy currents in a conducting test specimen. Presence of discontinuities such as fasteners, seams, corrosion and cracks under fasteners cause bending of eddy currents resulting in a normal component of flux. An MOI sensor made of garnet with easy axis of magnetization perpendicular to the sample is placed above the foil. If linearly polarized light is incident normally on the garnet sensor, the plane of polarization of light is rotated. When the reflected light is viewed through the analyzer, local occurrence of normal magnetic flux is seen as a "dark" or "light" area in the magneto-optic image depending on the direction of magnetization.

A major source of noise and degradation in MOI data is due to magnetic domain boundaries in the garnet sensor. The domain boundaries appear as small, filament-like structures in MO images (Figure 3 (a)) and can severely mask presence of cracks and corrosion. The presence of this textured background in MO images due to domain structures makes detection of cracks and corrosion difficult. This paper presents signal-processing methods for eliminating the noise due to domain boundaries and enhancing signals due to objects of interest. Two separate approaches are presented, namely, static and motion based. Two algorithms using the static approach are presented. The first algorithm uses the continuous wavelet transform (CWT) to separate out the noise, which is at a smaller scale, from crack signals at larger scales. An alternate static method, Threshold Averaging Filter, uses a combination of conventional and morphological image processing algorithms for eliminating the noise. The second approach, however, is dynamic, and uses the fact that, as the MO imager is moved over the sample, the noise due to domain boundaries in the sensor remains static relative to the sensor whereas objects and cracks in the specimen are in relative motion. A motion detection algorithm can thus separate out the defect signals from noise. Although the motion based approach can prove to be very effective, the results can be sensitive to the direction of scan.

The rest of this paper is organized as follows. Sections 2, 3 and 4 describe the three algorithms to noise mitigation that are proposed in this paper. Section 5 discusses the results of applying the algorithms to MOI data and compares the results of the three proposed techniques. Finally, Section 6 draws conclusions from the results and suggests avenues for future work.

2 Scale-based Filtering

Scale-based filtering takes advantage of the fact that the noise due to domain boundaries is at a smaller scale than the signal due to the objects of interest. Thus, by selecting all objects at an optimal scale, noise due to the domain boundaries can be easily eliminated.

The continuous wavelet transform ([5]) is a well-known approach to separating objects at different scales in images. The 2D continuous wavelet transform is given by [6]

$$W(\sigma,\tau_1,\tau_2) = \frac{1}{\sqrt{\sigma}}\int\int f(x,y)\psi^*(\frac{x-\tau_1}{\sigma},\frac{y-\tau_2}{\sigma})dxdy \quad (1)$$

where $f(x,y)$ is the original image, σ is the scale parameter, τ_1 and τ_2 are translation parameters, ψ is the mother wavelet function, and * represents complex conjugation. For a fixed scale, the wavelet coefficients can be computed in the frequency domain:

$$W(\sigma,\omega_1,\omega_2) = \sqrt{\sigma}F(\omega_1,\omega_2)\Psi^*(\sigma\omega_1,\sigma\omega_2) \quad (2)$$

where ω_1 and ω_2 are frequency variables, and F and Ψ are the 2D Fourier transforms of f and ψ respectively.

The major issues with this approach are the selection of the wavelet and the optimal scale for the image under investigation. In this study, the optimal scales for MO images were selected using purely subjective criteria, while the Gaussian function given by

$$G(x,y) = (\frac{1}{\sigma})\exp(-\frac{(x/\sigma)^2+(y/\sigma)^2}{2}). \quad (3)$$

was chosen as the wavelet. The resultant image is sharpened using a high-pass filter.

3 Threshold Averaging Filter

This algorithm first uses a nonlinear function to map the MO image's grayscale pixel values into a fixed range. In a typical magneto-optic image, the pixel intensity values range from 0 to some maximum value, I_{max}, that varies from image to image. In this study, the nonlinear mapping, accomplished via the tanh(.) function, maps the varying range of pixel values to the interval [0,1] as

$$g(x,y) = \tanh(\alpha * f(x,y)) \quad (4)$$

where $f(x,y)$ is the original MO image and α is an appropriate scaling factor [7]. This mapping enhances the contrast in the image and allows the development of filtering algorithms to be independent of the source of the MO data. Using subjective fidelity criteria [5], the optimal value of α was determined to be equal to $1/I_{max}$ for this application.

A simple moving-average filter can then be used to eliminate the high-frequency noise corresponding to the edges of the domain boundaries while preserving the lower frequency elements. The moving-average filter [5] replaces a given pixel by the average of the pixels in a given window:

$$h(x,y) = \frac{1}{P} \sum_{(i,j)\in S} (g(x_i,y_j)) \tag{5}$$

where S is the set of points in the neighborhood of (x,y) including (x,y) and P is the total number of points in the neighborhood. However, this technique degrades the edges of objects in the image, resulting in blurring the image. This is caused by a lack of discrimination between object pixels and background/noise pixels. Threshold Averaging Filter is a modified version of the moving average filter with edge preservation property [5]. The Threshold-Averaging filter with Edge Preservation removes the high frequency noise while preserving the crispness of the edges. The filter discriminates between background and object pixels by the use of a threshold value. This filter only averages over the pixels that are close in intensity value to the center pixel and are in the window. Mathematically, this can be written as

$$h(x,y) = \frac{1}{P} \sum_{i=0}^{n} \sum_{j=0}^{n} (g(x_i,y_j)) \tag{6}$$

where

$$\| (g(x_i,y_j) - g(x,y)) \| \leq T,\ (x_i,x_j) \in S \tag{7}$$

and P is the number of pixels that satisfy these conditions. The Threshold-Averaging filter selects only pixels in the window with similar intensities as the center pixel (x,y) that is being replaced. This allows the pixels along an object's edge to retain their sharpness. In MO images, a threshold value of 0.2 was determined to be optimal. The window size and the number of algorithm iterations can be adjusted to maximize the performance of the filter.

The result of the Threshold-Averaging stage contains many high intensity noise pixels along object edges. To eliminate these pixels, a morphological top-hat filtering mask operation followed by median filtering [8] was used. The morphological top-hat filtering operation consists of subtracting a morphologically opened image from the original image. Opening consists of eroding an image and then dilating the eroded image. Erosion of X by B is defined as:

$$(X \otimes B)(s,t) = \min\{X(s+x,t+y) - B(x,y) \mid (s+x),(t+y) \in D_x; (x,y) \in D_b\} \tag{8}$$

where X is the object and B is the structuring element, and D_x and D_b are the domains of X and B. Dilation of X by B can be defined as

$$(X \oplus B)(s,t) = \max\{X(s-x,t-y) + B(x,y) \mid (s-x),(t-y) \in D_x; (x,y) \in D_b\}. \tag{9}$$

The morphological top-hat filtering is then expressed as

$$Y = X - (X \otimes B) \oplus B. \tag{10}$$

Y is then used as a binary mask for the Threshold-Averaged data. The resulting image is then median filtered using a 7×7 window producing an image with minimal noise and sharp edges. The overall schematic of the NLM algorithm is summarized in Figure 1.

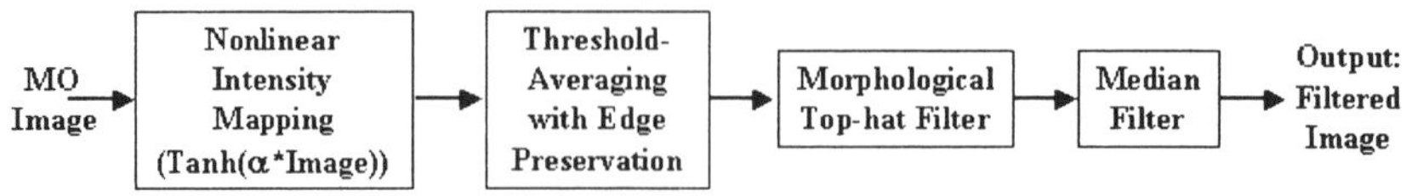

Figure 1: Threshold averaging filter with edge presenvation.

4 Motion-based Filtering

4.1 Additive Frame Subtraction (AFS)

The motion based filtering approach for MO images is based on the fact that the background noise (domain noise) remains static with respect to objects when the sensor is moved since the domain noise is generated within the sensor. Therefore the filtering of MO images for suppressing domain noise is equivalent to segmenting moving objects from a static background. Most segmentation techniques for dynamic-scene analysis are based on the detection of pixel-intensity changes in a frame sequence [9]. Thus, a simple subtraction of the current frame from previous frames will remove most of the background, while retaining images from rivets and cracks. The difference image $D_i(x,y)$ is obtained by:

$$D_i(x,y) = f_{n-i}(x,y) - f_n(x,y) \tag{11}$$

where $f_n(x,y)$ is the current frame and $f_{n-i}(x,y)$ is the frame obtained i instances earlier.

The difference image above also removes any overlap between frames in the object of interest. Complete information about the object of interest can be recovered by using multiple difference images. This is achieved by subtracting the current frame from several previous frames, and combining the difference images using a MAX operation:

$$S_n(x,y) = \max_{1 \leq i \leq w} (D_i(x,y)). \tag{12}$$

Here, w represents the number of difference images combined within the MAX operation. Each pixel of $S_n(x,y)$ is the result of the MAX operation on $\{D_1(x,y), ..., D_w(x,y)\}$. This procedure is referred to as Additive Frame Subtraction (AFS). In practice, S_n can be computed recursively using

$$S_n(x,y) = \max(S_{n-1}, D_n(x,y)), \; S_1 = D_1, \; 2 \leq n \leq w \tag{13}$$

A major issue in Additive Frame Subtraction is the selection of the optimal value of w. If w is too small, the algorithm will fail to extract the entire object from the MOI sequence. Conversely, if w is too large, the algorithm becomes computationally expensive . Two possible techniques for choosing w are presented below.

1. **Fixed w**: Let v be the scanning velocity (in inches/sec), s the object size along the scanning direction(inches), and r the frame rate(frames/sec). The relationship between w, v, s, r can be expressed as

$$w \geq \frac{sr}{v} \tag{14}$$

Using equation (14), w can be predefined if the scanning velocity is kept constant, since s and r are constant.

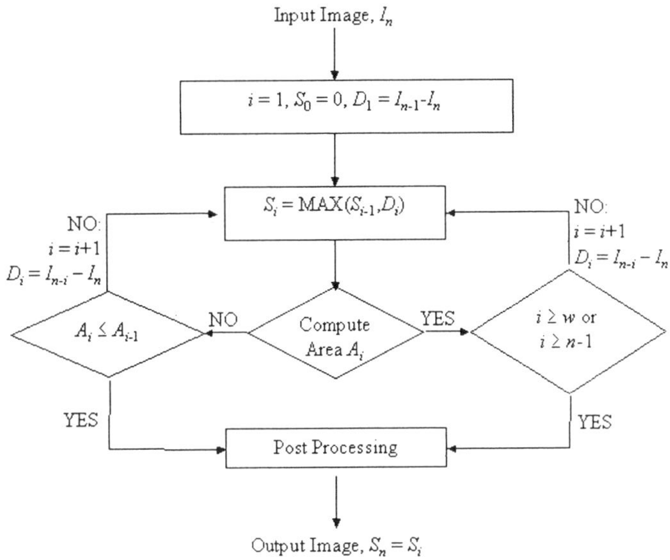

Figure 2: Motion-based filtering.

2. **Adaptive *w***: An alternative approach is to calculate the area of the regions with non-zero intensity in the difference image, $D_i(x,y)$. Typically, the area of $D_i(x,y)$ increases with i and reaches a maximum when the object in the 2 frames is completely separated. The value of i for which the area reaches a maximum can be selected as the optimal estimate of w.

4.2 *Postprocessing*

Post processing is performed to enhance the image contrast and remove any remaining noise. Median filtering [5], followed by an intensity stretching operation, is performed to map the intensity distribution of $S_n(x,y)$ to the range [0, 255]. Next, thresholding is done to remove any remaining noise in the intensity stretched image. The threshold value is obtained by Otsu's method [10], and lies between 50 and 80 in typical MO images.

The overall approach for the motion-based filtering algorithm is summarized in Figure 2.

5 Results And Discussion

The proposed algorithms were applied to several MO images and a sample of the results are presented in this section. Figure 3 presents the results of the three algorithms. Figure 3 (a) shows the original MO image, where the noise due to domain boundaries is clearly visible. The dark objects in the center are rivets with fastener. The distortion in the rivet signal on the right is due to the presence of a crack under the fastener. Figure 3 (b) shows the result

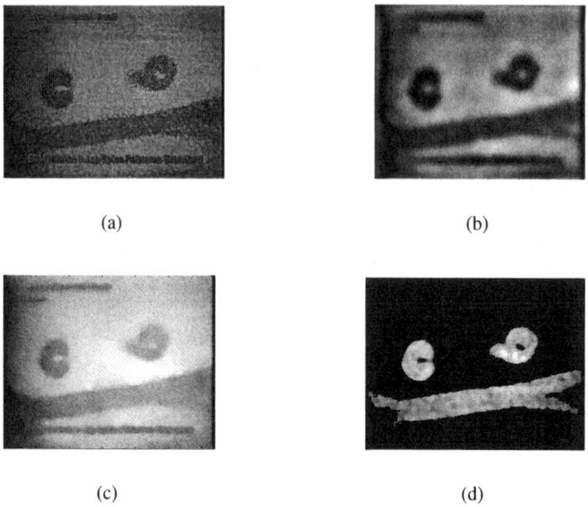

Figure 3: (a) Original image (b) Scale-based Filtering (c) Threshold Averaging Filter (d) Motion-based Filtering.

of scale-based filtering, Figure 3 (c) shows the filtering result using the Threshold Averaging Filter algorithm using a window size of 5×5 pixels, and Figure 3 (d) presents the results of motion-based filtering. Similar results are presented in Figure 4 for another MO image of a sample with corrosion.

The results presented indicate that the resulting image in all cases has better signal-to-noise ratio and better contrast with respect to the object of interest. While the proposed approaches result in a similar amount of reduction in the noise due to domain walls, each approach has its advantages and limitations. Motion-based filtering is seen to be most effective in totally removing static domain noise. However, this method is sensitive to the scan direction and can lose objects (such as seams) that extend parallel to the scan direction. The static approaches (Threshold Averaging Filter and scale-based filtering) do not suffer from this drawback. The static approaches tend to work best for MO images with geometrical structures such as rivets. However, in the case of corrosion patches, these algorithms are less effective. Although the CWT offers a one-step technique for eliminating noise, the resulting images appear blurred. The Threshold Averaging Filter is a multistage approach but provides a noise free image that is relatively sharper than the CWT technique. Thus, selecting the appropriate method for reducing noise due to domain walls is critical.

6 Conclusions

The presence of noise from domain walls in the MO sensor can significantly affect the interpretation of MO images. Three different algorithms – a static Threshold Averaging Filter algorithm, a Scale-based algorithm using the wavelet transform, and a Motion-based filtering algorithm – for the removal of such noise were presented, and compared.

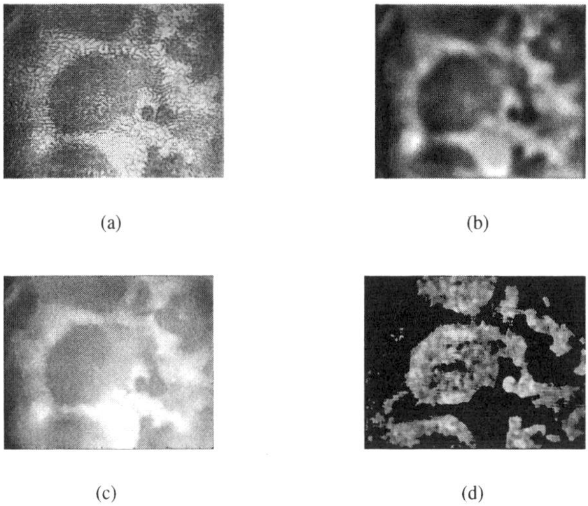

Figure 4: (a) Original image (b) Scale-based Filtering (c) Threshold Averaging Filter (d) Motion-based Filtering.

Although motion based filtering is very effective, it is sensitive to the scan direction and velocity. Static filters operate independently on each frame and hence do not use information available in the video sequence.

Evaluation of the proposed algorithms using additional data is under way. Hardware implementation of all three approaches for real-time processing is also in progress. Such a system can be directly connected to the MO imager, and process the data as it is acquired. Finally, the use of classification algorithms to identify regions with cracks is under investigation.

References

[1] G. L. Fitzpatrick, "Flaw Imaging in Ferrous and Nonferrous Materials Using Magneto-Optic Visualization," U.S. Patents 4,625,167, 4,755,752, 5,053,704 and associated foreign patents, and other patents pending.

[2] "Flaw Imaging in Nonferromagnetic Conductors using Magneto-Optic Detection Methods," (National Science Foundation Final Report), Principal Investigator: Gerald L. Fitzpatrick, Grant Number ECS-86-04228, 1989.

[3] G. L. Fitzpatrick, "Imaging Near Surface Flaws in Ferromagnetic Materials Using Magneto-Optic Detectors," *Review of Progress in Quantitative Nondestructive Evaluation*, D.O. Thompson and D.E. Chimenti (Eds.), Vol. 4, pp. , 1985.

[4] G. L. Fitzpatrick, "Imaging Near-Surface Flaws in Ferromagnetic Materials," *Proc. Eleventh World Conference on Nondestructive Testing*, ASNT, 1985.

[5] R. Gonzalez and Woods, *Digital Image Processing*, Prentice Hall, Upper Saddle River, NJ, 2002.

[6] S. Mallat, *A wavelet tour of signal processing*, Academic Press, 1998.

[7] J. Lim and S. S. Udpa, "Defect Extraction in MOI Image Based on Wavelet Packet Transform and Morphological Technique," *Review of Progress in Quantitative Nondestructive Evaluation*, D. O. Thompson and D. E. Chimenti (Eds.), Vol. 20, pp. 611-618, 2001.

[8] N. Eua-Anant, L. Udpa and J. Chao, "Morphological Processing for Crack Detection in Eddy Current Images of Jet Engine Disks," *Review of Progress in Quantative Nondestructive Evaluation*, (Ed. D. O. Thompson and D. Chimenti), Vol. 18, 1999, pp. 751-758 .

[9] L. G. Shapiro and G. C. Stockman, *Computer Vision*, Prentice Hall, NJ, 2001.

[10] N. Otsu, "A threshold selection method from gray-level histograms", *IEEE Trans. Syst.*, SMC-9, 1979, 62-66.

Electromagnetic Nondestructive Evaluation (VIII)
T. Sollier et al. (Eds.)
IOS Press, 2004

Crack Detection in Subsurface Layers of Riveted Aluminium Structures

Maxim MOROZOV, Guglielmo RUBINACCI, Antonello TAMBURRINO, Salvatore VENTRE, Fabio VILLONE
Associaz. EURATOM/ENEA/CREATE, DAEIMI, Università degli Studi di Cassino
Via Di Biasio 43, I-03043 Cassino (FR), Italy

Abstract. This paper presents recent results of eddy current inspection for tiny cracks growing out from a fastener hole with inserted rivet in the second layer of Aluminium-to-Aluminium structures. The experimental data are compared with results of a numerical simulation aimed at solving the inverse problem.

1. Introduction

The present paper continues on improving aircraft maintenance procedure by introducing non-destructive technique based on measurement of the magnetic perturbation field of eddy currents as stated in the previous paper [1]. The relevant eddy current probe contains the fluxset sensor which offers high sensitivity to deep subsurface defects at low frequencies of inspection [2].

The objective of the work is the development of a reliable method of detection and reconstruction of small cracks hidden under rivet heads. The work mainly focuses on the second-layer cracks in riveted aluminium-to-aluminium structures. Samples containing artificial cracks growing out from the fastener hole constitute inspection cases under examination. Experimental data are compared with results of numerical simulation in order to assess the possibility of solving the inverse problem.

The paper is organized as follows. In section 2 the experimental apparatus and inspection cases are described in details. Section 3 briefly recalls the numerical model used for the simulations. Measurement results are compared with computations in section 4. Section 5 contains conclusions and a summary of planned future developments.

2. Experimental set-up and samples

The eddy current probe includes an excitation coil and fluxset sensor centred below its base, as shown in Fig.1. Parameters of the eddy current probe are summarised in table 1. The fluxset sensor presented in Fig.2 consists of a ribbon shaped magnetic core made of annealed metallic glass, driving and pick-up solenoids wound on the core. The sensitive axis of the sensor is directed along its core. The core is periodically saturated by triangular field produced by the driving solenoid at frequency at least twice as high as frequency of the measured field. A symmetric voltage is induced in the pick-up solenoid in zero field. When the flow of eddy currents is distorted by a defect the resulting magnetic field causes an asymmetry of the pick-up voltage.

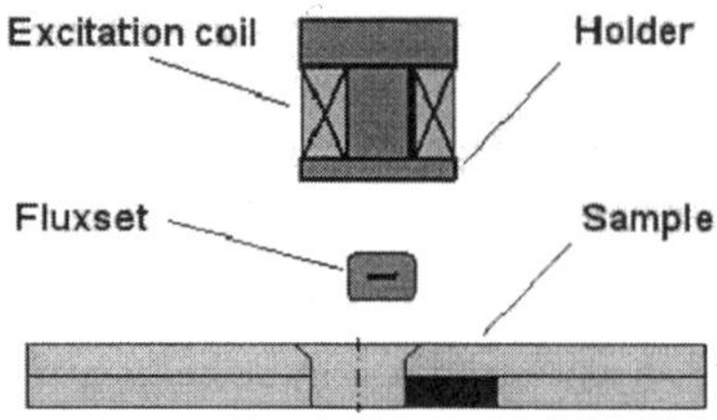

Fig.1 EC probe and sample outline

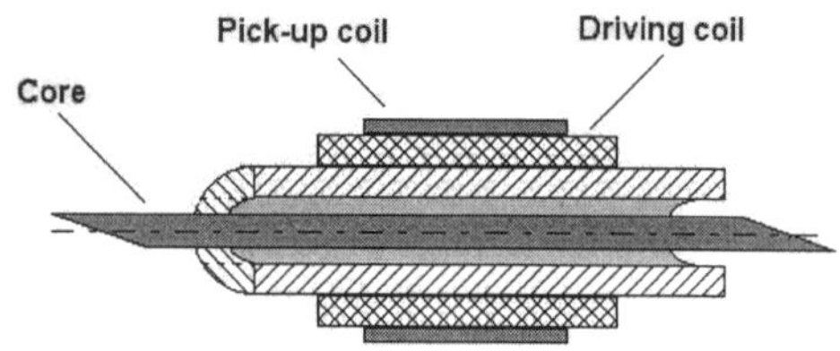

Fig.2 Fluxset sensor arrangement

The experimental set-up is sketched in Fig.3. The excitation coil of the probe is fed with sinusoidal current from the waveform generator G_1 at frequency f_{exc}. The generator G_2 produces triangular driving current at frequency $f_{dr} \geq 2f_{exc}$. A general-purpose data acquisition system furnished with analog-to-digital converter is used to readout pickup signal. The key parameters of the data acquisition system are its sampling rate (500 MS/s) and number of waveform points captured in one cycle (1.E+5). The acquired waveforms are passed to a workstation via GPIB interface (standard IEEE 488.2). The probe response is extracted from acquired waveforms using digital signal processing algorithm whose block-diagram is presented in Fig.4. The algorithm fulfils two major steps: first, the eddy current signal expressed in units of time (since the fluxset sensor operates in the time domain) is extracted from the acquired pick-up voltage waveform through pulse position detection; second, the demodulated signal is multiplied by rectangular reference waveform with frequency f_{exc}, both in phase and with 90 deg. phase shift with respect to the excitation current in order to obtain real and imaginary parts of the signal, respectively. The obtained response is linearly proportional to the eddy current field [1].

The samples represent aluminium-to-aluminium double layer sandwiches with fastener holes and inserted rivets. The layers are electrically insulated. General samples characteristics are listed in table 2. The samples contain EDM notches originating from the holes. The artificial notches simulate real cracks and for simplicity will be referred to as "cracks" in the further text. The inspection cases of the primary interest are cracks lying in the second layer. Upper side cracks are referred to as ID (inner defects) and reverse side cracks are referred to as OD (outer defects). These acronyms are followed by the defect depth expressed as percentage of the sample thickness (i.g. ID50%).

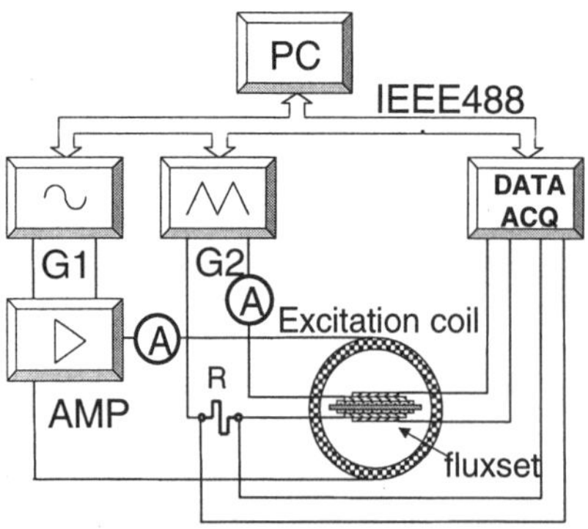

Fig.3 Outline of the experimental set-up

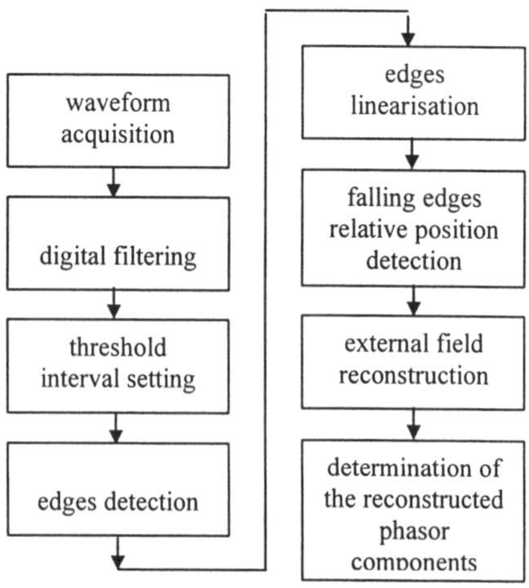

Fig.4 Digital signal processing algorithm

Table 1. Eddy current testing parameters

excitation coil	length	20.0mm
	inner diameter	10.0mm
	outer diameter	11.5mm
	number of turns	100
	lift-off	~0.8 mm
excitation frequency		998 Hz
excitation current (sinusoidal)		0.5 A
fluxset sensor lift-off		~0.5 mm

Table 2. General samples characteristics

plate dimensions		200 mm×200mm
plate thickness		2 mm
hole diameter		4 mm
rivet	head diameter	6 mm
	head height	1 mm
	leg diameter	4 mm
material		aluminium 2024T3
conductivity of plate / rivet		18 MS/m
EDM notch thickness		≅ 0.1 m

3. Simulation environment

The experimental results obtained with the apparatus briefly described in the previous section have been compared with some numerical simulations. This is done first of all to validate the procedure of quantitative calibration of fluxset. In fact, this calibration is done a homogeneous external field, but we can question whether the calibration constant is the same in the presence of a field pattern due to defects and/or holes since the sensitive region of the sensor is estimated to be 1-2 millimetres. This is particularly significant since in the past it has been demonstrated that fluxset in fact measures an average of the field in its core [3], and this can make the calibration procedure quite complicated [4].

Secondly, having a simulation tool able to reliably and efficiently solve the so called "direct" problem (given the excitation and the crack geometry, estimate the measurements provided by fluxset), we can tackle also the solution of the so called "inverse" problem (given the measurements provided by fluxset, find the crack geometry).

The numerical formulation used is described in [5] and here briefly recalled. The starting point is an integral formulation in terms of a two-component electric vector potential [6]; this choice has a number of favourable consequences that we recall here. Using an integral formulation allows us only to discretize the conducting domains where the eddy currents are induced (i.e., neither the air nor the exciting coils), automatically enforcing regularity conditions at infinity. The introduction of the electric vector potential **T** such that the current density **J** is $\mathbf{J}=\nabla\times\mathbf{T}$, guarantees that **J** is solenoidal. Finally, the choice of the two-component gauge minimizes the number of discrete unknowns required.

The equations to be solved are the standard eddy current equations in the frequency domain. We express the electric field as:

$$\mathbf{E} = -\mathrm{j}\,\omega\,\mathbf{A} - \nabla\varphi \tag{1}$$

φ is the scalar electric potential and **A** is the divergence-free magnetic vector potential given by:

$$\mathbf{A}(\mathbf{x},t) = \frac{\mu_0}{4\pi}\int_{V_c}\frac{\mathbf{J}(\mathbf{x}',t)}{|\mathbf{x}-\mathbf{x}'|}dV' + \mathbf{A}_0(\mathbf{x},t) \tag{2}$$

where V_c is the conducting domain, μ_0 is the magnetic permeability of the vacuum and $\mathbf{A}_0$ is the contribution of the external current density. The magnetic flux density is obtained as $\mathbf{B}=\nabla\times\mathbf{A}$.

From the numerical point of view, the formulation is solved using finite elements: a mesh of the conducting domain is given, and an edge element basis functions $\mathbf{N}_k$ is introduced for $\mathbf{T}$:

$$\mathbf{T}=\sum_k I_k \mathbf{N}_k \quad \Rightarrow \quad \mathbf{J}=\sum_k I_k \nabla\times\mathbf{N}_k \tag{3}$$

On the one hand, the choice of edge elements allows us to enforce the right continuity conditions of the various electromagnetic quantities; on the other hand, their properties are fully exploited both for the gauge and boundary conditions imposition.

The electric constitutive equation is imposed in weak form using Galerkin approach:

$$\int_{V_c} \nabla\times\mathbf{N}_k \cdot (\eta\mathbf{J} + j\omega\,\mathbf{A})dV = 0 \quad \forall\mathbf{N}_k \tag{4}$$

η is the resistivity; using (2) this results in:

$$(\mathbf{R}+j\omega\mathbf{L})\,\mathbf{I} = \mathbf{U} \tag{5}$$

$\mathbf{I} = \{I_k\}$, $\mathbf{U} = \{U_k\}$ and

$$\mathrm{L}_{ij} = \frac{\mu_0}{4\pi}\int_{V_c}\int_{V_c} \frac{\nabla\times\mathbf{N}_i(\mathbf{x})\cdot\nabla\times\mathbf{N}_j(\mathbf{x}')}{|\mathbf{x}-\mathbf{x}'|}dVdV' \tag{6}$$

$$\mathrm{R}_{ij} = \int_{V_c} \nabla\times\mathbf{N}_i \cdot \eta\nabla\times\mathbf{N}_j dV \tag{7}$$

$$\mathrm{U}_i = -\int_{V_c} \nabla\times\mathbf{N}_i \cdot j\omega\,\mathbf{A}_0\, dV \tag{8}$$

Supposing that the crack has a negligible thickness, it can be schematised as a surface (not necessarily planar), discretized via a set of finite element facets (defect pixels), where the normal component of the current density must vanish. In order to reduce the computational load, and exploiting linearity, we use superposition: the total current density is the sum of the solution computed in absence of crack (unperturbed solution $\mathbf{J}_0$) plus the perturbation $\delta\mathbf{J}$ due to the presence of the defect.

$$\mathbf{J} = \mathbf{J}_0 + \delta\mathbf{J} \tag{9}$$

In particular, on the crack surface, since the total current density normal component must be zero, we impose that

$$\mathbf{J}\cdot\mathbf{n} = 0 \Rightarrow \delta\mathbf{J}\cdot\mathbf{n} = -\mathbf{J}_0\cdot\mathbf{n}, \tag{10}$$

This approach offers the great advantage that $\mathbf{J}_0$ can be calculated either analytically, or numerically using the scheme described above on a mesh that does not depend on the crack geometry. Conversely, when solving for $\delta\mathbf{J}$ the mesh must account for the crack only, so that the mesh refinement is required only close to the crack, regardless the position of the exciting source. Due to the properties of edge elements, the set $\delta\mathbf{G}$ of perturbation currents crossing the crack facets (that must be equal the unperturbed currents $-\mathbf{G}_0$) is:

$$\delta \mathbf{G} = \mathbf{P}\, \delta \mathbf{I} \tag{11}$$

where $\delta\mathbf{I}$ are the coefficients of the expansion of $\delta\mathbf{J}$ in terms of edge elements, and $\mathbf{P}$ is a (m,n) sub-matrix of the edge-facet incidence matrix with coefficients 0, +1 or -1. The degrees of freedom of the edge element expansions are in fact related to the line integrals of $\mathbf{T}$ along the edges, and the circulation of $\mathbf{T}$ along a closed line gives the total current (flux of $\nabla\times\mathbf{T}$) linked with the line. We then make a change of variables:

$$\delta \mathbf{I} = \mathbf{K}\, \delta \mathbf{X} + \mathbf{P}^{+}\, \delta \mathbf{G} \tag{12}$$

where $\mathbf{K}$ is a (n,n-m) matrix given by an orthonormal basis for the null space of $\mathbf{P}$, $\mathbf{P}^{+}$ is the pseudo-inverse of $\mathbf{P}$, and $\delta\mathbf{X}$ is an auxiliary variable.

Galerkin's procedure in terms of these new variables yields:

$$\mathbf{K}^{T}\mathbf{Z}\mathbf{K}\, \delta \mathbf{X} = \mathbf{K}^{T}\mathbf{Z}\, \mathbf{P}^{+}\, \mathbf{G}_0 \tag{13}$$

solving which we have $\delta\mathbf{X}$, and hence $\delta\mathbf{I}$ from (12).

Knowing $\delta\mathbf{I}$ we calculate the magnetic flux density perturbation at given points $\delta\mathbf{B} = \{\delta B_k\}$ simply via a multiplication for a suitable matrix coming out of the Biot-Savart law:

$$\delta \mathbf{B} = \mathbf{F}\, \delta \mathbf{I} \tag{14}$$

Once the perturbation field is known, the total field is obtained simply adding up the unperturbed field.

The finite elements meshes of 3968 and 3328 elements, used for the unperturbed and perturbed solutions, respectively, are shown in Fig. 5a and Fig. 5b. The plates are assumed to be electrically insulated with the rivet being in poor electrical contact with the plates.

4. Results

The eddy current testing was performed with parameters given in table 1. The probe moved above the test piece along the central line containing defect. Fig. 6 presents the probe response to a fastener hole with rivet without cracks. There is a slight discrepancy between experimental and computational results. Taking in account that results for higher testing frequency (5kHz) presented in paper [1] were in a satisfactory agreement

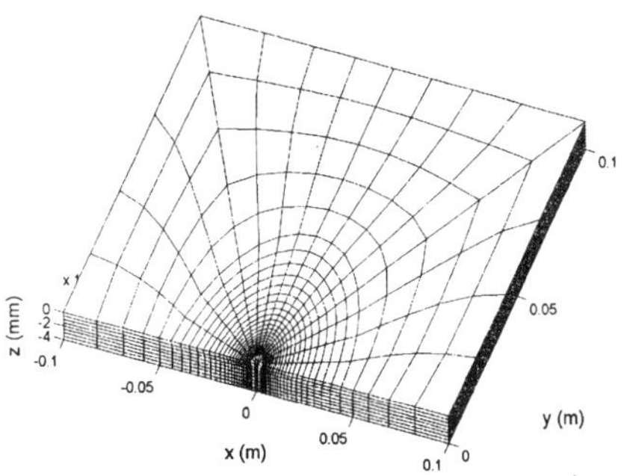

(a)

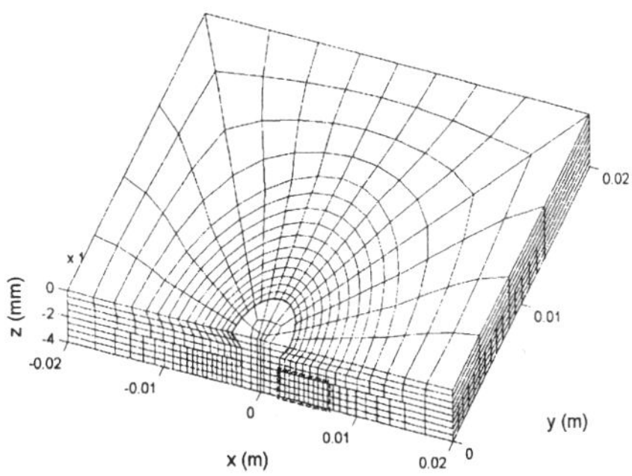

(b)

Fig. 5 Finite elements mesh applied for (a) unperturbed and (b) crack solutions, respectively

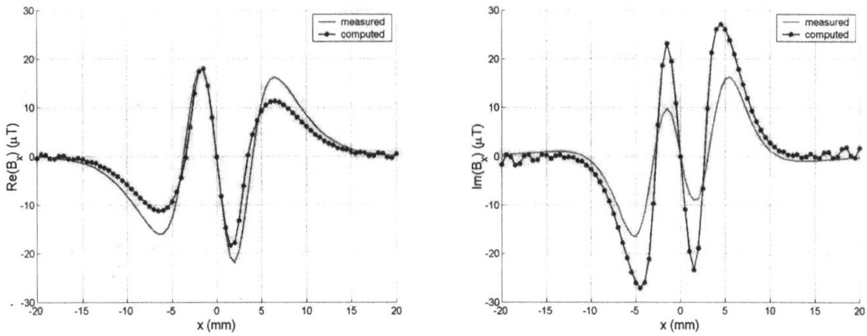

Fig. 6 Response to a fastener hole with rivet, measurements vs. simulation:
left: real component and right: imaginary component

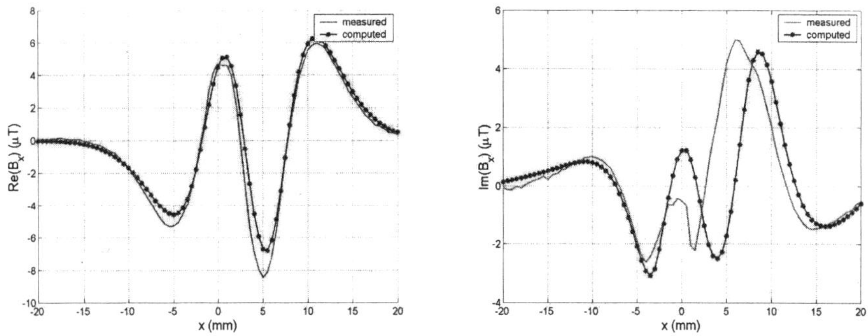

Fig. 7 Crack contribution of the response to an ID100% 5 mm long crack in the 2nd layer originating from a fastener hole with rivet, measurements vs. simulation: left: real component and right: imaginary component

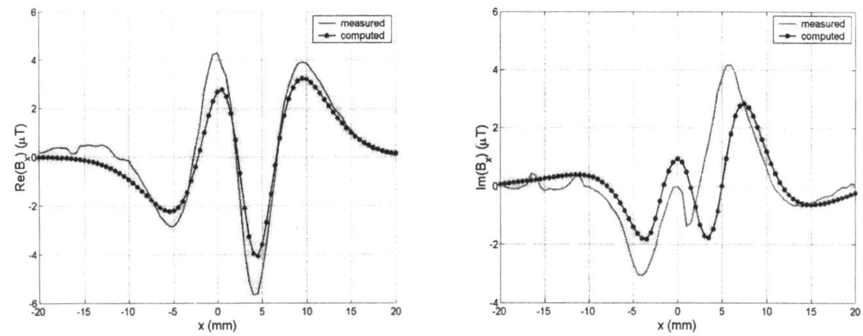

Fig. 8 Crack contribution of the response to an ID100% 3 mm long crack in the 2nd layer originating from a fastener hole with rivet, measurements vs. simulation: left: real component and right: imaginary component

it is possible to assume that respective computational error might be due to inappropriate mesh discretisation since the eddy currents distribution is more diffuse at lower frequencies and the contribution of finer elements near the hole where the eddy current changes are most drastic might be distorted by more coarse elements in outskirts. On the other hand further mesh refinement will increase computational cost significantly.

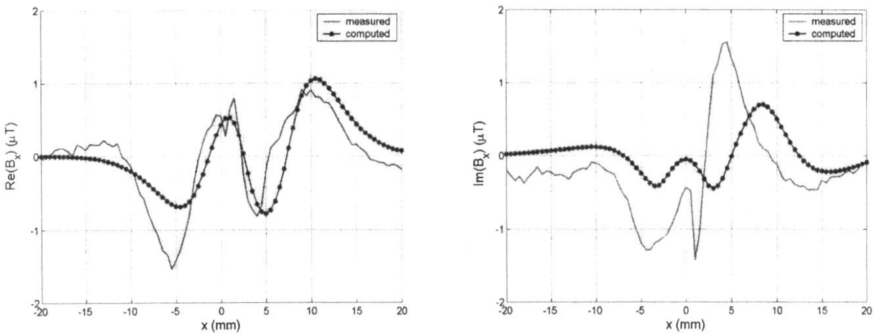

Fig. 9 Crack contribution of the response to an ID50% 5 mm long crack in the 2nd layer originating from a fastener hole with rivet, measurements vs. simulation: left: real component and right: imaginary component

There might also be a measurement error due to uncertainty of the lift-off of the excitation coil and fluxset sensor.

Figs. 7-9 present crack contributions of the responses to defects of various configuration in the second layer. The crack contribution signal is defined as the difference between the signal coming from a hole with a rivet and the signal for the same structure including a crack. The crack contribution, obtained as a difference between two large signals, is thus very sensitive to noise. Measured crack contributions are strongly affected by noise since they are difference between two much higher signals, that is crack originating from hole and only hole. Reasonably good agreement between experimental results and simulations comes from the uniformity of the discretisation of the solution region in the region of the crack.

5. Conclusions and outlook

Analysis of both experimental and numerical results and their comparison suggests the following conclusions:

- the discrepancy in the unperturbed numerical solution (only hole and rivet) mainly in the imaginary component, which is dominant at low frequencies, might be due to inappropriate mesh discretisation;
- better agreement on the crack contributions is obtained since the perturbed current is computed in only a small uniform region;
- measured crack contributions are subject to noise since they are result of subtraction of two much higher signals, that is hole with rivet and crack compared to hole and rivet.

Therefore the future development should address:

- computing L and R matrices in double precision;
- application of parallel computing;
- filtration of the hole and rivet signal from the measured responses.

Acknowledgements

This work is supported by the Italian MIUR and ASI, with the contribution the European Union and ALENIA.

References

1. M. Morozov, G. Rubinacci, A. Tamburrino, S Ventre, Fabio Villone, 'Aeronautical structures examination using fluxset type ECT probe' , Proceedings of the 8th International Workshop on Electromagnetic Nondestructive Evaluation ENDE-2002, Saarbrücken, Germany, June 12-14, 2002, in press
2. A. Gasparics, C.S. Darocsi, G. Vertesy, J. Pavo, "Improvement of ECT Probes Based on FLUXSET Type Magnetic Field Sensor", Electromagnetic Nondestructive Evaluation (II), R. Albanese et al. (Eds.), IOS press, pp. 146-151 (1998)
3. R. Albanese, G. Rubinacci, A. Tamburrino, F. Villone, "Phenomenological Approaches Based on an Integral Formulation for Forward and Inverse Problems in Eddy Current Testing", *Int. J. of Applied Electromagnetics and Mechanics,* Vol. 12, No. 3-4/2000, pp. 115-137
4. J. Pavo, O. Biro, K. Hollhaus, "Calculation of the field map from the measurement data of the fluxset sensor", *COMPEL*, Vol. 20, no. 2, pp. 404- 416 (2001)
5. R. Albanese, G. Rubinacci, F. Villone, "An integral computational model for crack simulation and detection via eddy currents", *J. of Comp. Phys.*, Vol. 152, pp. 736-755 (1999)
6. R. Albanese, G. Rubinacci, "Finite element methods for the solution of 3D eddy current problems",*Advances in Imaging and Electron Physics*, Vol. 102, pp. 1-86 (1998)

Electromagnetic Nondestructive Evaluation (VIII)
T. Sollier et al. (Eds.)
IOS Press, 2004

CODECI, a new system for the inspection of surface breaking flaws based on eddy current array probe and high resolution CCD camera

Thierry SOLLIER [(1)], Boguslaw LORECKI [(2)], Olivier GOUPILLON[(3)]
(1) CEA SACLAY, SISC Bâtiment 611, 91191 Gif-sur-Yvette, France
(2) CYBERNETIX, Rue Albert Einstein, BP 94,13382 Marseille, France
(3) EUROCOPTER, Aéroport Marseille-provence, 13725 Marignane Cedex, France

Abstract: a new device has been developed for the inspection of surface breaking flaws in non-ferrous materials. This device includes an array of eddy current (EC) coils, a high-resolution camera, a light source and an encoder. The probe is scanned manually over the work specimen. The data streaming from the camera and the EC array probe are simultaneously processed and compared. False calls are rejected and remaining flaws are characterized precisely by their length, width and depth.

1. Introduction

The inspection of many cast or machined mechanical pieces is performed by the dye penetrant testing technique. This technique gives a general view of potential flaws breaking the surface but can hardly give accurate information, if any, on the depth of the flaw. Moreover one needs to clean carefully the test specimen to obtain reliable results and this operation is time consuming. EUROCOPTER is a major helicopter manufacturer and uses dye penetrant testing for the inspection of several critical pieces. EUROCOPTER, CEA and CYBERNETIX worked together in the frame of a joint national program (CODECI) to develop an alternative technique for the inspection of critical parts such as gear casing.

The aim of the new system is to provide a complete and reliable description of potential flaws (length, width, depth) in real time. The end user could thus quickly decide to accept, reject or repair the piece and save time on inspection and manufacturing cycles. The main feature of the CODECI system is to combine in real time two different techniques (digitalized visual inspection and eddy current testing, ECT) with signal processing. The optical imaging device is developed by CYBERNETIX and gives the location, length and width of the flaws at the surface. The EC array probe system is developed by CEA and is used in conjunction with image data processing to evaluate the depth of the flaws.

2. Description of the system

The CODECI system is based on a real time PC platform, and accepts several kinds of duplex probes (optical and EC). A general view of the system is given in figure 1.

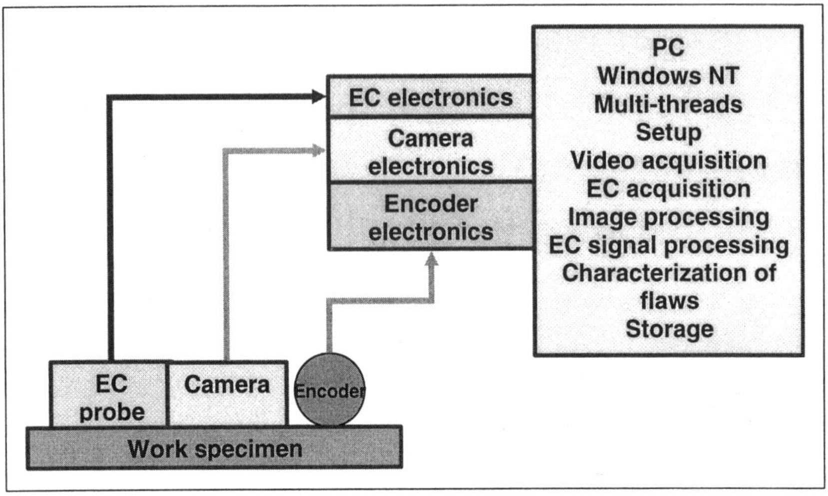

figure 1: General architecture of the CODECI system

In this project, the requirements of the end-user are to detect and characterize a flaw with a minimum length of 1 mm and a minimum depth of 0.2 mm in light alloys used in aeronautics (aluminum, magnesium and titanium based alloys). In order to meet these requirements, a specific probe has been developed.

2.1 EC array probe

The EC array probe is made of 32 ferrite core coils assembled on 2 rows. The outer diameter of the coils is 1 mm and the inner diameter is 0.4 mm. The spatial sampling of EC data is 0.5 mm for both the multiplexing axis (X) and the scanning axis (Y). The small diameter of the ferrite core gives a high resolution suitable for the detection and characterization of small surface breaking flaws. The resolution of the coil at –6 dB for a frequency of 700 kHz is 0.8 mm. This resolution was evaluated for an EDM notch 2 mm long, 0.2 mm deep and 0.1 mm width, scanned perpendicularly to its main axis with a sampling pace of 0.1 mm (figure 2). The EC signal obtained for two nearby EDM notches is shown in figure 3. The notches have a length of 1 mm, a depth of 0.1 mm and are separated by a gap of 2 mm. The detection of the two notches is clear and could be achieved up to a gap of approximately 1.2 mm (corresponding to the coil's resolution at –12 dB). The EC coils are operated in the impedance mode in a Wheatstone bridge. A reference coil, within the probe, is used to balance the bridge. This mode of operation was preferred to the separate transmit/receive mode in order to have an isotropic response of the probe with respect to the orientation of the flaw. That point was part of the end-user specifications. The width of the covered area swept by the probe is 16 mm and the multiplexing sequence is given in figure 4.

2.2 EC instrumentation

The EC hardware is based on a specific board plugged in a PC slot and an external multiplexer. The EC board hosts a signal generator, a programmable Wheatstone bridge, a signal amplifier and a DSP. The DSP unit performs the EC demodulation numerically.

A specific multiplexer with 32 channels has been developed for the EC array probe technology. The multiplexing frequency ranges from a few Hz to 1 kHz. The 32 coils are thus scanned at a maximum recurrence of 30 Hz and, considering a sampling rate of 0.5 mm during the probe displacement, the nominal speed of the probe is 15 mm/s. This velocity is convenient for acquisitions performed manually but can be increased, if needed, by adding more EC boards into the PC.

The size of the coils and the sampling rate are defined upon the end-user specifications for characterization performances, but could be adapted to various configurations resulting from different inspection speeds.

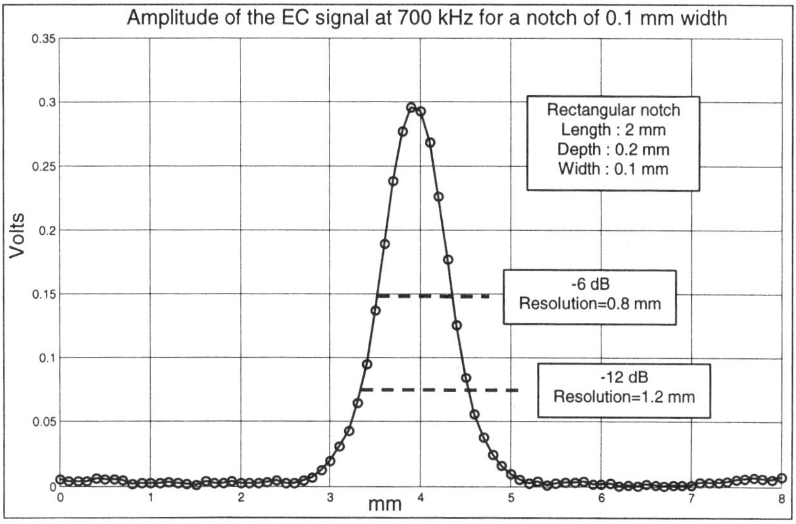

figure 2 : Evaluation of the coil resolution for a frequency of 700 kHz

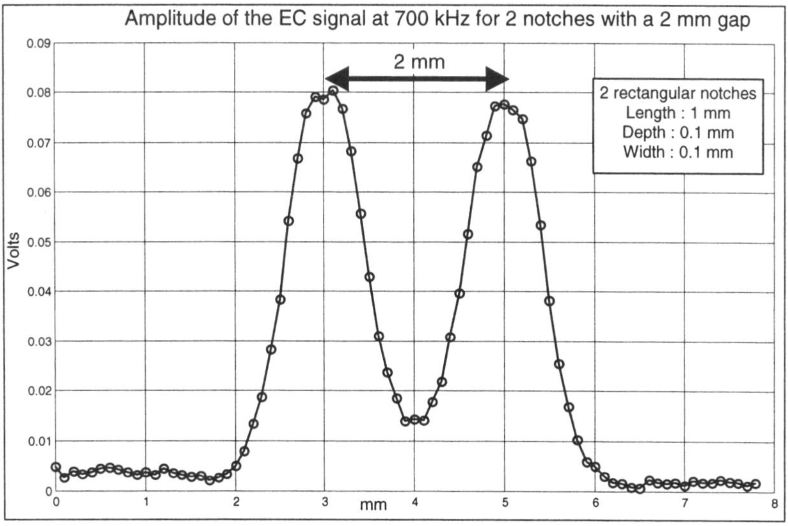

figure 3 : Coil response for 2 notches with a gap of 2 mm

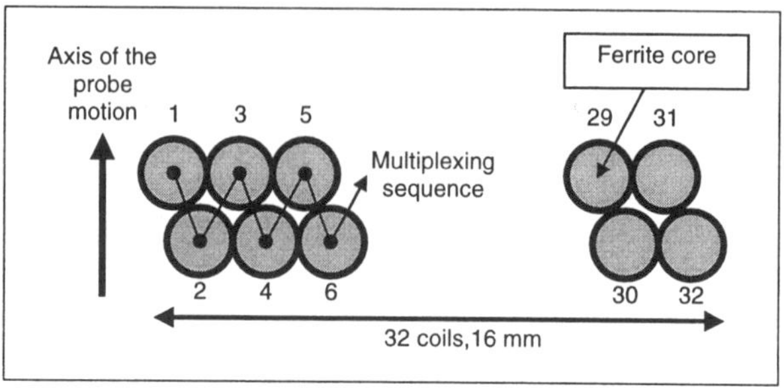

figure 4: EC array probe configuration and multiplexing sequence

2.3 CODECI probe

The CODECI probe includes a miniaturized and high-resolution camera (pixel ≈ 15 μm), a homogeneous illumination of the test specimen by LEDs with a specific diffusive cone, an encoder and the EC array probe. A general view of the probe is given in figure 5.

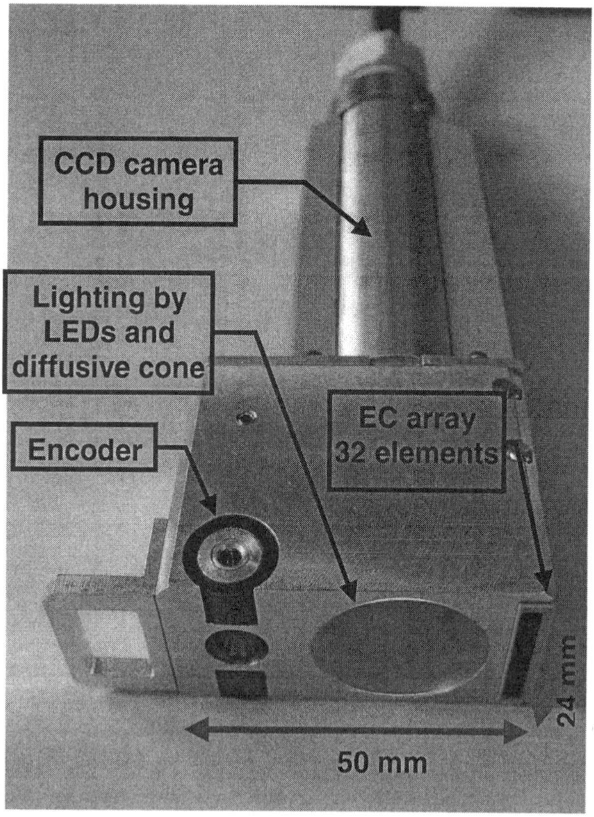

figure 5: General view of the CODECI probe

2.4 Software architecture

Several processes (or threads) are running simultaneously on the remote PC, mainly: digital image acquisition, digital image processing, EC acquisition, EC signal processing, overall characterization mixing the two techniques and real time display. The detection algorithm for both vision and ECT is based on a threshold value. The digital image processing gives the location, extension and width of flaws at the surface. The EC data processing gives the amplitude and phase of the EC signal. The overall characterization process combines optical and ECT data. An intersection on flaw location is done on binary images and thus excludes any artifacts such as scratches.

3. Characterization of the flaws

The aim of the system is to give the location, length, width and depth of the flaws as accurately as possible, in real time and without limitation of the acquisition time. In the case of a hand-held system, real time was set to one second for the overall characterization. This delay was considered fully acceptable for the operator.

The real time analysis is however a stringent constraint for the system, and influences mainly the methodology of the flaws characterization. The optical image gives quickly and with a great accuracy (±100 μm) the dimensions of the flaw at the surface of the work piece. The depth is then evaluated from EC data. The proposed method relies on the measurement of the maximum of the impedance variation as the probe scans the flaw. Given the length of the flaw at the surface from optical measurement, the depth is evaluated by comparison to simulated signals stored in a look-up table. In figure 6, the impedance variation is computed and plotted against the depth for various lengths of rectangular notches. The figure shows the drastic and well-known influence of the flaw length upon the impedance variation. It can be noticed that, even for a coil with an outer diameter of 1 mm, the impedance variation for a given depth is significantly modified when the length of the flaw varies from 1 mm to 4 mm.

The length of the flaws could also be estimated from EC signal but the resolution and bandwidth of the CCD camera are clearly many times better than the one reasonably achievable with the EC array probe.

4. Results

The CODECI system has been evaluated in laboratory on various alloys used in the aeronautics industry (aluminum, titanium and magnesium based alloys).

For aluminum-based alloys, two frequencies were used simultaneously: 300 kHz and 700 kHz. The corresponding operating points in the normalized impedance plane are given in figure 7. The normalized impedance plane gives the reactance and inductance of the coil above an aluminum slab normalized by the coil's inductance in air [1] The selection of these frequencies gives a good coupling between the coil and the slab (R/X_0 is high) and a good SNR for the instrumentation (high values of impedance changes). The depth of flaws is evaluated for each frequency and then averaged in order to increase the overall accuracy.

An example of the system output is given in figure 8for EDM notches of 1 mm length and various depths for a work specimen of aluminum alloy. The detection is good and the accuracy of the depth evaluation is about 0.2 mm.

The system has also been tested in the field by the end-user and has proved to be operational but limited to planar surfaces due to the size and rigidity of the probe. The detection and characterization of flaws near the edge of a work specimen is limited to a distance of 3 mm from the edge. A general view of the probe in position over a helicopter gear housing is given in figure 9.

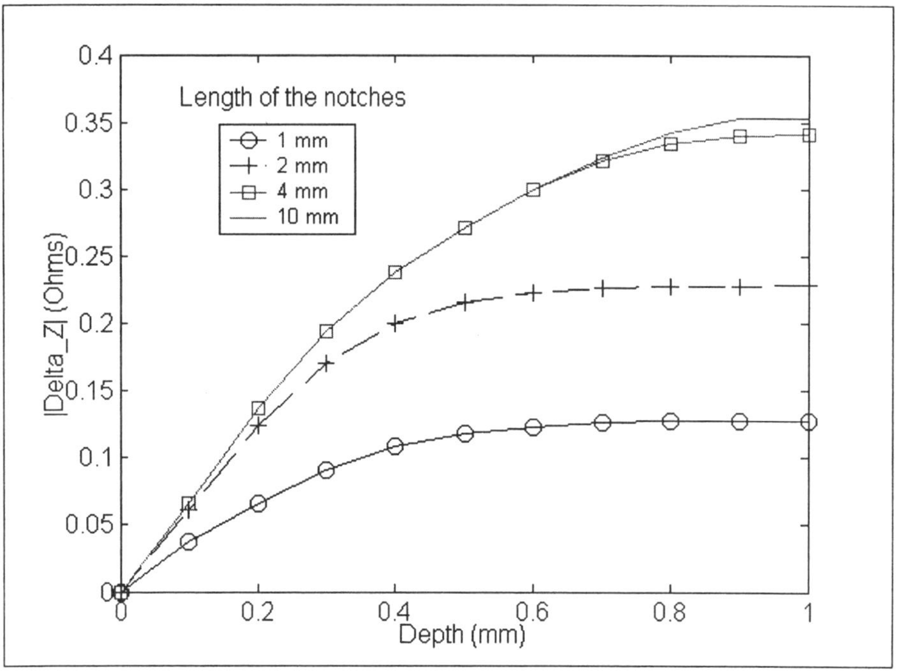

figure 6: Maximum of the variation of the coil impedance for a rectangular EDM notch. The test frequency is 300 kHz and the material is aluminum. The notches have a width of 0.2mm, a length of {1, 2, 4 , 10 mm} and a depth ranging from 0.1 to 1 mm.

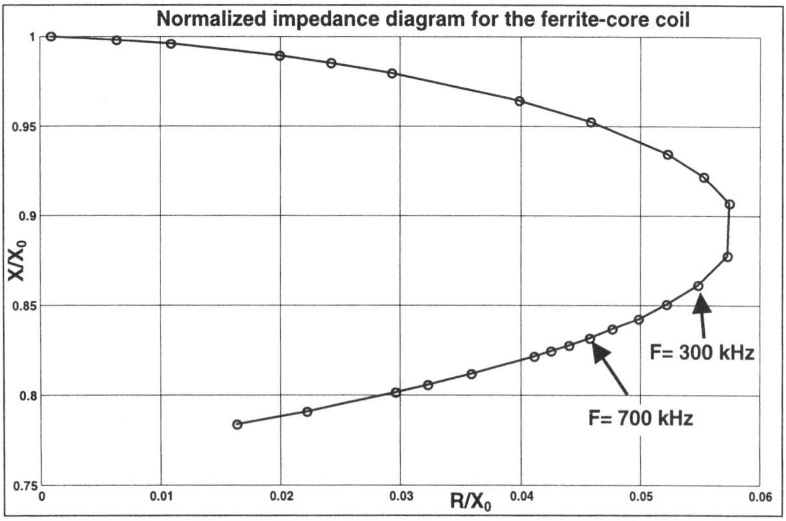

figure 7 : normalized impedance of the ferrite-core probe above an aluminum slab

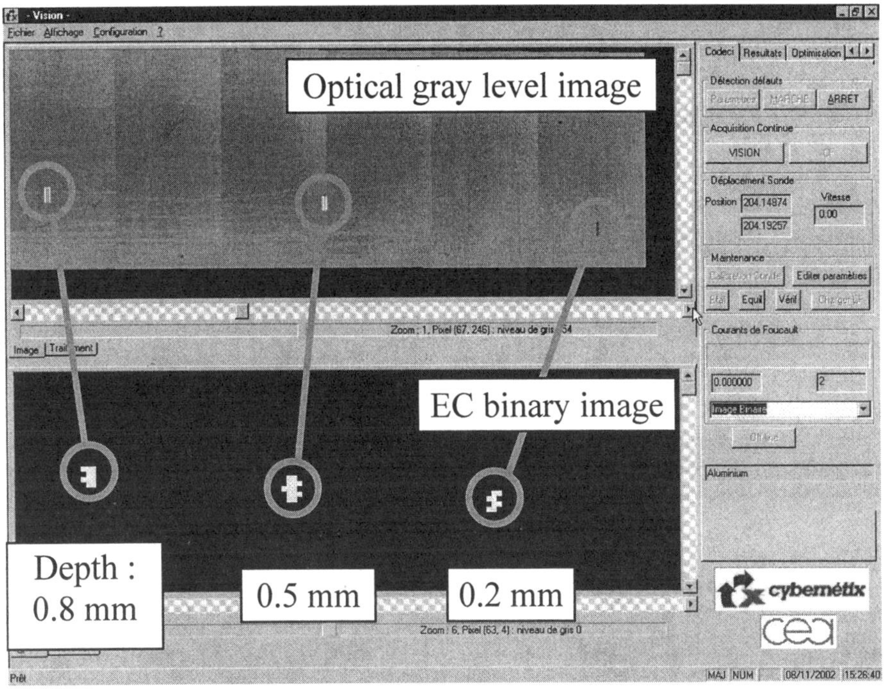

figure 8: Detection of EDM flaws of 1 mm length in an aluminum alloy

figure 9: Evaluation of the CODECI system in the field over an helicopter gear housing

5. Conclusions

A new inspection system based on EC array probe and high-resolution camera was developed and tested in laboratory and in industrial environment for real time non-destructive applications. It was shown that mixing both techniques enhanced the sizing ability and offered more overall accuracy than each technique considered solely. The use of simulated EC signals was the key point for a real time evaluation of the depth based on the measurement of the impedance variations as coils pass over flaws, knowing accurately their size from optical means.

Up to now, the system is operational for flaws with regular shape located on flat or weakly curved surfaces. The sizing accuracy is approximately 100 μm for the length and 0.2 mm for the depth of the flaws. The probe can be swept over the work specimen at a speed up to 15 mm/s.

Future works should address more complex inspection configurations such as network of cracks, geometrical singularities (flaws located near edges or holes) and more complex geometries for the work specimen.

References

[1] F. Buvat et al., A fast model of eddy current ferrite-core probes for NDE, ENDE'2003, published in these Proceedings.

Electromagnetic Nondestructive Evaluation (VIII)
T. Sollier et al. (Eds.)
IOS Press, 2004

Evaluation of the Effect of Sodium in Steam Generator Tubes using Remote Field

Ovidiu MIHALACHE, Kunihiko YOKOYAMA, Masashi UEDA, Takuya YAMASHITA
International Cooperation and Technology Development Center,
ISI System Development Group,
Japan Nuclear Cycle Development Institute, 1, Shiraki, Tsuruga-shi, Fukui-ken, Japan

Abstract. This paper characterizes, using the remote field technique, the influence of the sodium (Na layer and Na patches) on the signal arising from outer defects in steam generator tubes of a fast breeder reactor. The effect of the sodium is evaluated using a simulation tool based on a finite element code, for several cases of outer defects in steam generator tubes. The results from the numerical simulations are compared with the experimental measurements in a limited number of cases. Optimization of the remote field detection system improves the detection of defects in steam generator tubes in the presence of sodium.

Introduction

Heat exchanger tubes are among the critical systems in a nuclear power plants. The main function of the steam generator (SG) tubes is to transfer heat from the reactor core to the secondary side to produce steam, which drives the electrical turbine. In a sodium-cooled fast breeder reactor, only the walls of the heat exchanger tubes separate sodium and water. The high reactivity of sodium with water makes the integrity of steam generator tubes a critical component with respect to the safety of fast breeder reactors.

The remote field eddy current technique (RFEC) is an important non-destructive testing tool [1]. It is used mainly to detect corrosion in ferromagnetic conducting tubes and pipes. The method was also shown to be effective in the inspection of non-magnetic tubes [2]. This technique is based on the indirect electromagnetic coupling at low frequency between an excitation and a detection coil. The energy flow penetrates twice through the tube wall, making the method equally sensitive to both inner and outer tube discontinuities. Compared with a classical eddy current testing system, the remote field detection area is located far from the excitation coil, usually at a distance between two to three tube diameters. Until now, several approaches have been used to model the remote field: theoretical solutions [3], which are limited to axisymmetric geometries and axisymmetric numerical models. A three dimensional numerical model is very cumbersome [4] and accurate results could be computed until now only for non-magnetic steam generator tubes [2].

In a fast breeder reactor that uses liquid sodium (Na) as coolant, the signal from the deposits of sodium and sodium layer located outside of the steam generator tubes can mask or modify a defect signal.

In the present paper we estimate the effect of sodium in the in-service inspection of steam generators tubes using remote field technique. The simulation results, obtained using an axisymmetric finite element code [5], are compared with experimental measurements.

The code is used to evaluate the sodium patch and sodium layer signal for various outer defects (OD) in SG tubes. Based on a trial and error approach, a remote field device is optimized in order to increase both the defect signal and the signal/noise ratio.

1. Experimental measurements and comparison with numerical simulations of sodium signal in steam generator tubes using remote field eddy current

Sodium heats the steam generator tubes containing high-pressure water in a sodium-cooled fast breeder reactor. Due to the high reactivity of sodium with oxygen and water their contact between these elements should be avoided at all costs. A small leak in the evaporator tube leads to the formation of sodium hydroxide, which is corrosive. Large defects leading to a leak can result in interaction between sodium and water causing the generation of hydrogen, which is explosive.

Therefore, in-service inspection to evaluate the steam generator tubes integrity is necessary. For non-magnetic steam generator tubes, eddy current testing is fast and reliable. If steam generator tubes are made of ferromagnetic materials, detection of outer defects is difficult using classical eddy current testing, due to the limited penetration of eddy currents. The remote field eddy current technique is suitable for detection of outer defects in ferromagnetic tubes. The remote field detector signal can be digitized and recorded for off-line automated software analysis. However, several challenges exist when using this method: the sodium on the external surface of steam generator tubes can modify the defects signals. An additional complication arises due to the lack of information about the location of sodium patch and thickness of the sodium layer present on the outer surface.

An experiment was set up to measure the influence of sodium on the defect signal when the RFEC technique is used. A 400 mm tube length was cut from a ferromagnetic steam generator. In the middle part of the tube, an electrical discharged machined (EDM) defect was created. The defect is circumferential with a 20% through wall depth and 5 mm long. The tube was placed in a container filled with liquid Na at around 300^0 C and kept for several hours (see Figure 1a), until a small Na layer was deposited on the external surface of the tube. The whole experiment took place in a closed chamber filled with argon at higher pressure in order to prevent oxygen contamination. The oxygen fraction in the chamber was maintained continuously at less than 1 ppm. After the sodium was deposited, the SG tube temperature was lowered to room temperature and the external surface of the tube was protected with a plastic pouch filled with argon. Figure 1b shows a close view of the defect

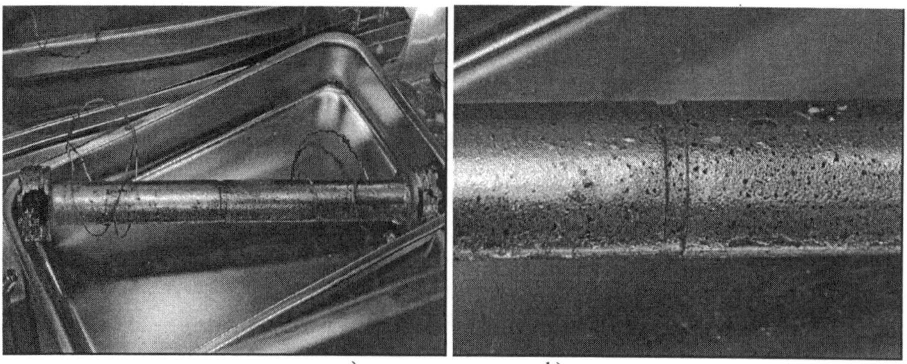

a) b)

Figure 1. a) Cut part from a steam generator tube covered with sodium; b) Close view of defect area

Figure 2. Close view of tube defect-free zone

area. A thin sodium layer was also deposited inside the defect. Figure 2 shows a close-up view of the tube surface without defect. It can be noticed that small sodium patches are located mainly in the lower part of the tube. Also, the sodium layer surface is far from smooth. Small sodium dots, patches and islands, each with different thickness, pepper the surface.

The RFEC signal from the defect (OD20% groove, length 5mm) was measured at 150 and 250 Hz before and after sodium was deposited (see Figure 3a). It can be seen that at both frequencies, the sodium layer does not significantly affect the defect signal. The phase of the signal remains unchanged but its amplitude decreases slightly. A finite element simulation based on a 2D axisymmetric code [5] shows the influence of the sodium layer in Figure 3b. In the simulation, the sodium layer has the same conductivity (23.8×10^6 S/m) and thickness (0.05 mm) along the tube. The sodium layer covers the bottom of the defect as well as the defect walls. The SG tube has the following parameters: 24.2 mm inner diameter, 31.8 mm outer diameter, 3.5×10^6 S/m electrical conductivity and 100 relative magnetic permeability. The simulation results agree very well with experimental measurements both in phase and amplitude at several excitation frequencies. In addition, consistent with experimental observation, the simulations predict a reduction in the amplitude of defect signal when a thin sodium layer covers the tube.

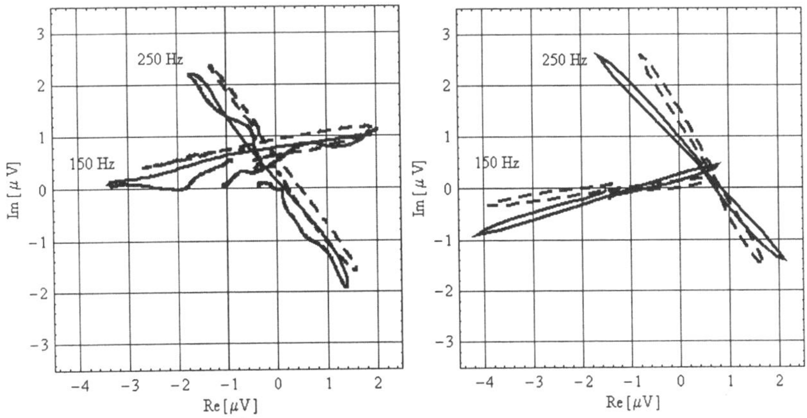

a) b)

Figure 3. OD 20% groove RFEC signal – without Na layer, --- with Na layer;
a) measurement data; b) simulation results

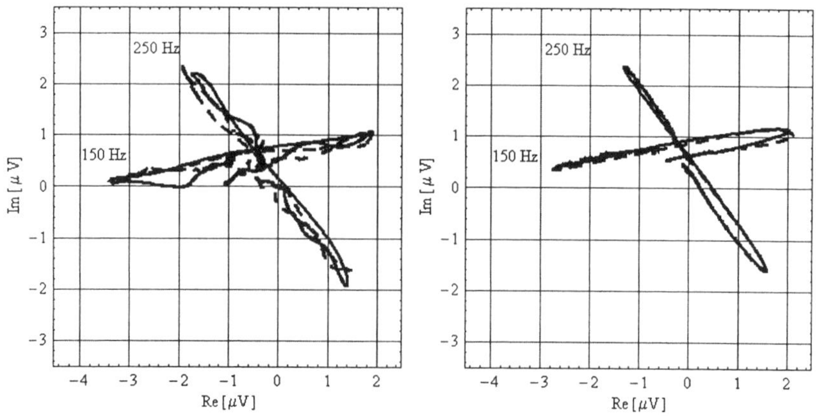

a) b)

Figure 4. OD 20% groove RF signal – 1st measurement, --- 2nd measurement; a) without Na; b) with Na layer

The 2-dimensional axisymmetric model can predict accurately the effect of the sodium layer only when a very fine mesh is used in all zones between the detection and excitation coils. The finite element mesh consisted of 220,000 linear triangular cells.

Figure 4 shows two separate measurements from the same defect (OD 20% deep, 5 mm long) with and without the sodium layer. It can be seen that the RFEC signal is consistent.

2. Simulations of Na patch, layer and band signal

Measuring the sodium signal using RFEC in steam generator tubes is difficult and only a small number of cases can be experimentally simulated. We have seen before that it is possible to model the influence of sodium on the defect signal using finite element techniques. Several cases were considered: a) when the defect is filled with sodium; b) when the defect is under a patch of sodium; c) when the defect is under a full circumferential band of sodium. In the second case, an approximate solution was constructed, for circumferential defects based on the following linear approximation:

$$S_{\text{Na drop+defect}^*} = \frac{l_\varphi}{2\pi r} S_{\text{Na band+defect}^*} + \left(1 - \frac{l_\varphi}{2\pi r}\right) \cdot S_{\text{defect}^*} \tag{1}$$

where r is the tube radius, l_φ is the patch length in the circumferential direction, $S_{defect*}$ is the signal from a defect filled with sodium and with tube covered with a sodium layer and $S_{Na\_band+defect*}$ is the signal from a defect filled with sodium and under a circumferential sodium band and, the tube is covered with sodium layer.

For circumferential outer defects with a small length (0.5 mm) and depths between 20%tw and 50%tw the filling of defect with liquid sodium will slightly modify the defect signal (see Figure 5a,c). If the defect is long, for example 10 mm, the RFEC signal decreases of 30% to 40% from the no-fill case (see Figure 5b). However, in this case, the simulations show that the defect is detectable if the defect signal has high amplitude. The operating frequency is 150 Hz.

The next case considers when the defect is under a small Na patch. Several conditions were simulated: when the Na layer is 0.1 and 0.3 mm thick. The sizes of the Na patches in the simulations were chosen to be: 8 mm axial length, 8 mm circumferential length, 3 mm high and 20 mm axial length, 5 mm circumferential length, 10 mm high. The Na patch signal and the signal from a circumferential OD 20%, length 0.5 mm under Na patch are compared in Figure 6 and Figure 7. The excitation frequency is 150 Hz. In all cases, the amplitudes of Na patches signal are about half the amplitude of the signal due to defects under Na patches. The simulations show that even outer defects with small openings masked by Na patches can be detected using RFEC.

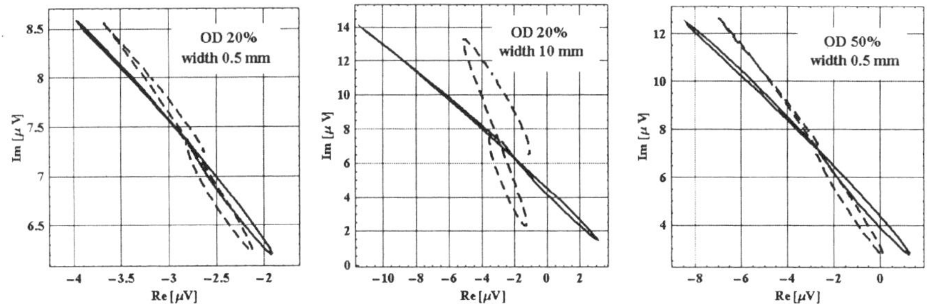

a) b) c)

Figure 5. Comparison of defect signal and defect filled with Na signal.
OD grooved RFEC signal with Na layer, - grooves not filled with Na, .--- grooves filled with Na

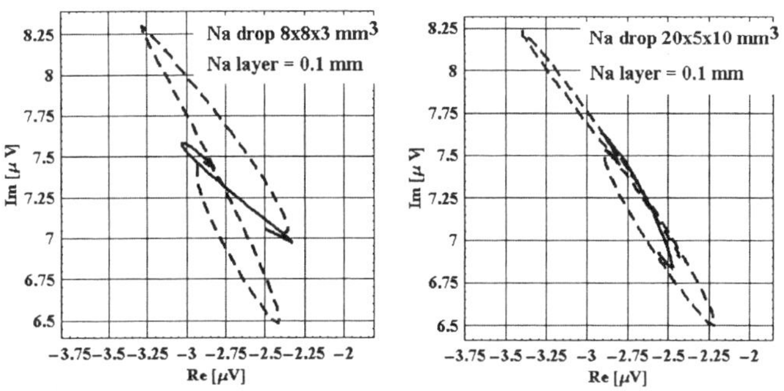

Figure 6. Comparison of the Na patch signal (continuous line) and defect under the Na patch signal (dashed line) for OD20%, length 0.5 mm. Na layer thickness is equal to 0.1 mm.

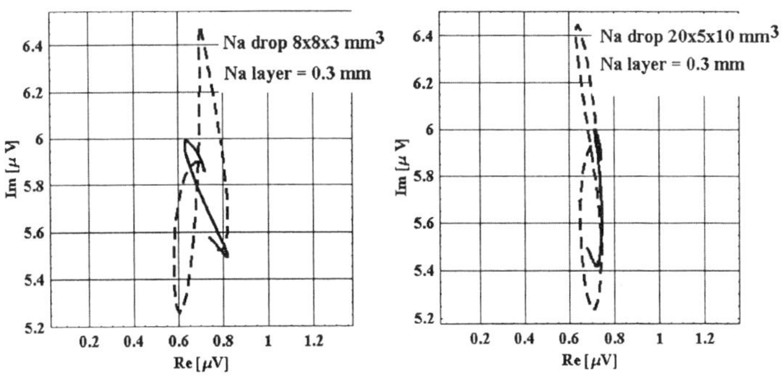

Figure 7. Comparison of the Na patch signal (continuous line) and defect under the Na patch signal (dashed line) for OD20%, length 0.5 mm. Na layer thickness is equal to 0.1 mm.

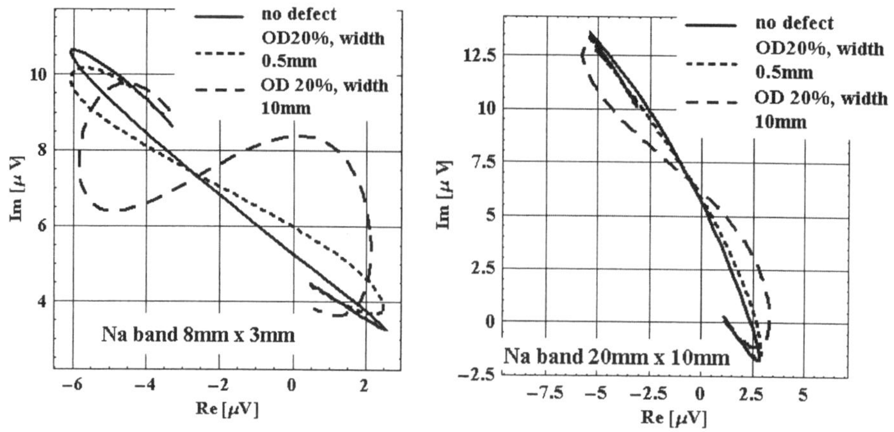

Figure 8. Comparison of the Na patch signal (continuous line), OD20%, 0.5 mm long defect under Na band signal (dot line) and OD20%, 10 mm long defect under Na band signal (dashed line). Na layer thickness is equal to 0.1 mm.

The most difficult situation is when the outer defect is masked by a circumferential Na band. The simulation results (see Figure 8) show that it is possible to detect outer defects with small opening if the Na band size is 3 mm high and 8 mm long. For a "higher" Na patch (20 mm long and 10 mm high), only wide defects (length 10 mm) can be seen using RFEC. However, in practical situations, a Na band can appear only in the vicinity of weld areas (where steam generators tubes are connected) or near the support plate. In order to eliminate the Na band signal, a specific signal analysis, based on multi-frequency technique, should be used.

3. Optimization of the remote field system

As was seen in the previous section, the presence of sodium band on the outer surface of steam generator tubes might hide the defect signal. The thickness of the Na layer, the locations and positions of sodium patches and Na island-layers usually are unknown. In the RFEC signal there are several sources of noise arising due to vibrations of coil probes, probe velocity, tube curvature, electronic noise and others. However, in the next analysis the authors focused only on the perturbation due to Na located outside the SG tubes. The "sodium perturbation" represents how the defect RFEC signal is affected by the presence of sodium. There are several difficulties in defining "sodium perturbation". For example, when Na fills the defect, if the "filling" is considered a perturbation then this perturbation exists only if the defect is present. There is no pure "Na filling" noise in the absence of a defect. The situation is different in the case of Na patch, where the Na patch signal can be computed in the absence of the defect. The third situation appears in the case of Na layer. It is possible to calculate only the Na layer signal but the amplitude of this signal is almost zero since is will be only a point in the complex signal representation.

To define the "sodium perturbation" the following conventions will be adopted in the present paper: a) if the noise arises due to a Na patch, the noise signal is defined as Na patch signal; b) when the noise is present due to the Na layer or Na filling the defect, the noise signal is defined as follows:

$$Signal_{perturbation} = Signal_{defect+perturbation} - Signal_{defect} \quad (2)$$

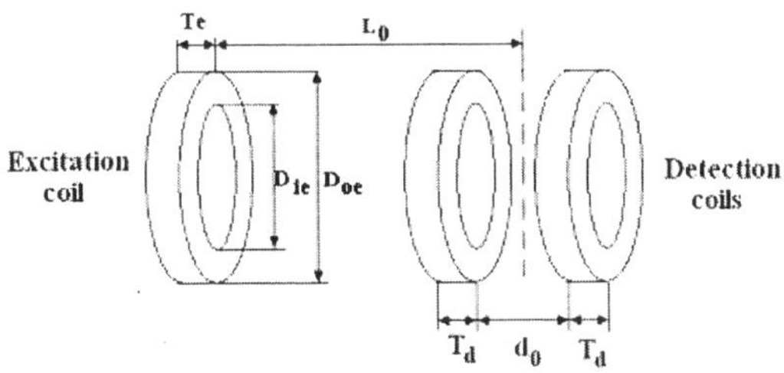

Figure 9. Parameters to be optimized in the remote field device

Table 1. Remote field device parameters

PARAMETERS	Initial RF configuration	Optimized RF configuration1	Optimized RF configuration 2
Distance excitation-detection (L_0)	120 mm	60 mm	60 mm
Distance between detection coils (d_0)	3 mm	11 mm	1 mm
Detection length (T_d)	3 mm	5 mm	3 mm
Excitation length (T_e)	8 mm	8 mm	8 mm

The signal/noise (S/N) ratio is defined in all cases as:

$$S/N = \frac{Signal_{defect}}{Signal_{perturbation}} \tag{3}$$

The RFEC device has been optimized in order to increase both the outer defect signal and signal/noise (S/N) ratio. The noise was considered to be only due to "sodium perturbation". Several parameters of the remote field device can be varied (see Figure 9): a) the distance between excitation and detection coil (60 to 140 mm); b) excitation and detection length (1 to 9 mm); c) distance between detection coils (1 to 30 mm); d) excitation frequency (50 to 1000 Hz).

A trial and error approach was employed. The defect signal and the S/N ratio were simulated in every specific situation: a) Na layer (0.3 mm thick); b) Na filling the defect; c) Na patch d) any combination of effects from a) to c). In order to evaluate the detectability of defects, responses from various outer defects signals were computed, with defects depth ranging from 20% to 50% through wall, and defect lengths varying from 0.5 to 10 mm.

The remote field parameters are given in Table 1 for the initial configuration and for

Table 2 Amplitude of signal from defects at 250 Hz excitation frequency

	Initial RF configuration	Optimized RF configuration 1	Optimized RF configuration 2
SIGNAL FROM DEFECT	Ampl. [μV]	Ampl. [μV]	Ampl. [μV]
OD 20%, length 0.5 mm	2.3	9.5 (310%)	4.2 (81%)
OD 20%, length 5 mm	7.8	34.2 (340%)	14 (81%)
OD 20%, length 0.5 mm Na layer	1.4	6.8 (370%)	3 (110%)
OD 20%, length 0.5 mm Na layer+Na filled	1.3	5.8 (350%)	2.7 (110%)
OD 20%, length 5 mm Na layer+Na filled	3.4	15 (350%)	7.2 (110%)
OD 20%, length 0.5 mm Na layer+Na filled+Na band	3.4	25 (630%)	7 (98%)
OD 20%, length 5 mm Na layer+Na filled+Na band	3.1	23 (640%)	6.2 (100%)

Table 3. Signal/Noise ratio

	Initial RF configuration	Optimized RF configuration 1	Optimized RF configuration 2
Signal=OD 20%, length 0.5 mm Noise=Na layer	1.7	2.2 (30%)	2.2 (30%)
Signal=OD 20%, length 0.5 mm Noise=Na layer+Na filled	1.5	1.7 (17%)	1.8 (22%)
Signal=OD 20%, length 5 mm Noise=Na layer+Na filled	1.2	1.3 (5%)	1.3 (12%)
Signal=OD 20%, length 0.5 mm Noise=Na layer+Na filled+Na band	0.5	0.2 (-36%)	0.4 (-7.1%)
Signal=OD 20%, length 5 mm Noise=Na layer+Na filled+Na band	0.8	0.7 (-17%)	0.8 (-2.2%)

the two final optimized configurations. Table 2 and Table 3 are presented the amplitude of the defect signal (with and without noise) and signal/noise ratio respectively. During optimizations the thickness of Na layer was increased from 0.05 mm to 0.3 mm and correspondingly the 20% OD (length 0.5 mm) RFEC signal amplitude decreased from 2.3μV to 1.4μV (see Table 2). Also, it was observed that for thin defects there is a small influence from the filling with sodium on the RFEC signal. In the optimized configurations 1 and 2, the defect signal increases from 300% to 600% and from 80% to 100% respectively. However, in the optimized configuration 1 and 2, the signal/noise ratio when the defect is under a Na band decreases by 20% to 30% and by 2% to 8% respectively. In the case when only a Na layer covers the outer surface of SG tubes and Na fills the defect, the S/N ratios for both optimized configurations increase between 5% and 30%.

4. Conclusions

The effect of sodium on the RFEC signal from outer defects in steam generator tubes was shown to be properly modeled by a simulation tool based on the finite element method. Experimental measurements of the Na-layer effect agreed very well with simulation results. The effect of Na patch and when Na fills the defect was computed. Considering the signal due to Na as a perturbation signal, the remote field device was optimized in order to decrease the sodium influence. The optimization resulted in the prediction of a proper operation point of the remote field system, in which the amplitude of the defect signal was shown to increase from 300% to 600%, and the signal to noise ratio, up to about 30%. However, in the case when there is a Na band the S/N ratio decreases up to 36%.

References

[1]. D. L. Atherton, S. Sullivan and M. Daly, A remote field eddy current tool for inspecting nuclear reactor pressure tubes, *Br. J. Nondestructive Testing* **30** (1988), 22-28.

[2]. H. Fukutomi, T. Takagi and M. Nishikawa, Remote field eddy current technique applied to non-magnetic steam generators tubes, *NDT&E International* **34** (2001), 17-23.

[3] N. Ida and M. Raugi, Theoretical Models for the Remote Field Effect, *IEEE Transactions on Magnetics*, **29** (1993), 1356-1358.

[4] S. Sharma, Y. Sun and L.Udpa, Finite Element Modeling for Simulating Remote Field Flaw Detection in Aircraft Frame, *IEEE Transactions on Magnetics* **35** (1999) 1750-1753.

[5] O Mihalache, Advanced Remote Field Computational Analysis of Steam Generators Tubes, *The 8th International Workshop on Electromagnetic Nondestructive Evaluation*, Saarbrucken, Germany, June 12-14 (2002).

Electromagnetic Nondestructive Evaluation (VIII)
T. Sollier et al. (Eds.)
IOS Press, 2004

Improvement of Ultrasonic Flaw Images by Electromagnetic Acoustic Transducer

Eiji MATSUMOTO and Hideaki FUNAOKA
Department of Energy Conversion Science, Kyoto University, Kyoto 606-8501, Japan

Abstract. This paper attempts to improve the ultrasonic flaw image by an Electro-Magnetic Acoustic Transducer of Lorentz type on the basis of numerical simulation. In order to obtain clearer images of internal flaws, several techniques of waveform processing and ultrasonic imaging are examined. It is found that the so-called ALOK method is the most effective one for ultrasonic flaw imaging by the EMAT.

1. Introduction

An ultrasonic wave is influenced by the geometrical, mechanical and physical conditions of the material. Ultrasonic Non-Destructive Evaluation is widely applied to detection of flaws, measurements of material properties, geometries, residual stresses etc. UT-NDE refers to the variation of the wave characteristics during propagation, and it enables us to obtain important information from the interior of the object by a portable apparatus. In general, transmission and reception of ultrasonic waves are achieved by a transducer between the mechanical and the electromagnetic fields. Recently, EMATs have been expected to have several advantages with respect to conventional transducers using piezoelectric or magnetostrictive materials. An EMAT consists of an exiting coil and magnets and utilizes the electromagnetic force and the electromagnetic induction for transmission and reception of mechanical waves. Thus it can be applied to metals without contact, and many kinds of waves such as longitudinal, transverse and surface waves can be generated by arranging the coil and the magnets. A Lorenz type of EMAT utilizes the Lorentz force due to the eddy current under the static magnetic field by the magnets, cf. [1]. For the Lorentz type of EMATs, many theoretical studies have been done and we shall refer to only some examples. Ludwig et al. [2] and Thompson [3] studied the basic principle, Ogi et al. [4] numerically analyzed transmission of ultrasonic waves on the basis of FEM, and Mitsuda et al. [5] analyzed both transmitting and receiving processes including propagation of ultrasonic waves. Funaoka et al. [6] simulated ultrasonic inspection of flaws by the EMAT. In this paper, we attempt to visualize the internal flaw by the EMAT, and examine several techniques of waveform processing and ultrasonic imaging to improve the flaw image.

2. Basic Equation of Ultrasonic Inspection by EMAT

Figure 1 shows the sketch of the cross section of an EMAT consisting of an elliptic coil and a couple of permanent magnets. When a dynamic electric current is supplied to the coil, the eddy current is generated near the surface of the conductive specimen. Then the electromagnetic fields around the EMAT are governed by

$$\Delta \boldsymbol{A} - \mu\sigma\left(\nabla\phi + \frac{\partial \boldsymbol{A}}{\partial t}\right) - \mu \boldsymbol{J}_0 = 0\,,\ \nabla \cdot \boldsymbol{J}_e = -\nabla \cdot \sigma\left(\nabla\phi + \frac{\partial \boldsymbol{A}}{\partial t}\right) = 0\,, \tag{1}$$

where $\boldsymbol{A}$ is the vector potential, ϕ the scalar potential, $\boldsymbol{J}_0$ the forced current density in the coil and $\boldsymbol{J}_e$ the eddy current density in the specimen. Then the magnetic flux density $\boldsymbol{B}$ and the electric field $\boldsymbol{E}$ are given by

$$\boldsymbol{B} = \nabla \times \boldsymbol{A}\,,\ \boldsymbol{E} = -\frac{\partial \boldsymbol{A}}{\partial t} - \nabla\phi\,. \tag{2}$$

Under the magnetic flux density $\boldsymbol{B}$, the free charges in the eddy current are subjected to the Lorentz force, which is expressed per unit volume in the form

$$\boldsymbol{F} = \boldsymbol{J}_e \times \boldsymbol{B}\,. \tag{3}$$

When a high frequency current is supplied to the coil, the eddy current and the Lorentz force are localised near the surface of the specimen. Then a mechanical wave is created there and propagates into the specimen. If the specimen is an isotropic homogeneous elastic material, propagation of the elastic wave is governed by the equation of motion as

$$\rho\frac{\partial^2 u_i}{\partial t^2} = \frac{\partial \sigma_{ij}}{\partial x_j} + F_i\,. \tag{4}$$

Here double indices in a term are summed from 1 to 3, $\boldsymbol{u}$ is the displacement, ρ the material density, $\boldsymbol{\sigma}$ and $\boldsymbol{\varepsilon}$ the stress and the strain tensors defined by the displacement derivatives

$$\sigma_{ij} = \lambda \cdot \varepsilon_{kk}\delta_{ij} + 2\mu\varepsilon_{ij}\,,\ \varepsilon_{ij} = \frac{1}{2}\left(\frac{\partial u_i}{\partial x_j} + \frac{\partial u_j}{\partial x_i}\right), \tag{5}$$

where λ and μ are the Lame's elastic constants. The elastic wave is reflected from the boundaries and the internal flaws and the reflected waves reach the surface faced by the EMAT.

Oscillation of the free charges due to the reflected wave generates the eddy current under the static bias magnetic field. Then the current in the specimen for the receiving process is expressed as

$$\boldsymbol{J} = \sigma\left(\boldsymbol{E} + \frac{\partial \boldsymbol{u}}{\partial t} \times \boldsymbol{B}_0\right). \tag{6}$$

In general, the dynamic magnetic fields induced by the eddy currents in the transmitting and the receiving processes are very small compared with the static magnetic field due to the magnets. Hence we may identify the magnetic flux density for the receiving process with the static magnetic flux density $\boldsymbol{B}_0$ because it appears only in the term multiplied by small quantity $\boldsymbol{v}$. The current (6) plays a similar role to the forced current in the coil for the transmitting process. That is, for the receiving process of the EMAT, instead of (1) the vector and the scalar potentials are governed by

$$\Delta \boldsymbol{A} - \mu\sigma\left(\nabla\phi + \frac{\partial \boldsymbol{A}}{\partial t} + \frac{\partial \boldsymbol{u}}{\partial t} \times \boldsymbol{B}_0\right) = 0\,,\ \nabla \cdot \left[\sigma\left(\nabla\phi + \frac{\partial \boldsymbol{A}}{\partial t} + \frac{\partial \boldsymbol{u}}{\partial t} \times \boldsymbol{B}_0\right)\right] = 0\,. \tag{7}$$

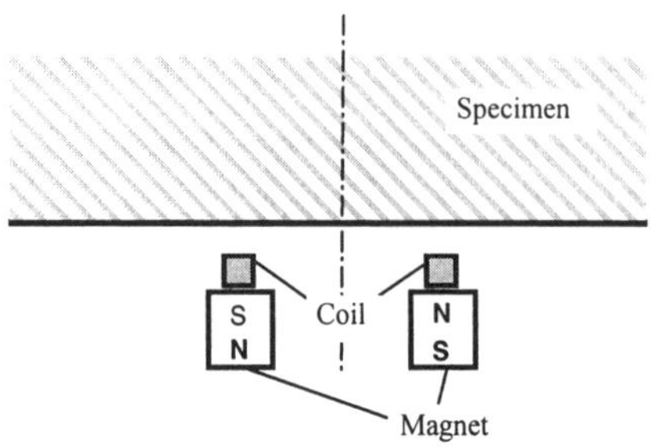

Fig.1 Sketch of cross-section of EMAT

Through the electromagnetic induction, the perturbed magnetic field gives rise to the voltage variation in the coil, which is the received signal of the reflected wave.

3. Numerical Simulation of Ultrasonic Inspection of Flaws by EMAT

Let us assume that the electromagnetic fields around the EMAT are two-dimensional, and let x and y be the two-dimensional coordinates in the plane of Fig.1. Then the vector potential and the electric current are expressed in the forms

$$\boldsymbol{A} = (0, 0, A), \ \boldsymbol{J} = (0, 0, J), \ \boldsymbol{J}_e = (0, 0, J_e). \tag{8}$$

We divide the electromagnetic fields into the large static fields caused by the permanent magnets and the small dynamic fields caused by the source coil and the wave motion. For analysis of the static magnetic field, we employ FEM with first-order rectangular elements. The length of the minimum edge of a rectangular element is 0.5mm. Each permanent magnet has a square cross-section with 5mm edges, and two magnets are located at 15mm distance. The flux density in the permanent magnets is 1.0T. We analyze the magnetic field in the 60mm×50mm rectangular region around the magnets. In this case, there is no current and hence the scalar potential vanishes. It is easy to see that the equation (1) for the component A of the vector potential reduces to a Laplace equation. The calculated equipotential lines of A are shown in Fig. 2.

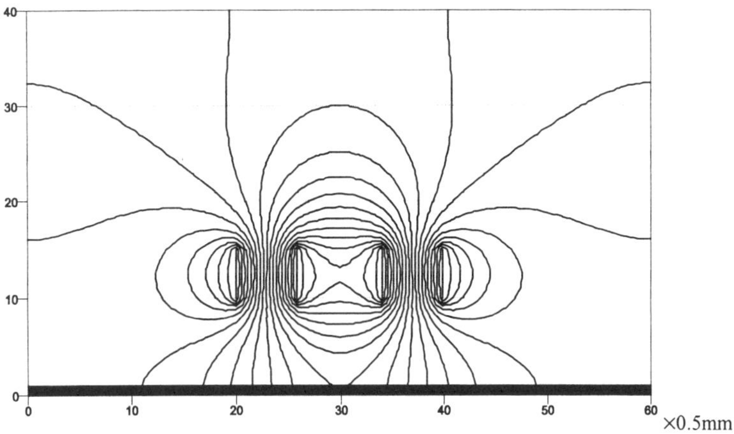

Fig.2 Equipotential lines of the static magnetic field due to the permanent magnets

Next we consider the dynamic magnetic field generated by the coil of the EMAT. In this case, the vector potential component A and the scalar potential ϕ are governed by the system of partial differential equations (1). We use smaller finite elements than the case of the static analysis in the region containing the specimen of aluminum with rectangular cross-section 30mm×10mm. The exiting coil is made of copper wire with rectangular cross-section 1mm×1mm and located over the permanent magnets with 1.0mm lift-off from the center of the specimen surface, see Fig.1 The material parameters used in the calculation are as follows. The permeabilities of the environment (the air), aluminum and copper are assumed to be equal to that of the vacuum 1.26×10^{-6}H/m, the conductivities of aluminum and copper are, respectively, 3.64×10^{7}S/m and 5.99×10^{7}S/m, and the environment is assumed to be an insulator. We supply one cycle of sinusoidal current to the coil such that the direction of the current in each wire is opposite to each other. The current density has the amplitude 1.00×10^{7}A/m^2 and the frequency is 100kHz. Figure 3 shows the distribution of the induced eddy current density at the instance of its maximum.

We observe that the eddy current is locally distributed near the surface of the specimen faced to the EMAT coil.

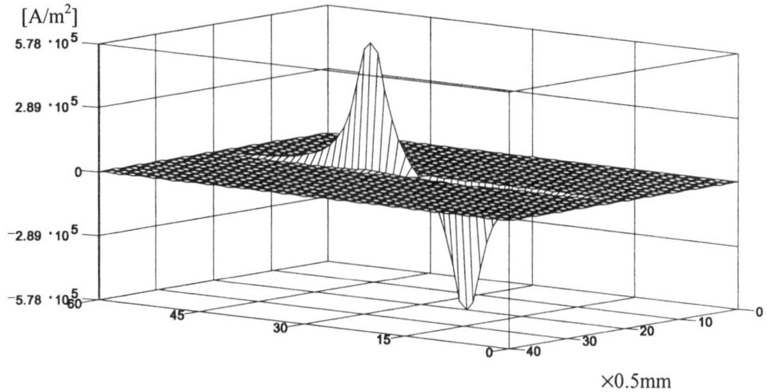

Fig.3 Distribution of eddy current by EMAT coil in conductive material

Since the electromagnetic fields are two-dimensional, the magnetic flux density $\boldsymbol{B}$ and the Lorentz force $\boldsymbol{F}$ can be expressed in the form

$$\boldsymbol{B} = \left(B_x, B_y, 0\right),\ \boldsymbol{F} = \left(F_x, F_y, 0\right) = \left(-J_e B_y, J_e B_x, 0\right). \tag{9}$$

Thus the Lorentz force lies in the x–y plane. If the axis y is set at the center of the EMAT, the distribution of the Lorentz force has a symmetry with respect to the y–axis because the magnetic field and the current have also symmetries, refer to Fig.1. In fact, since J_e and B_y are antisymmetric and B_x is symmetric with respect to the y–axis, from (3) F_x and F_y become, respectively, symmetric and antisymmetric. Figure 4 shows the calculated components of the Lorentz force near the surface of the specimen. It is found that each component of the Lorentz force has two peaks corresponding to each pair of wire and magnet and the maximum strength of the horizontal component F_x is larger than the one of the vertical component F_y. That is, the EMAT shown in Fig.1 mainly excites the transverse wave. Since the eddy current and the Lorentz force are localized within the skin depth of the dynamic magnetic field, an elastic wave is excited near the boundary surface facing the EMAT.

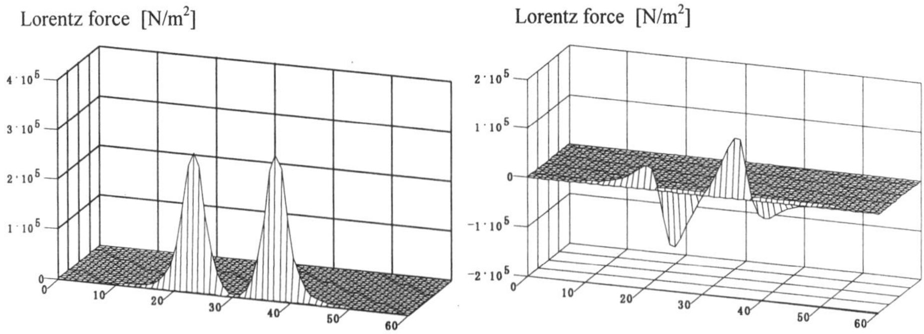

(a) Lorentz force parallel to specimen surface

(b) Lorentz force normal to specimen surface

Fig.4 Distribution of Lorentz force near specimen surface

Now we analyze the wave motion in the specimen which is caused by the above Lorentz force. Suppose that the specimen has the rectangular cross section 300mm×50mm and a crack with 10mm×2mm cross-section is located at its centre. For the numerical calculation of elastic waves, we employ the finite difference method with higher order formula for the time domain. Figure 5 shows the wave fields around the crack at time 12μs. The longitudinal and the transverse components of the wave field are defined by the derivatives of the displacement $\nabla \cdot \boldsymbol{u}$ and $\nabla \times \boldsymbol{u}$, respectively. We see that the transverse wave is reflected from the crack surface and diffracts beyond the crack. We see also that the longitudinal wave is reflected from the crack edges by the collision of the transverse wave. That is, a mode conversion from the transverse to the longitudinal waves takes place at the crack edges.

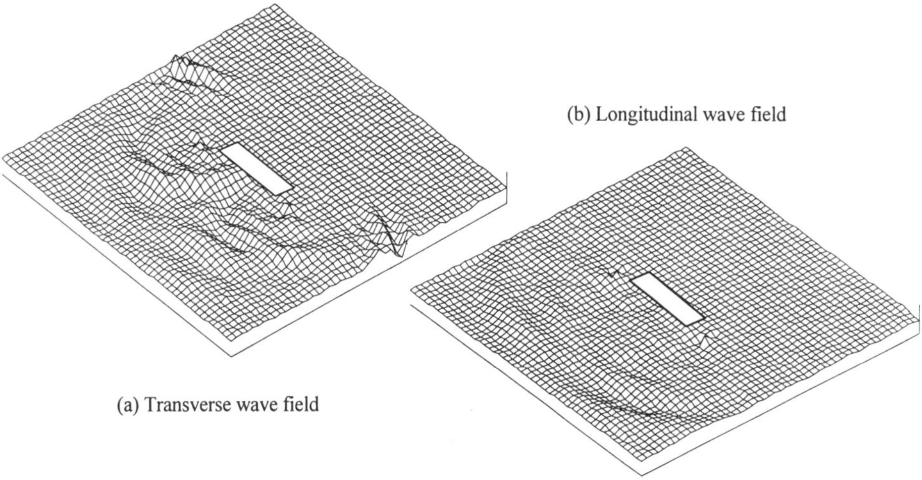

Fig.5 Wave field around a crack

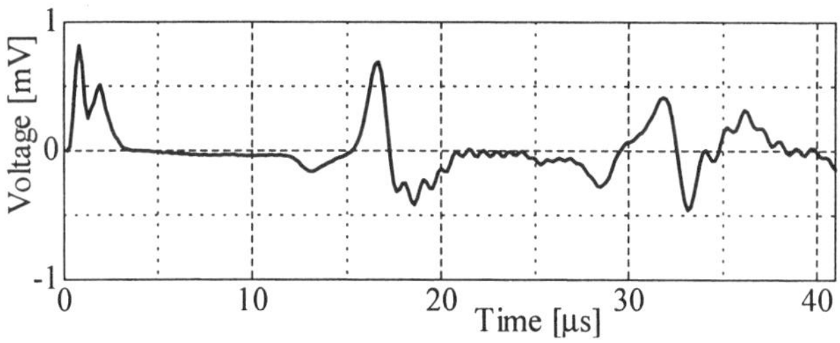

Fig.6 Received waveform in EMAT

If the displacement $\boldsymbol{u}$ is obtained as a function of the time and the coordinates, the electromagnetic fields in the receiving process of the EMAT can be calculated from (7). The variation of the voltage in the coil, i.e., the received waveform by the EMAT, is given in Fig.6. The first peak indicates the incident wave and the second one at 16.5 μs and the third one at 33.0 μs indicate, respectively, the reflected transverse waves from the crack surface and the opposite boundary surface. As discussed in the above, the EMAT also excites the longitudinal wave though the component F_y of the Lorentz force. Although the amplitude of the induced longitudinal wave is not small, the signal of the longitudinal wave does not clearly appear in Fig. 6. It may reflect the fact that the EMAT is not sensitive to the longitudinal motion of the specimen. These characteristics of the EMAT are convenient to ultrasonic inspection because the flying time of the wave can be easily converted to the distance by means of the speed of the transverse wave.

4. Improvement of Ultrasonic Flaw Image by EMAT

The received waveform itself has important information in the interior of the material (A-scan), but the exact position and the shape of the crack cannot be estimated only from the waveform received at a point. We then scan the EMAT linearly along the specimen surface by keeping the constant lift-off. In Fig. 7, the received waveform at a position is expressed by the vertical line at each horizontal position with the shading corresponding to its time variation. This is the original ultrasonic image of the flaw (B-scan). To obtain the image, we reduced the number of the data and applied the smoothing processing to each waveform by taking the moving average. This image has many ghost patterns caused by the multi-reflections, inappropriate imaging method, etc, and the pattern of the crack is much larger than the exact size.

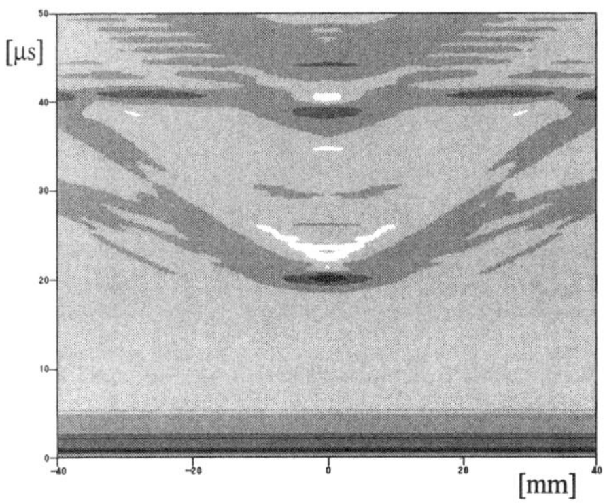

Fig.7 Original ultrasonic crack image

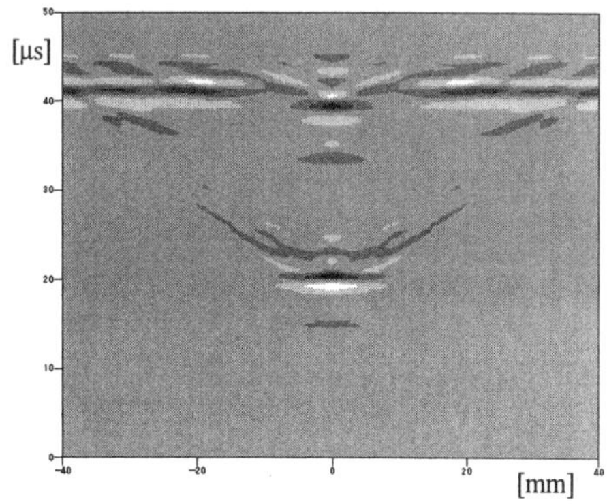

Fig.8 Crack image by deconvolution

As shown in Fig. 6, the pulse of the incident wave has a time width larger than the crack width converted by the speed of the transverse wave. In order to obtain a more accurate flight time of the wave, let us perform the deconvolution to the received waveform. Let $f(t)$ and $g(t)$ be the incident and the reflected waveforms. When the incident wave is reflected from a discrete number of reflection points, the received waveform will be expressed as

$$g(t) = a_1 f(t-\tau_1) + a_2 f(t-\tau_2) + \cdots \quad , \tag{10}$$

where a_i is the reflection coefficient of the i-th reflection point and τ_i is the flight time of the reflected wave from it. More generally, by increasing the number of the reflection points to infinity, we may express $g(t)$ in terms of the convolution of $f(t)$ in the form

$$g(t) = \int_{-\infty}^{+\infty} a(\tau) f(t-\tau) d\tau , \tag{11}$$

where $a(\tau)$ is the local reflection coefficient of the point in the distance corresponding to the flying time τ of the transverse wave. The Fourier transform of (10) becomes

$$G(\omega) = A(\omega) F(\omega) . \tag{12}$$

Suppose that from the experiment or the simulation the waveforms of the incident and the reflected waves are obtained. Then the distribution of the reflection coefficient $a(\tau)$ can be reconstructed by the inverse Fourier transform of the ratio $G(\omega)/F(\omega)$ between the Fourier spectra of the reflected and the incident waves as

$$a(t) = \int_{-\infty}^{+\infty} \frac{G(\omega)}{F(\omega)} e^{i\omega t} d\omega . \tag{13}$$

We applied FFT algorithm to numerical calculation of the Fourier and the inverse Fourier transforms in deconvolution. Figure 8 shows the image after deconvolution of the received waveforms. As expected, the image of the crack becomes sharper. The shadows growing from the ends of the crack towards inclined directions are caused by the assumption that the transmitted and the received waves are plane waves travelling in the normal direction to the boundary (B-Scan).

In fact the incident wave radiates out form the EMAT and it is also reflected in all directions from the crack edges. Let us regard the incident and the reflected waves as cylindrical waves. Then a pulse in the received waveform after deconvolution indicates that the corresponding reflection point may be located on the circle with the centre of the EMAT and the radius of the half flight time of the pulse. The real reflection point will also lie on the circles of possible reflection points for other positions of the EMAT. In order to illustrate the real reflection point, we accumulate such reflection coefficients at each position of the plate. Figure 9 is thus obtained image of the crack (SAFT-Method, refer to [7]). We see that the inclined shadows from the crack edges disappear but the crack pattern is still larger than the exact size.

In the above SAFT-Method, we have assumed that the wave is a cylindrical wave with the same amplitude in all the direction. The transmitted waves, however, have the directivity such that the normal wave field has the maximum amplitude, and the sensitivity of the EMAT has also a similar directivity. Thus, taking into account such directivities (ALOK method, refer to [7]), we reach the results shown in Fig. 10, where the size of the crack pattern becomes close to the exact one.

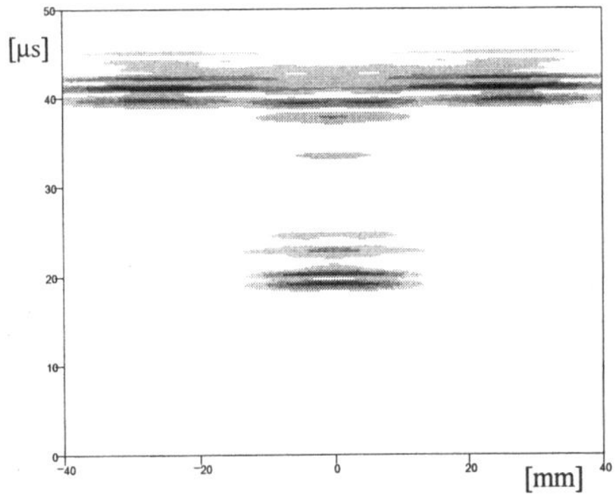

Fig. 9 Crack image by SAFT method

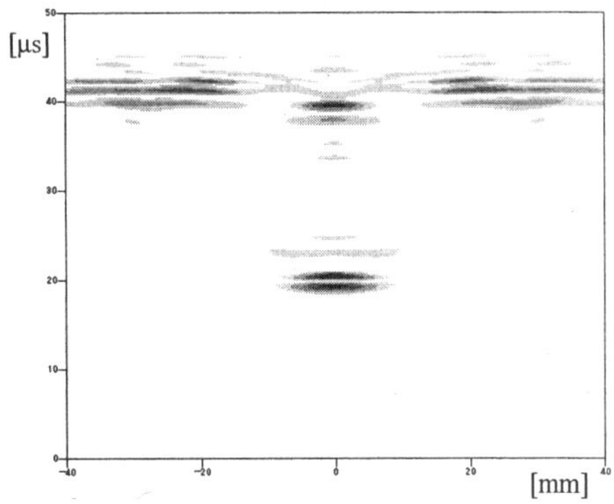

Fig. 10 Crack image by ALOK method

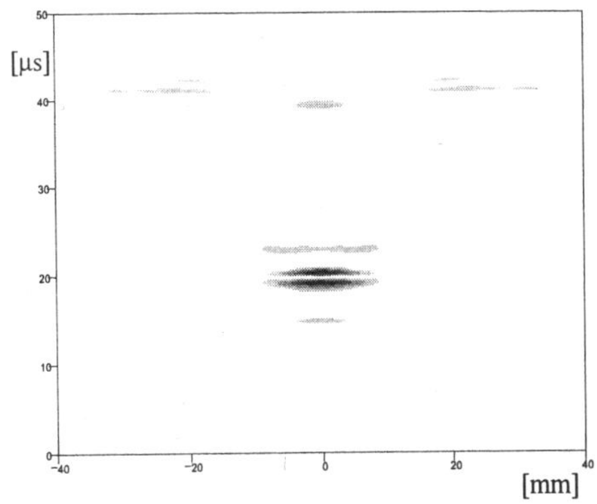

Fig. 11 Crack image by ALOK method with damping

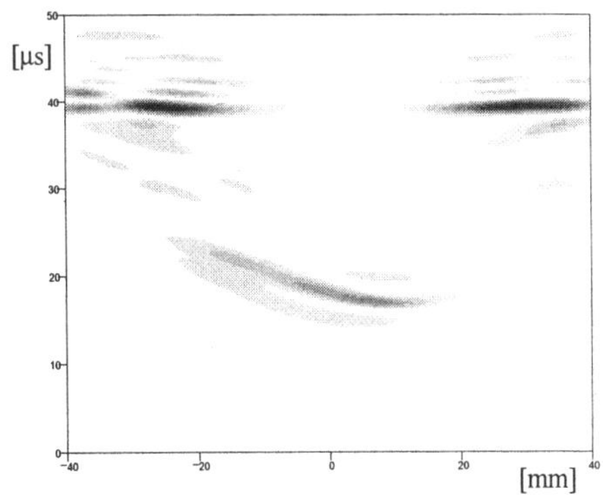

Fig. 12 Image of inclined crack by ALOK method

The actual waves in the material have always a damping during propagation caused by many effects such as viscosity, thermoelastic effect, ferromagnetic effect and micro eddy current. The damping of the waves has both advantages and disadvantages in ultrasonic nondestructive evaluation. The most serious disadvantage is that the reflected waves do not have enough amplitude for accurate evaluation. On the other hand, when there is a strong reflection point near the surface, the ghost patterns by multiple reflections can be reduced by the damping of the wave. Such damping can be simulated by multiplying the received waveform by a time decaying function. Figure 11 shows the crack image by use of the waveforms with damping, where the patterns of the second reflection and the other boundary almost disappear.

In order to show an unsuccessful example of the most effective ALOK method discussed in the above, we consider a crack located in an inclined direction with respect to the specimen surface. The image obtained is shown in Fig.12, where we see that the pattern of the crack is curved. In this case, the reflected wave propagates strongly in a specific direction, which brings errors in estimation of crack location because such directional distribution of reflected waves cannot be assumed in advance. Thus it will be a common disadvantage of the methods using a single transducer.

It should be noted that in all ultrasonic images low-pass and high-pass filters have been applied to the received waveforms in advance to eliminate the high frequency noise due to discrete calculations and the low-frequency noise from the translation motion of the specimen. In order to pick up the incident and the reflected waveforms from the received waveform, we employed a smooth rectangular window whose high frequency spectra can be cut by the low-pass filter.

5. Conclusions

The following conclusions have been reached in this paper.

1. Simulation method of ultrasonic inspection is presented based on FEM and finite difference method.
2. Although longitudinal waves are transmitted by the EMAT and reflected by the mode conversion from the crack edges, the received waveform mainly consists of the signals of transverse waves.
3. Several techniques of waveform processing and ultrasonic imaging are examined to improve the flaw image. It is found that ALOK method is the most effective one for the present EMAT.

References

[1] H. Ogi and M. Hirao, Line-Focusing of Ultrasonic SV Wave by Electromagnetic Acoustic Transducer, *J. Acoust. Soc. Am.* **103** (1998) 2411-2415.

[2] R. Ludwig and X-W. Dai, Numerical Simulation of Electromagnetic Acoustic Transducer in the Time Domain, *J. Appl. Phys.* **69** (1991) 89-98.

[3] R. B. Thompson, Physical Acoustics 19, Academic Press, New York, 1988.

[4] H. Ogi, M. Hirao, K. Minoura and H. Fukuoka, Quasi-Nonlinear Analysis of Lorentz-Type EMAT by Finite Element Method, *Trans. JSME*, 61A (1995) 638-645.

[5] T. Mitsuda and E. Matsumoto, Numerical Analysis of Ultrasonic Inspection using Electromagnetic Acoustic Transducer, *Studies in Applied Electromagnetics and Mechanics* 23 (2002) 60-67.

[6] H. Funaoka, T. Mitsuda and E. Matsumoto, Simulation of Ultrasonic Flaw Detection by Electromagnetic Acoustic Transducer, *JSAEM Studies in Applied Electromagnetic and Mechanics*, **14** (2002) 285-292.

[7] J. Krautkrämer and H. Krautkrämer, Ultrasonic Testing of Materials, Springer-Verlag, Berlin, 1990.

Electromagnetic Nondestructive Evaluation (VIII)
T. Sollier et al. (Eds.)
IOS Press, 2004

Eddy Current Testing of Stress Loaded Steel Samples

Tomasz CHADY, Ryszard SIKORA, Stanisław GRATKOWSKI,
Bartosz BORKOWSKI, Marcin CARYK, Grzegorz PSUJ
Technical University of Szczecin, ul. Sikorskiego 37, 70-313 Szczecin, Poland

Abstract. The purpose of this paper is to present the results of stress damage observation using an eddy current sensor. Parameters suitable for quantitative stress evaluation are also proposed. The experimental verification of the developed technique is done and the selected results are presented.

1. Introduction

Defects, stress and fatigue damages are causing changes of materials' magnetic and electric properties. Therefore, various electromagnetic techniques are used for the evaluation of ferromagnetic materials degradation. The method based on magnetic leakage flux observations finds widespread use in a great number of applications. It is based on a detection of leakage fields, caused by a change in the reluctance of ferromagnetic materials, which are excited with a DC field. The leakage magnetic fields are measured by scanning the surface of the specimen with some kind of a flux density sensitive transducer (e.g. GMR sensor) [1]. The main disadvantages of this method are: dependence of measurements on magnetic history of the samples, necessity of preliminary treatments (e.g. magnetization, demagnetization) of the material under testing. An eddy current method is less sensitive to these parameters and therefore, authors propose to use this method for evaluation of stress loaded ferromagnetic samples.

2. Sensor and measuring system description

Construction of the probe is shown in Fig. 1. It contains four excitation coils (*A*, *B*, *C*, *D*) and a single search coil (*S*) mounted on a ferrite core (*F*).

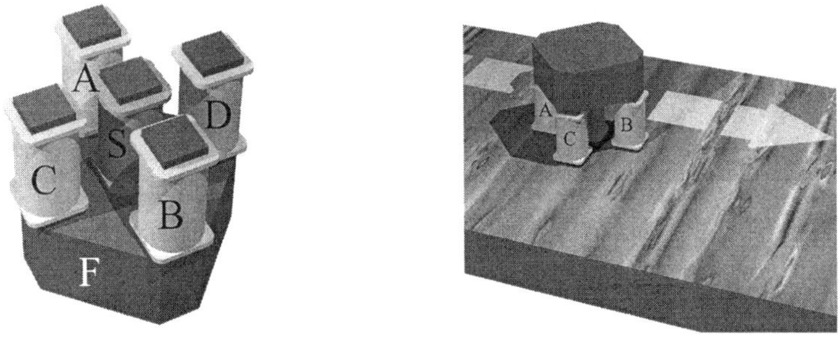

Fig. 1. Views of the eddy current probe (*A*,*B*,*C*,*D* - excitation coils, *S* - search coil, *F* - ferrite core)

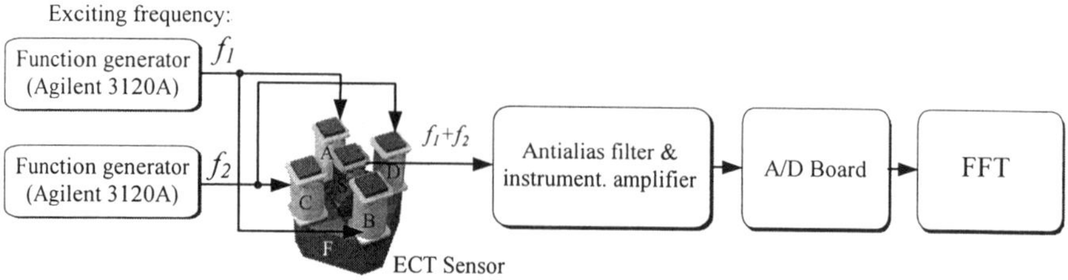

Fig. 2. Block scheme of the measuring system.

The probe construction is the same as presented in [2]. The main difference is the size of the sensor. In case of examination of stress loaded samples the most important factor is a high spatial resolution and sensitivity of the transducer. Therefore, taking into account optimisation results achieved in [2] the distance between transducer's columns was set to 1.5 mm. The other dimensions (having minor influence on probe performance) were assumed in this way, that proportions of the original transducer [2] were preserved. The height of the core is 7 mm and the total width is 6 mm. The core was manufactured using a high quality ferrite block obtained from MAGNETICS^{R}.

The search coil has 100 turns (ϕ 0.04 mm), and each of the four exciting coils has 25 turns (ϕ 0.2 mm). The sensor has two sensitivity axes. Excitation coils *A* and *B*, which belong to one axis of sensitivity are energized with frequency f_1, while the coils *C* and *D* are supplied with a signal having frequency f_2. The search coil *S* measures a mixed signal that contains both components. The signal is amplified by the instrumentation amplifier and filtered by a low pas analog anti-alias filter. The analog signal is converted into digital form by an A/D converter (20 MSamp/s, 16 bit resolution). Finally, Fast Fourier Transform is utilized to separate the frequency components. System designed in this way allows us to detect features having different orientations during a single scan. Workability of the system was already verified in case of the INCONEL plates' examination [3]. In this paper, authors propose to use the system for inspection of ferromagnetic materials.

3. Description of the samples

All the experiments were carried out for seven planar specimens made of low alloy steel (SS400). The dimensions of the specimens were as follows: thickness 2.17 mm, width 30 mm, total length 270 mm. The shape of the specimens is shown in Fig. 3. Each sample was tensile deformed in the longitudinal direction at a room temperature by an Instrom type machine.

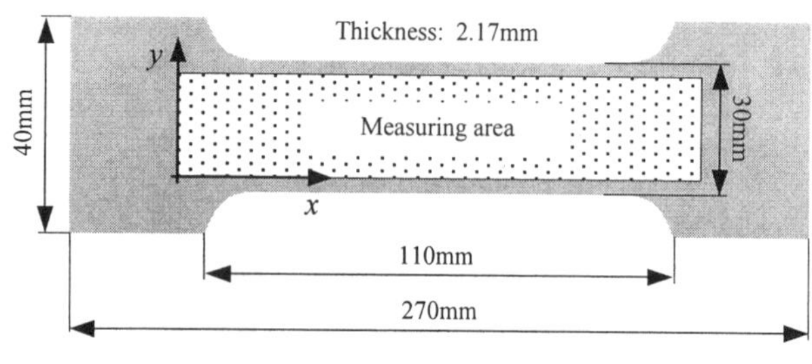

Fig. 3. Dimensions of specimen used for experiments (material SS400).

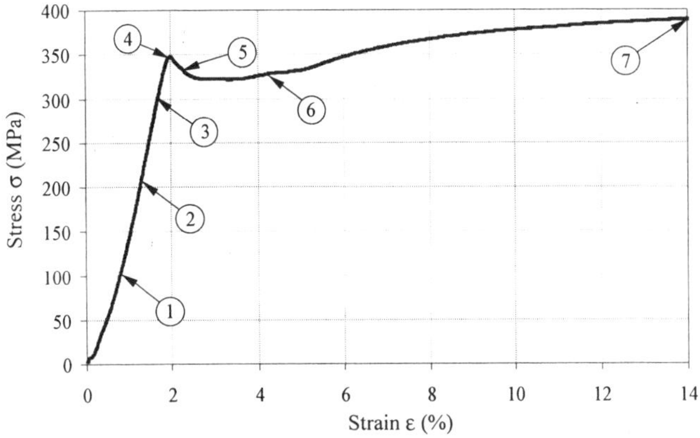

Fig. 4. Stress-strain curves of SS400 samples. The curves are marked with labels corresponding to the index of each sample.

The strain-stress curves for all seven samples are shown in Fig. 4. The yield of this material was around 350 MPa. The residual strain ε were between 0.7% and 14% while the maximum tensile stress was 389 MPa.

The samples SS400-1, SS400-2 and SS400-3 are loaded with the maximum stress of 103.2 MPa, 206.4 MPa and 300 MPa respectively, which is lower than the yield stress. The sample SS400-4 was tensile deformed exactly up to the yield point. A stress-strain history of the remaining three samples (SS400-5, SS400-6 and SS400-7) has passed the yield. Such selection of the samples enable us observation of the material changes caused by different levels of the applied stress.

While it was not necessary, all the samples were demagnetised before measurements. Results of experiments show (Fig. 5), that the magnetic history of the samples has no influence on the measurements.

4. Conditions of the experiments

The sensor was scanned along the *x*-axis (the longitudinal direction) in steps of 0.5 mm and along the *y*-axis in steps of 1mm. The scanning area was 180mm x 23mm (Fig. 3). The liftoff (the gap between the specimen and the sensor) was measured to be 0.2 mm.

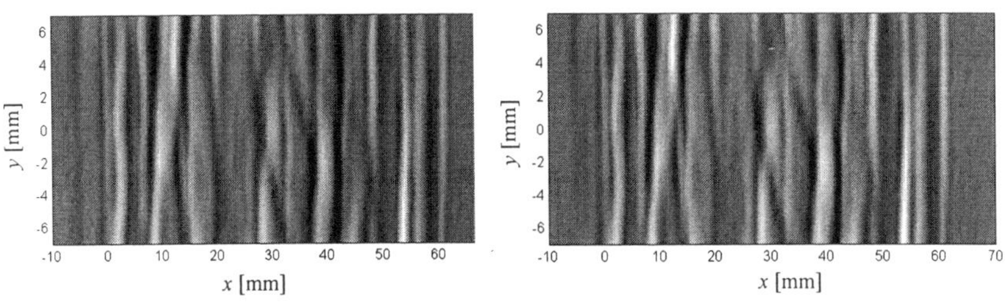

Fig. 5. Comparison of a signal obtained for the specimen after demagnetisation and a signal obtained for the same specimen after magnetization.

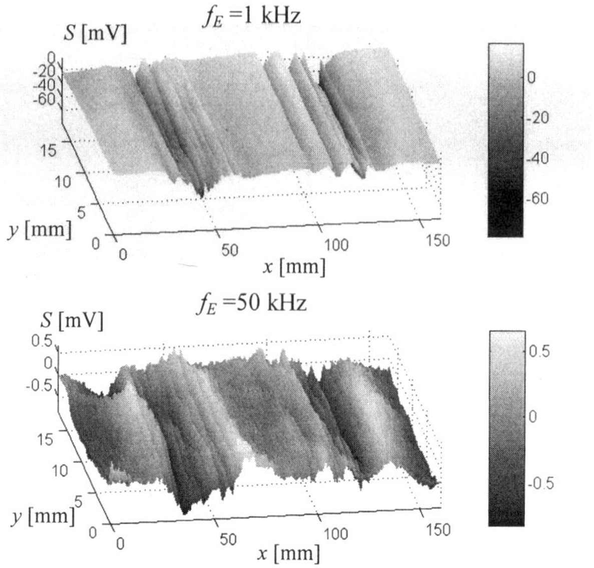

Fig. 6. Results of measurements for the tensile-loaded sample SS400-4 obtained for two different testing frequencies (f_E = 1 kHz, 50 kHz). Two-dimensional plots of S (mV) - the voltage signal obtained from the ECT transducer.

First experiments were carried out in order to select an optimal testing frequency. Therefore, a set of measurements using different excitation frequencies (f_E = 100 Hz ÷ 50 kHz) was done. The signals were very similar for all frequencies from the mentioned range. As an example signals measured for a frequency of 1 kHz and 50 kHz are shown in Fig. 6. The main difference was amplitude of a usable signal. It affects also the value of signal to noise ratio (SNR). The SNR is defined as a ratio of the signal changes caused by the material degradation to the amplitude measured for unloaded sample.

The maximum value of SNR was achieved for the frequency of 1 kHz and therefore this frequency was selected as a testing frequency for channel A-B and the frequency of 1.5 kHz for channel C-D. A difference between signal frequencies (0.5 kHz) is necessary in order to separate signals coming from the channels. The difference could be reduced but it will cause more complicated analysis. The excitation current is 30 mA.

5. Results of the measurements

The two-dimensional results of the measurements (the voltage signal obtained from the search coil - S [mV]) of tensile-loaded steel samples are shown in Fig. 7. The gray levels of the images correspond to the amplitude of the signal. Plots of the signals made in individual scales enable us observation of fine details in the case of the low loaded samples (the sample SS400-1, SS400-2 and SS400-3).

The effects of stress influence can be easily detected starting from the sample SS400-3. Detection of such small changes of the signals were impossible when all of them were plotted in the same scale. As apparently seen in Fig. 7, the maximum amplitude of the measured signal depends on the amount of applied stress. This relation will be discussed in the end of this section. Signals obtained from the eddy current sensors are blurred and the fine elements or edges are usually hidden. Therefore, the gradient filter, was applied. The results of filtering (a spatial derivative of the signal with respect to the x-coordinate) are shown in Fig. 8.

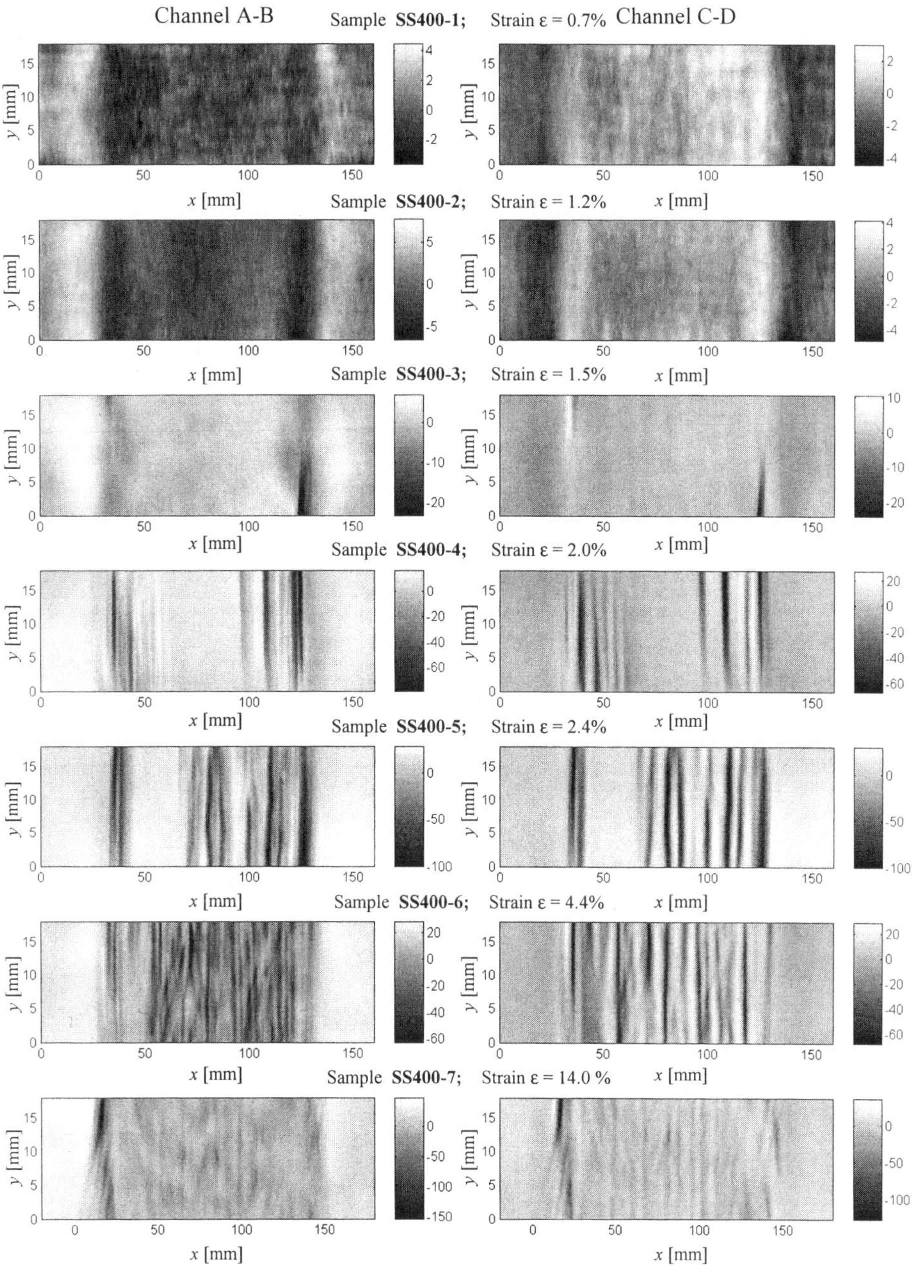

Fig. 7. Two-dimensional results of the measurements (signal obtained from the search coil - S (mV)) of tensile-loaded steel samples. The gray levels of the images correspond to the value of the signal. Each signal is plotted in individual gray scale.

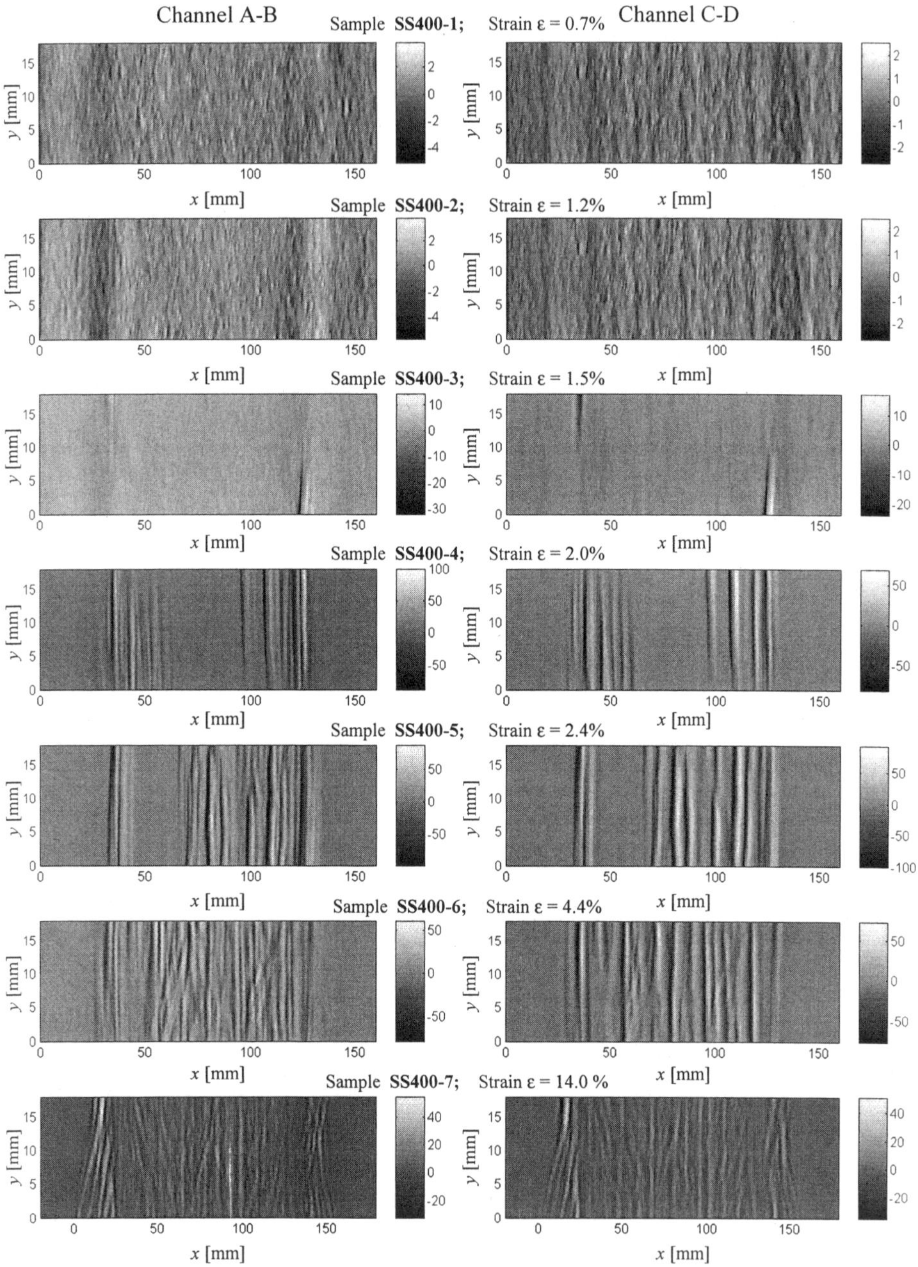

Fig. 8. Two-dimensional results of the measurements (of the signal S with respect to the x-coordinate $(\partial U(x,y)/\partial x)$) of tensile-loaded steel samples. The gray levels of the images correspond to the value of the signal's derivative. Each signal is plotted in individual gray scale.

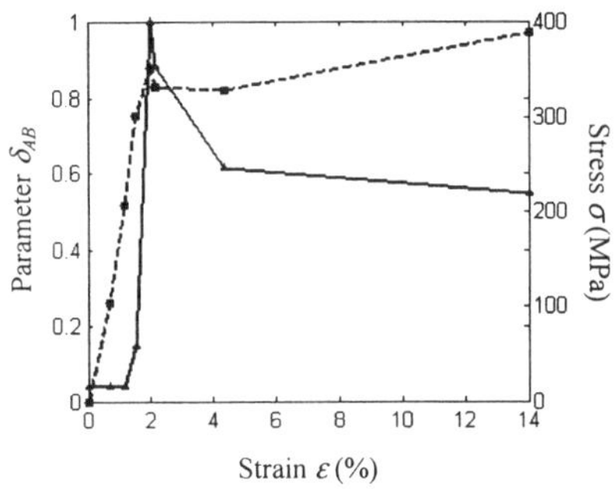

Fig. 9. Plot of parameter δ_{AB} calculated for all samples; signal measured from channel A-B; dashed line represents stress-strain curve

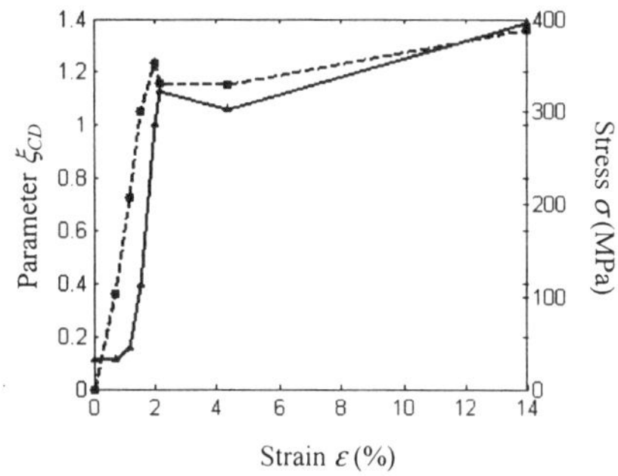

Fig. 10. Plot of parameter ξ_{CD} calculated for all samples; signal measured from channel C-D; dashed line represents stress-strain curve

One could observe that the abrupt changes of the signal are now more visible. Also in this case there is a strong coincidence between amount of the applied stress and the maximum amplitude of the derivative. It is visible that the signals from channel A-B contains more details, while the signal peaks measured from channel C-D have higher amplitude in case of relatively small elongation (SS400-3). Additionally, in practical applications the direction of elongation is unknown, and therefore both channels should be used. Some features of the measured signal and its derivative are directly related to the material changes caused by the stress. Two relative parameters (1), (2) were selected as valuable indices for the quantitative evaluation of the applied stress level.

$$\xi_{CD} = \min(S) / \min(S_{yield}) \quad (1)$$

$$\delta_{AB} = \max\left(\partial S / \partial x\right) / \max\left(\partial S_{yield} / \partial x\right) \quad (2)$$

where: S – is the signal obtained from the transducer, S_{yield} – is the signal measured when the sample is loaded to the yield.

The parameters calculated for all samples are shown in Fig. 9 and Fig. 10.

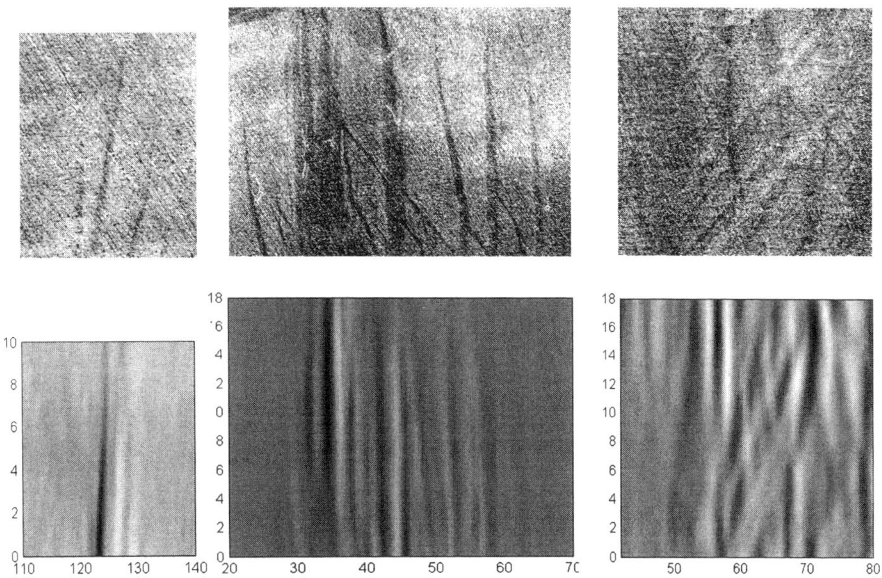

Fig. 11. Comparison of the selected signals measured for the sample SS400-3, SS400-4, SS400-6 and corresponding scans of the samples' surface (obtained using high resolution CCD sensor).

In order to provide a possibility to verify the measurements obtained from ECT sensor, high resolution scans of samples' surface were done. A CCD camera was used for this purpose. Signals obtained from ECT sensor and relevant scans are shown in Fig. 11. A coincidence between surface features and ECT signals can be observed. On the other hand, one could observe that the changes of ECT signals are not only due to the surface roughens but they mainly depend on deep material changes.

6. Conclusions

The presented results of experiments confirmed that proposed method could be used to detect and evaluate stress degradation in ferromagnetic materials. The main advantages of the method are: simplicity, high sensitivity to material degradation, very low cost of sensors, construction of matrix type sensor is possible, the proposed parameters of the signals enable quantitative evaluation of the applied stress, the results do not depend on magnetic history of the samples, preliminary treatment of the material under testing is unnecessary.

Acknowledgement

This study was supported in part by State Committee for Scientific Research, Poland, Grant no: 4T10C 003 22 and 4T10A 024 24

References

[1] T. Chady, Evaluation of Stress Loaded Steel Samples Using GMR Magnetic Field Sensor, IEEE Sensors Journal, Vol. 2, No. 5, October 2002, pp. 488-493.

[2] T. Chady, R. Sikora: "Optimization of Eddy Current Sensor for Multi-Frequency Systems", to be printed in IEEE Trans. Magn. April 2003

[3] T. Chady, R. Sikora: "Multi-frequency Eddy Current Probe Sensitive to a Flaw Orientation", The 8th International Workshop on Electromagnetic Nondestructive Evaluation, 2002 ENDE Short Papers, June 12-14, 2002, Saarbrucken, Germany, pp. 11.1-11.2

Electromagnetic Nondestructive Evaluation (VIII)
T. Sollier et al. (Eds.)
IOS Press, 2004

Minimally Invasive System for the Evaluation of Prosthetic Heart Valves

Shiu C. Chan, Mohachiro Oka*, Adam Downey, and Satish S. Udpa
Department of Electrical and Computer Engineering, Michigan State University, 2120 Engineering Building, East Lansing, MI 48824-1226, USA
**Department of Computer and Control Engineering, Oita National College of Technology, 1666 Maki, Oita, 870-0152, Japan*

Abstract. The Björk-Shiley Convexo-Concave (BSCC) prosthetic heart valve employs a pyrolytic carbon disc, held in place by two struts, as an occluder to block the flow of blood in one direction. The strut on the inlet side is integral to the suture ring while the strut on the outlet side is welded to the ring. Occasionally, after years of use, the weld at one of the two sites fractures, resulting in a single-leg separation (SLS) of the strut. A failure of the other weld would cause the disc to escape resulting in mortality. This paper describes the progress made to date in building a minimally invasive system for detecting SLS failures.

1. Introduction

The Björk-Shiley Convexo-Concave (BSCC) prosthetic heart valve (Figure 1) employs a pyrolytic carbon disc as an occluder to ensure unidirectional blood flow from the atrium to the ventricle of the heart. The disc is held in place by two struts (Figure 1). The strut on the inlet side is integral to the suture ring while the strut on the outlet side is tungsten inert gas welded to the ring. Occasionally, after years of use, the weld at one of the two sites fractures, resulting in a single-leg separation (SLS) of the strut [1]. A failure of the other weld would cause the disc to escape, resulting in mortality. Two different electromagnetic approaches for detecting SLS failures have been proposed to date [2, 3]. The first method relies on electromagnetic acoustic transduction (EMAT) while the second approach uses a gradiometer to assess the integrity of the valve. This paper describes the progress made to date in building the gradiometer-based system.

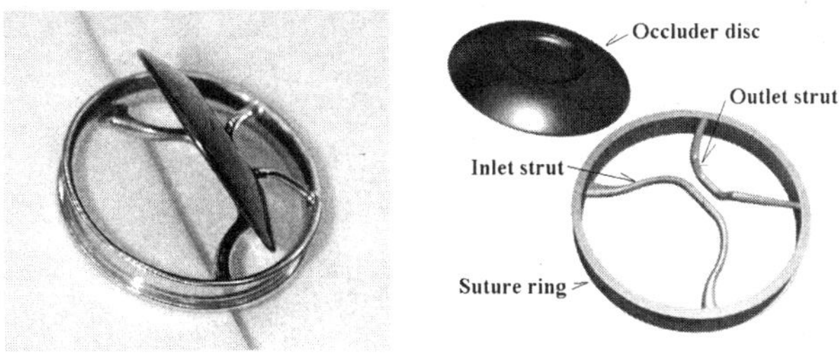

Figure 1 Björk-Shiley Convexo-Concave (BSCC) heart valve.

2. The Gradiometer-Based Approach

The gradiometer-based approach applies a uniform magnetic field over a volume enclosing the heart valve under test. A pair of large coils, arranged in a Helmholtz configuration, and excited by an alternating current source is employed to generate this uniform field. The coils are positioned such that the heart valve is positioned at the center of the coils and oriented in a manner that ensures that the suture ring is parallel to the applied field as shown in Figure 2. This orientation ensures that the eddy current induced in the suture ring, when subject to the time varying field, is minimal. Since the outlet strut is not parallel to the field, a current is induced in the strut, causing the field in the vicinity of the strut to be perturbed. The perturbation in the field is measured using a catheter-mounted gradiometer in the proximity of the strut. An SLS failure results a reduction in the outlet strut current and hence affects the degree of perturbation in the magnetic field. The gradiometer signal can, therefore, be used as a basis for detecting SLS failures.

3. System Components

Figure 3 shows a block diagram of the gradiometer-based detection system. The signal generator produces the excitation waveform which is amplified to drive the excitation coils. An impedance matching circuit (Figure 4) ensures that the total load impedance of the circuit and the excitation coils is matched to that of the amplifier output stage at the excitation frequency. The pair of excitation coils generates a uniform magnetic field around the heart valve during the test. Each excitation coil consists of 9 turns of 4 AWG (American wire gauge) wire and has a mean radius of 1 m. The coils are electrically connected in series and are spaced 1 m apart.

The gradiometer consists of three coaxial pancake coils (each wound with 16 turns of a 44 AWG wire) mounted on a nonmagnetic and electrically non-conducting spindle. The coils are labelled A, B and C, with coil A being closest to the tip of the gradiometer. The 2.5 mm diameter coils are spaced 1 mm apart, and each coil is 1-wire thick. Figure 5 shows a close-up view of the gradiometer. To improve robustness and sensitivity of the approach, differential measurements are made using coils A and C only. The signal from the gradiometer is applied to a lock-in amplifier (LIA) that uses the excitation signal as the reference signal to improve the signal-to-noise ratio.

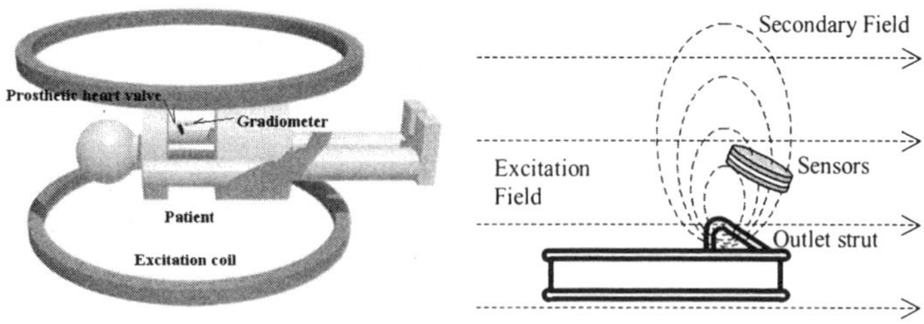

Figure 2 In the gradiometer-base approach, a pair of excitation coils are used to induce eddy currents in the outlet strut of the heart valve.

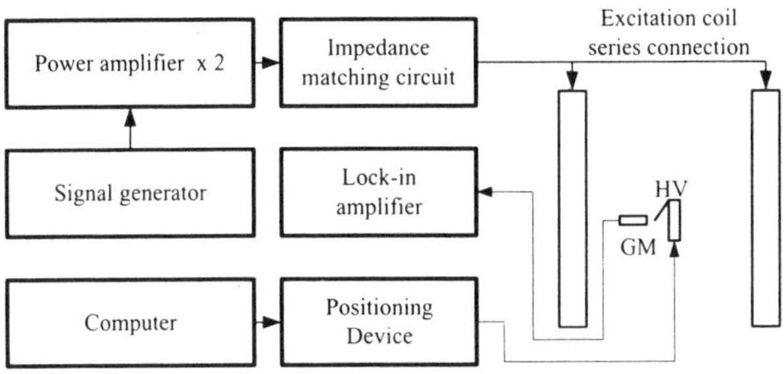

Figure 3 The system block diagram of the gradiometer-based detection approach.

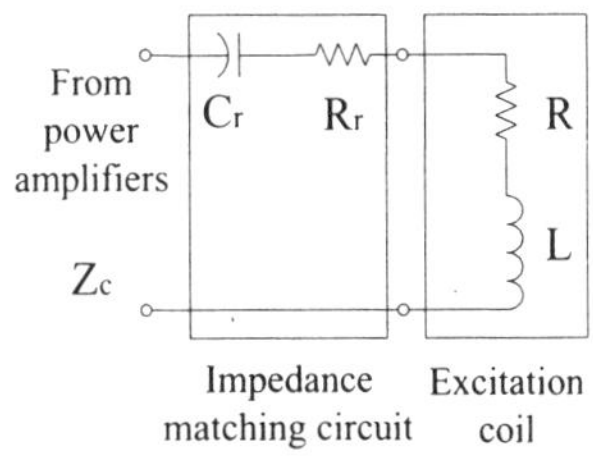

Figure 4 The impedance matching circuit consists of banks of power resistors and capacitors. The total impedance is adjusted to match the power amplifier's output impedance at the operating frequency.

Let the LIA input signal, P, be the difference between signals from coils A and C. Using the signal generator output as the reference signal, the LIA outputs are given by

$$X = P\cos(\theta) \quad (1)$$

$$Y = P\sin(\theta) \quad (2)$$

where θ is the phase angle between P and the reference signal. The outputs of the LIA are then analyzed to determine the condition of the heart valve.

4. Experimental Results and Analysis

Figure 6 shows the excitation coils and gradiometer configuration used in the experiments. The gradiometer is mounted at the centre of the diagonal crossbeam while the heart valve is mounted at the end of the radial beam. The field distribution predicted by a finite element simulation of the system (Figure 7) is shown in Figure 8. The simulation results indicate that the field is uniform at the center. Experimental test show that the variation in the magnetic flux density is less than 0.5% within a cylindrical volume of radius 0.25m and 0.5 m high.

The X and Y signals from the LIA are related to the gradiometer measurement of the field in the vicinity of the outlet strut and is a function of the location and orientation of the gradiometer in relation to the strut for a given excitation frequency. Figure 9 shows the |X|, |Y|, |P|, and |Phi| ($|\theta|$) signals from the LIA as a function of distance of the gradiometer from the plane of the outlet strut. In this figure, two tests each are performed on an SLS and IOS (intact outlet strut) valve. The excitation frequency is 45.4 kHz and the applied magnetic field is 0.6 gauss. From this figure, the peak |X| value is larger for a heart valve with an IOS than that with a SLS. The distinction between the two signals decreases with increased distance from the strut. For optimal sensitivity, the gradiometer should be placed within approximately 5mm from the outlet strut.

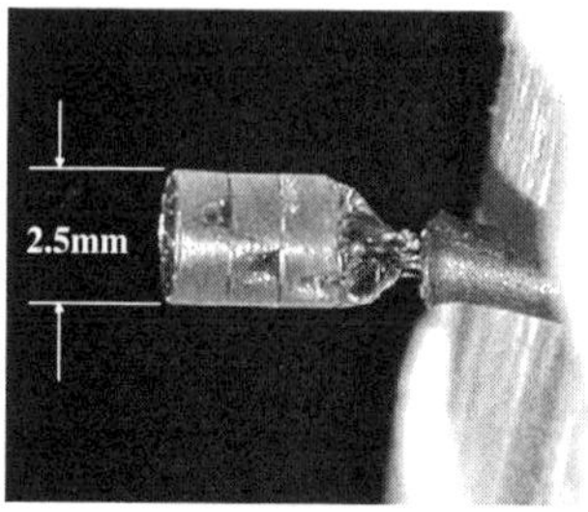

Figure 5 The gradiometer consists of three coaxially-mounted pancake coils.

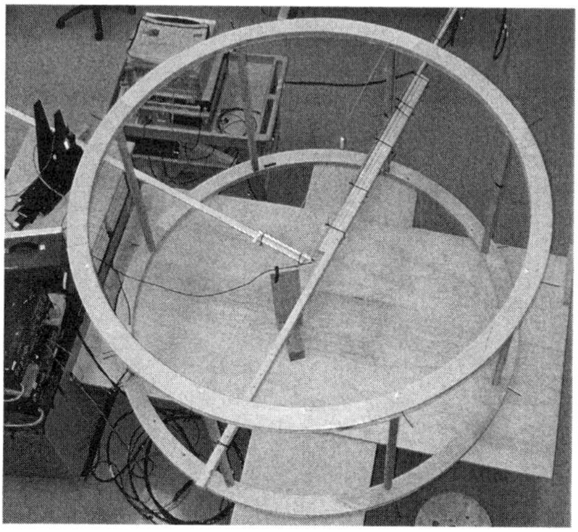

Figure 6 The excitation coils arranged in a Helmholtz configuration of 1m mean coil radius and separation.

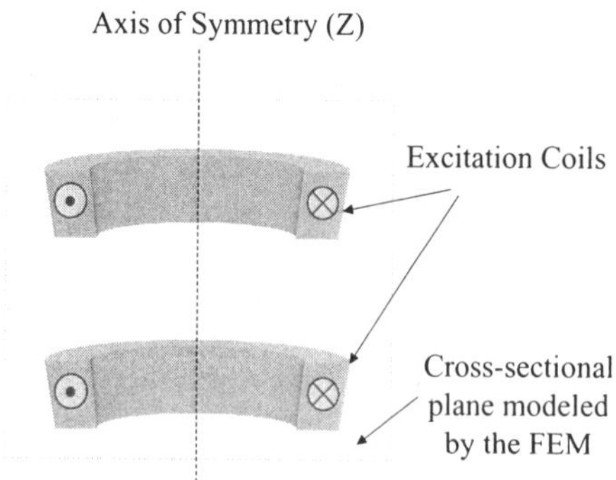

Figure 7 Cross-sectional plane of the excitation coils modeled by the FEM

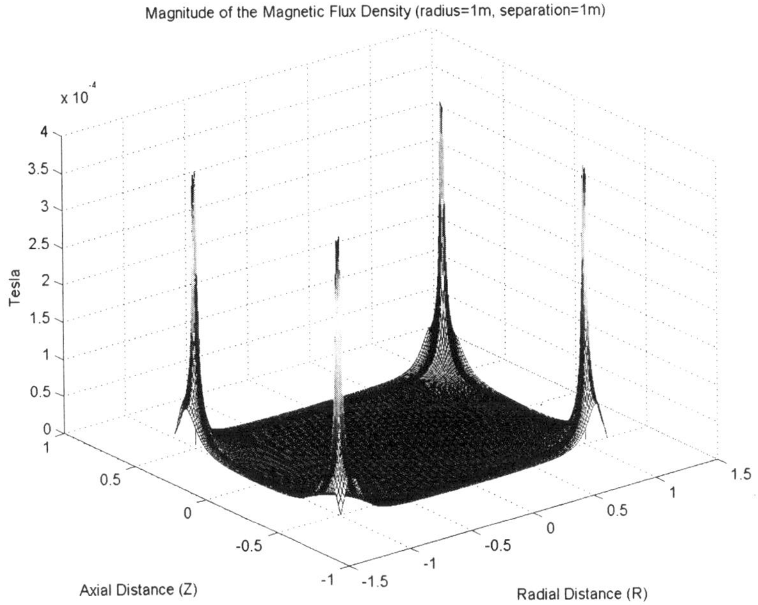

Figure 8 Finite element simulation results of the field produced by the Helmholtz coils. The center region of this plot shows the uniformity magnetic flux density in the region of interest. The applied current is 1.315A.

Figure 9 also shows the signal levels as a function of the distance between the gradiometer and the strut obtained from two trials of the experiment with IOS and SLS valves. In a clinical setting, the gradiometer is encased within a catheter sheath. The catheter is inserted into the heart and positioned close to the outlet strut. The position of the catheter is affected by the blood flow patterns within the heart and as a consequence is continually oscillating about its mean position with the ventricular systole and diastole. The movement of the gradiometer was modelled by fitting a polynomial, f_x, to the curves indicated in Figure 9 and the output of the gradiometer was calculated as a function of time assuming that the gradiometer oscillated at a frequency of 1.2 Hz (corresponding to heart rate of 72 beats/min). The maximum distance variation, L, is a function of the turbulence in the ventricle and the rigidity of the catheter.

Figures 10 and 11 show plots of |X| and |Y| as function of time for IOS and SLS valves for two trials of the experiment assuming L = 5mm. Figure 12 shows the corresponding plot when L = 1mm. In each of these figures, the "*" curve is the average of the estimates for the SLS cases and the "+" curve is the average for the IOS cases. The estimated relative changes in |X| as a function of excitation frequency is shown in Figure 13 for 14.995, 38.21, and 45.44 kHz (L =5mm).

Figure 10 through Figure 13 suggest that, a large |X| and |Y| signal variation over time is an indication of an IOS valve whereas a small signal variation indicate an SLS valve. Figure 13 suggests that higher excitation frequencies improve the discrimination between IOS and SLS cases and results in a more robust system.

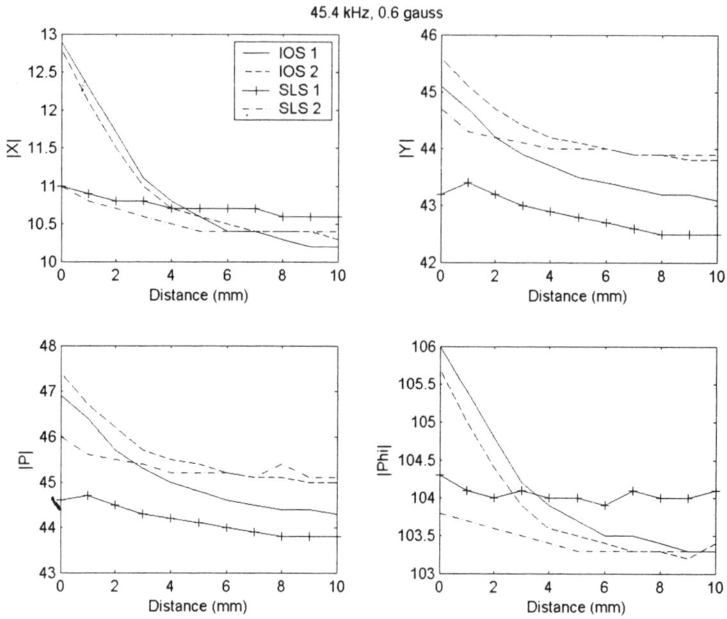

Figure 9 |X|, |Y|, |P|, and |Phi| (|θ|) signals from the LIA for two tests each of a SLS and IOS valve.

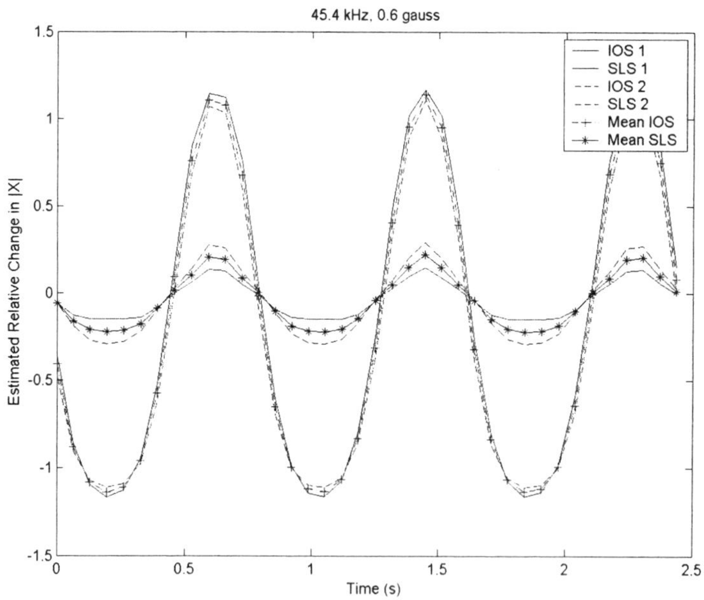

Figure 10 Estimated relative changes in |X| with respect to gradiometer-outlet strut distance changes over time for both IOS and SLS cases. L = 5mm.

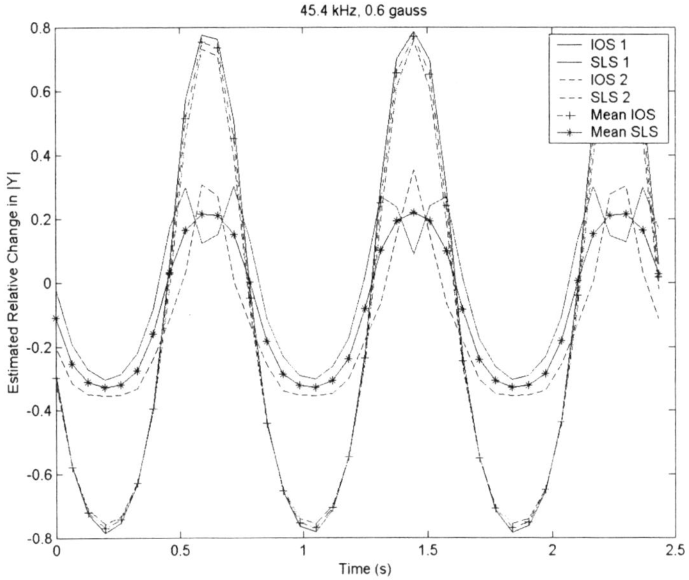

Figure 11 Estimated relative changes in |Y| with respect to gradiometer-outlet strut distance changes over time for both IOS and SLS cases. L = 5mm.

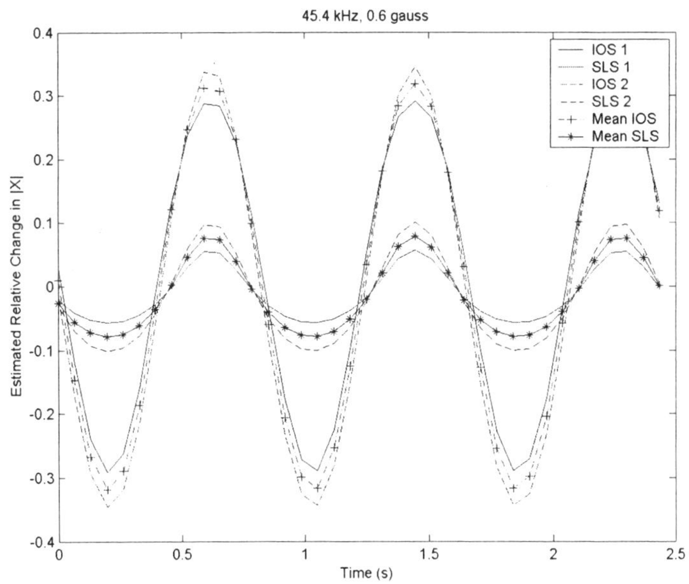

Figure 12 Estimated relative changes in |X| with respect to gradiometer-outlet strut distance changes over time for both IOS and SLS cases. L = 1mm

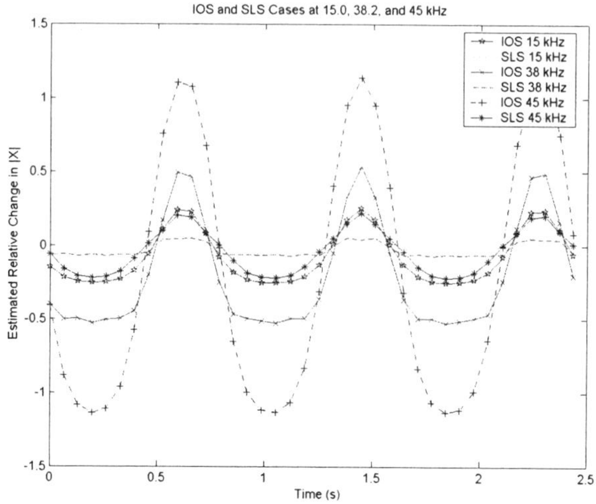

Figure 13 Estimated relative changes in |X| for IOS and SLS cases at 15.0, 38.2, and 45.4 kHz. L = 5mm.

5. Conclusions

A minimally invasive, gradiometer-based system for detecting SLS failures in BSCC heart valves has been described. A finite element model is used to estimate the region of uniform magnetic field between the excitation coils used in this approach. Results from the stationary experiments and simulated gradiometer movements show that the gradiometer-based approach is viable and shows considerable promise as a tool for detecting strut fractures.

Acknowledgement

This work is supported by the Bowling-Pfizer Settlement Fund. The support is gratefully acknowledged.

References

[1] J. G. Ghandler, J. L. Hirsch, W. W. ONeill, S. N. Oesterle, D. C. Miller, J. A. Kennedy and A. Faichney, Radiographic Detection of Single Strut Leg Separeation as a Putative Basis for Prophylactic Explantation of Bjork-Shiley Convexo-Concave Heart Valves, *World Journal of Surgery*, 20, No. 8, 953-960, (1996).

[2] S. C. Chan, Y. Li, L. Udpa, and S. S. Udpa, Electromagnetic Techniques for Detecting Strut Failures in Artificial Heart Valves, *Proceedings of the 8th International Workshop on Electromagnetic Nondestructive Evaluation*, Saarbrücken, Germany, 2002.

[3] F. Chen, S. Udpa, and Y. Sun, "Detection of Single Leg Separation Defects in Bjork-Shiley Convexo-Concave Prosthetic Heart Valves", *Proceedings of the 10th International Symposium on Applied Electromagnetics and Mechanics*, Tokyo, Japan, pp. 423-424, 2001.

Study on the wall thickness evaluation of hollow turbine blades

Yann LE BIHAN[1], Pierre-Yves JOUBERT[2]
[1] *LGEP, UMR 8507 CNRS – SUPELEC – Universités Paris XI et VI, Plateau de Moulon, 91192 Gif-sur-Yvette cedex, France*
[2] *SATIE, UMR 8029 CNRS – ENS Cachan, 61 avenue du Président Wilson, 94234 Cachan cedex, France*

Abstract. The external wall thickness of the hollow blades of jet engines has to be systematically checked after manufacturing in order to guarantee the blade strength. The thickness evaluation is mainly disturbed by the presence of partitions in the blades. An EC sensor dedicated to this application has been developed. The authors present the definition of three EC measurement configurations. Their accuracy towards thickness evaluation taking into account the effect of the blade partitions is then tested and analysed.

1. Introduction

The high pressure (HP) turbine blades of advanced jet engines are complex parts exposed to severe thermal-structural loading conditions. They are hollow in order to allow their cooling by an air flowing. The thickness of the external wall of a HP blade is a critical parameter for the part strength which has to be systematically checked after manufacturing. This work is focused on the HP blade of the jet engine M88-2 manufactured by Snecma Moteurs (Figure 1.a). Its internal cooling structure consists in channels separated by partitions (Figure 1.b). The blade is built in a single-crystal nickel-based superalloy (AM1).

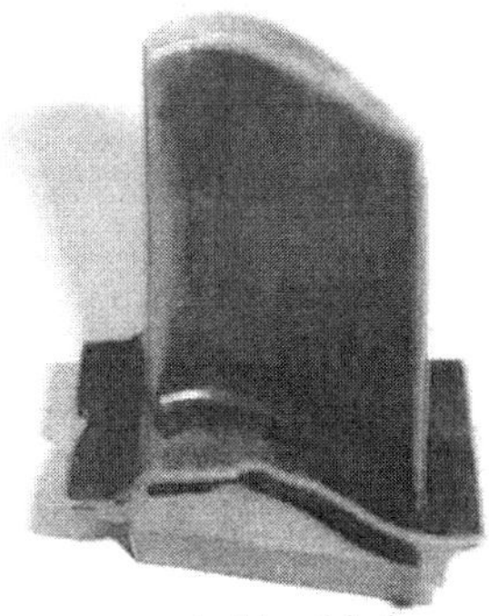

a: turbine blade

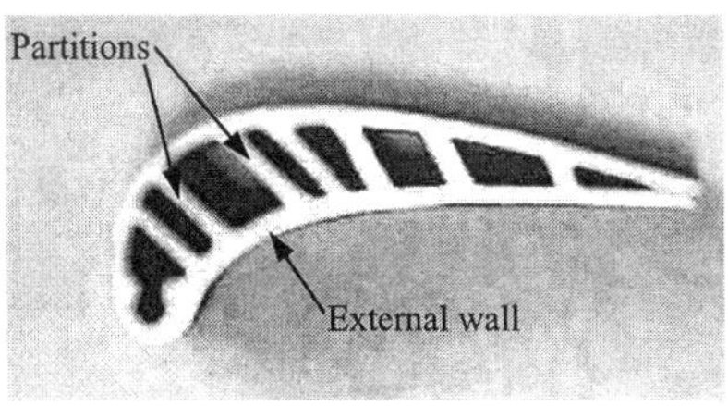

b: turbine blade cross section

Figure 1. Structure of the considered hollow blade

Considering an EC sensor positioned on the external wall of a blade (Figure 2), the sensor impedance depends on the wall thickness and the following main influence parameters:

- the possible presence of a partition in the vicinity of the sensor;
- the electrical conductivity of the blade material;
- the sensor/blade relative positioning, i.e. the sensor lift-off and tilt;
- the local curvature of the blade (concave or convex).

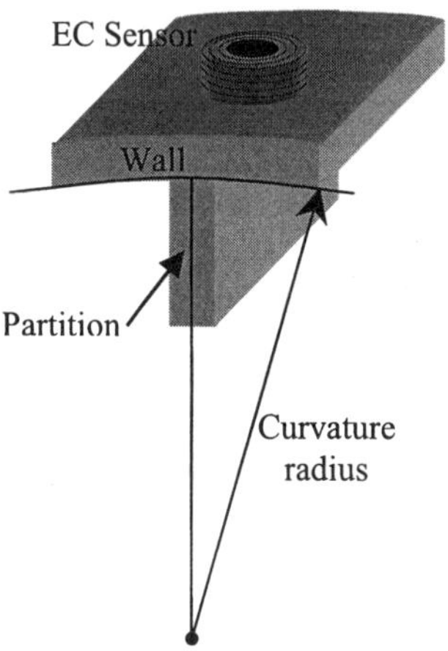

Figure 2. Local context of the EC measurement

According to previous experiments [1], the wall thickness evaluation of blades using conventional cup-core sensors is largely disturbed by the presence of the partitions. As a consequence, a sensor allowing to limit the effect of the partitions has been developed [2][3]. The structure of the sensor is shown in Figure 3. It consists in a pair of coils wound around each end of an U-shaped magnetic core. The coils are serially connected with the current flowing in phase opposition in such a way that the emitted excitation magnetic field flows from one pole to the other and shows a preferred direction. This property allows to reduce or increase the effect of the blade internal partitions on the EC measurements, depending on the field orientation with respect to the blade.

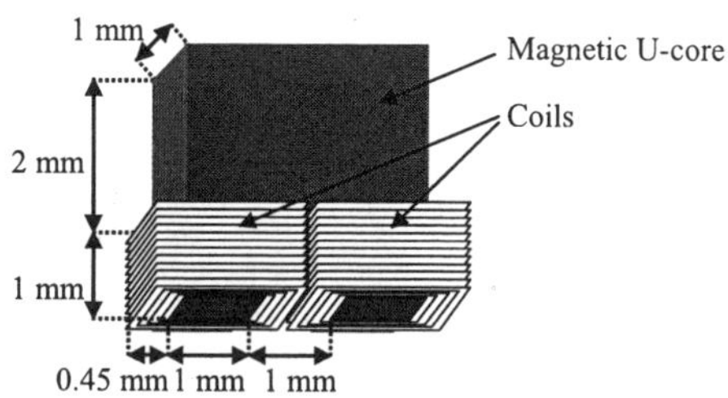

Figure 3. EC sensor structure

In this paper, the uncertainty of the wall thickness evaluation is studied taking into account the effect of the partitions.

2. EC measurement data in presence of partitions

Two operating modes of the sensor have been defined:

- the parallel mode (//), in which the sensor is oriented in such a way that the excitation field is parallel to the partition direction;
- the orthogonal mode (⊥) in which the sensor is oriented so that the excitation field is orthogonal to the partition direction.

A planar gauge was designed in order to test the effect of the partitions on the sensor impedance (Figure 4). The planar gauge is made of AM1 superalloy. The top of the gauge exhibits 10 steps of different thicknesses covering the thickness range of the blade wall. The bottom of the gauge shows ribs standing for the blade partitions and lying perpendicularly to the steps. Thus, the gauge allows to evaluate both wall thickness and partition effects on the EC sensor impedance.

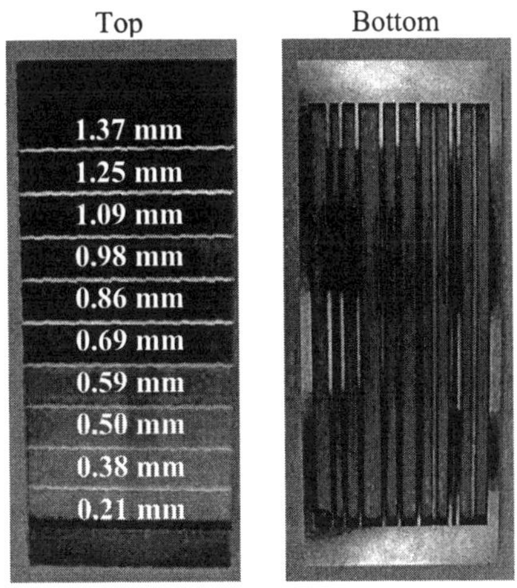

Figure 4. Planar gauge structure

EC measurements were realised on the 10 steps of the planar gauge. The measurement output quantity is the normalised impedance Z_n which is deduced from the sensor impedance by:

$$Z_n = \frac{Z - R_0}{X_0} = R_n + jX_n \tag{1}$$

where R_0 and X_0 denote respectively the resistance and reactance of the sensor in absence of conductive target. R_n and X_n are respectively the normalised resistance and reactance of the sensor.

As an example, Figure 5 shows the reactive part of the sensor impedance obtained at a 100 kHz excitation frequency for acquisitions achieved in both parallel and orthogonal measurement modes. As a result, if the sensor emits an excitation magnetic field parallel to the partition direction (parallel mode), the EC signal is weakly influenced by the partitions. On the contrary, if the sensor emits an excitation magnetic field orthogonal to the partition direction (orthogonal mode), the EC signal is highly influenced by the partitions. As a consequence, the wall thickness evaluation in presence of partitions was tested on the planar gauge for the three following measurement configurations:

- measurements in parallel mode at a single frequency (P1 configuration);

- measurements in parallel mode at two frequency (P2 configuration);
- measurements for both parallel and orthogonal modes, each one at a single frequency (01P1 configuration).

The first configuration requiring a single measurement at a single frequency is the simplest to implement and should provide a fair thickness evaluation because of the weak influence of the partitions on the EC signal. The combination of both parallel and orthogonal modes should allow to correct the partition influence on thickness evaluation thanks to the significant difference of influence of the partitions for these two modes. Nevertheless this configuration requires to perform a sensor rotation during the measurement (a double U-core structure could be considered later on). As an intermediate scheme, the utilisation of measurements in parallel mode using dual frequency should allow one to carry out the partition effect correction without sensor rotation.

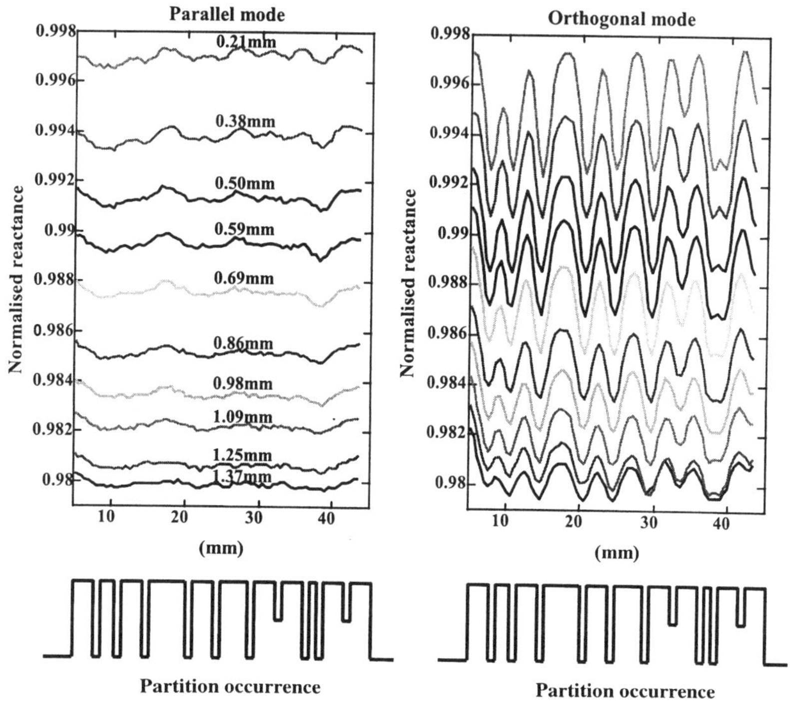

Figure 5. EC measurements on the planar gauge

3. Selection of the measurement configuration with respect to the partitions

3.1. Determination of the P1 configuration frequency

Since the partitions induce an impedance variation very similar to the thickness effect, the frequency of the P1 configuration was chosen in order to minimise the partition effect on the EC measurements. This was done by maximising the sensitivity ratio:

$$SR = \frac{\left|\Delta Z_n\right|_{thickness}}{\left|\Delta Z_n\right|_{partition}} \tag{2}$$

with $|\Delta Z_n|_{thickness}$ the modulus of the Z_n variation induced by a thickness variation and $|\Delta Z_n|_{partition}$ the modulus of the Z_n variation induced by a partition occurrence. *SR* was experimentally determined using EC measurements from the 0.5 mm and 0.59 mm steps which correspond to the nominal thickness range of the wall, and is shown in Figure 6 in function of the frequency.

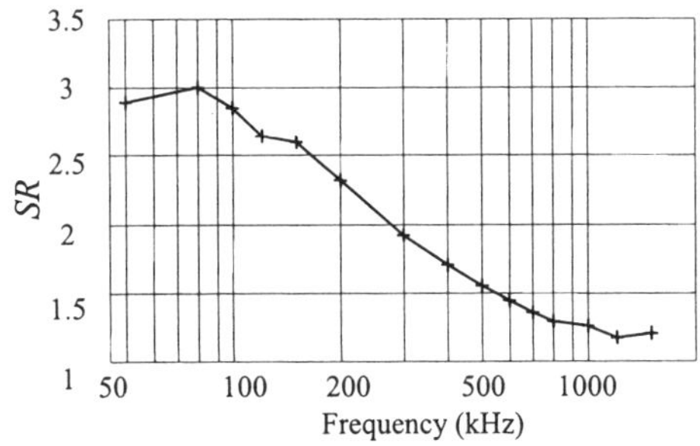

Figure 6. Sensitivity ratio vs. frequency

SR is maximum for a low excitation frequency. However a too low frequency leads to a poor thickness sensitivity since the EC magnitude is reduced. Consequently the excitation frequency was chosen at 100 kHz.

3.2. Determination of the P2 configuration frequencies

In order to allow the thickness evaluation in spite of the partition influence, the two excitation frequencies of the P2 configuration were chosen so as to maximise the complementarity of the information provided by each frequency on the thickness and the partitions. This was done by linearising locally the behaviour of the sensor. Let us define M the vector containing the four real variables issued from the P2 measurement and P the vector of the two considered actuating quantities:

$$M = \begin{pmatrix} R_{n1} \\ X_{n1} \\ R_{n2} \\ X_{n2} \end{pmatrix} \quad \text{and} \quad P = \begin{pmatrix} P_t \\ P_p \end{pmatrix} \tag{3}$$

where P_t denotes the thickness quantity and P_p the partition quantity. The relation linking M to P is nonlinear. Considering a small variation δP of the actuating quantity vector around an operating point leads to:

$$\delta M = J\,\delta P \quad \text{and} \quad J = \begin{pmatrix} \dfrac{\partial R_{n1}}{\partial P_t} & \dfrac{\partial R_{n1}}{\partial P_p} \\ \dfrac{\partial X_{n1}}{\partial P_t} & \dfrac{\partial X_{n1}}{\partial P_p} \\ \dfrac{\partial R_{n2}}{\partial P_t} & \dfrac{\partial R_{n2}}{\partial P_p} \\ \dfrac{\partial X_{n2}}{\partial P_t} & \dfrac{\partial X_{n2}}{\partial P_p} \end{pmatrix} \tag{4}$$

with δM the concomitant variation of the EC measurement vector and J the Jacobian matrix of the relation linking M to P, which depends on the operating point. The inverse problem consists then to estimate δP knowing δM. The least square solution is:

$$\delta \hat{P} = (J^t J)^{-1} J^t \delta M \tag{5}$$

The reliability of the inversion depends on the conditioning of the matrix $J^t J$, evaluated from its condition number, defined as [4]:

$$Cond(J^t J) = || J^t J || \, ||(J^t J)^{-1}|| \tag{6}$$

with $||.||$ a matrix norm. In this work the euclidean norm was used. The condition number is an indicator of the sensitivity of the solution towards perturbations and therefore of the possibility to separate the actuating quantities. The condition number is superior or equal to 1. The greater it is, the more the problem is ill-conditioned and the solution inaccurate. Since $J^t J$ is a symmetric positive matrix, its eigenvalues $\{\lambda_i\}$ are positives and:

$$Cond(J^t J) = \frac{max(\lambda_i)}{min(\lambda_i)} \tag{7}$$

The excitation frequencies f_1 and f_2 was chosen by selecting the set of frequencies giving the lowest condition number $Cond(J^t J)$. The Jacobian matrix was determined using EC data obtained from the 0.5 mm and 0.59 mm steps. The condition number is shown in Figure 7 vs. f_1 and f_2.

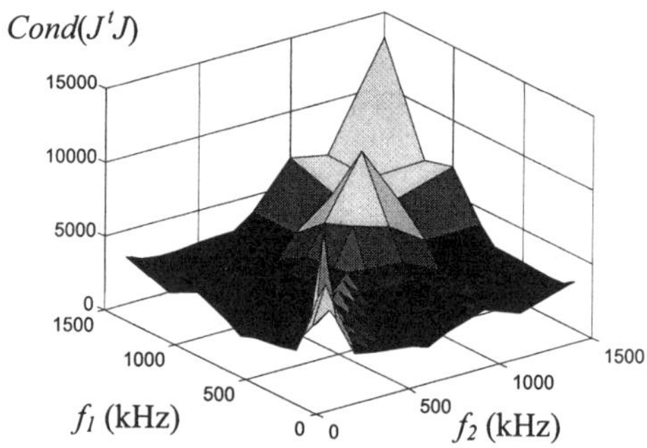

Figure 7. Condition number vs. f_1 and f_2

The condition number is minimised by the combination of a low frequency with a high frequency, which leads to f_1 = 100 kHz and f_2 = 500 kHz.

3.3. Determination of the O1P1 configuration frequencies

An identical scheme was implemented for the determination of the frequencies $f_{//}$ and $f_{\perp}$ of the O1P1 configuration. The measurement vector M and the Jacobian matrix J are then:

$$M = \begin{pmatrix} R_{n//} \\ X_{n//} \\ R_{n\perp} \\ X_{n\perp} \end{pmatrix} \quad \text{and} \quad J = \begin{pmatrix} \frac{\partial R_{n//}}{\partial P_t} & \frac{\partial R_{n//}}{\partial P_p} \\ \frac{\partial X_{n//}}{\partial P_t} & \frac{\partial X_{n//}}{\partial P_p} \\ \frac{\partial R_{n\perp}}{\partial P_t} & \frac{\partial R_{n\perp}}{\partial P_p} \\ \frac{\partial X_{n\perp}}{\partial P_t} & \frac{\partial X_{n\perp}}{\partial P_p} \end{pmatrix} \tag{8}$$

The evolution of the condition number relatively to $f_{//}$ and $f_{\perp}$ is shown in Figure 8:

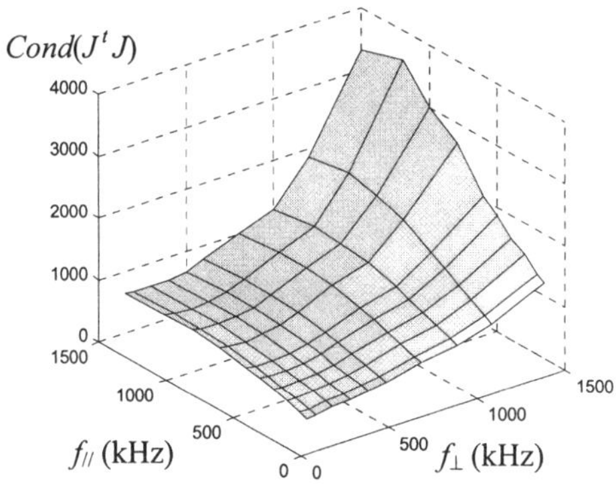

Figure 8. Condition number vs. $f_{//}$ and $f_{\perp}$

The condition number is minimized by using a low frequency excitation for both measurement modes. Taking into account that the excitation frequency should not be too low, the selected frequencies are: $f_{//}$ = 100 kHz and $f_{\perp}$ = 100 kHz.

4. Thickness evaluation performance for the three configurations

For each considered measurement configuration, the inversion of the EC data was carried out by using a behavioural inverse model. The used inverse model consists in a two layers multilayer perceptron neural network (NN) with logistic activation functions in the hidden layer and a linear activation function in the output layer. NN of this kind are actually universal and parsimonious approximators [5][6]. The NN is fed with the real and the imaginary parts of the normalised impedance whereas its scalar output provides the evaluated thickness. The NN internal parameters were adjusted via a training set obtained from the gauge. The training goal was to minimise the mean square error (*MSE*) between the NN output and the training set output. The Levenberg-Marquardt algorithm was implemented for the training [7]. The generalisation capability of the NN was tested according to the hold-out procedure by calculating the *MSE* obtained on a test set constituted of input/output data not included in the training set [8]. The aim of this method is to avoid NN over-parametrisation by selecting the NN size (number of neurons in the hidden layer) showing the lowest *MSE* on the test set. The training set uses data issued from eight of the ten gauge steps, the test set containing data from the two remaining steps. Table 1 shows the standard uncertainty (u) of the thickness evaluation obtained on the training and test sets for each previously defined configuration.

Table 1. Thickness evaluation results on the NN sets

	P1	P2	O1P1
Training set	$u = 12$ µm	$u = 11$ µm	$u = 11$ µm
Test set	$u = 11$ µm	$u = 12$ µm	$u = 10$ µm

The three considered NN models show a good generalisation capability since for each configuration the standard uncertainties on the training and test sets are similar. By comparison, the standard uncertainty obtained with a classical cup-core sensor on the same gauge is about 50 µm [1]. However, one can notice that the standard uncertainty is not significantly decreased by the combination of two frequencies (P2) or of two modes (O1P1) comparatively to the simple P1 configuration. A study of the different causes of error according to the GUM recommendations [9] has shown that the constancy of the thickness evaluation uncertainty is due to the thickness uncertainty of each step of the gauge. This one was characterised using geometrical measurements by a standard uncertainty of about 9 (m. The step thickness uncertainty is then the major limitation factor of the thickness evaluation accuracy for the considered test. Indeed, the uncertainty of the evaluation procedure cannot be lower than the uncertainty of the gauge used for the inverse model calibration. As a consequence, the simple measurement configuration P1 is sufficient since no benefit can be obtained from the combination of other measurements, except if a more complete and accurate calibration of the gauge is carried out.

Conclusion

The goal of the work is to devellop a method for the wall thickness evaluation of hollow turbine blades. This study shows the benefits that arise from the implementation of an oriented excitation field EC sensor since it reduces the partition disturbance. Indeed, experimental results show that the wall thickness evaluation uncertainty is significantly decreased in comparison to classical cup-core EC sensors. In addition, a multi-frequency excitation or a dual field orientation can be implemented, if a better thickness evaluation accuracy relatively to the partition effect is required and if the gauge calibration procedure allows to benefit from them. It must be noticed nevertheless that the accuracy can be then limited by other influence parameters which were not taken into account in the study.

Acknowledgements

This work is supported by Snecma Moteurs through a research contract.

References

[1] Y. Le Bihan, Design and characterisation of an eddy current device dedicated to the nondestructive evaluation of the wallthickness of hollow turbine blades. PhD thesis, ENS Cachan, 2000.

[2] Y. Le Bihan, P.-Y. Joubert and D. Placko, Wall thickness evaluation of single-crystal hollow blades by eddy current sensor, *NDT&E International*, **34** (2001) 363-368.

[3] Y. Le Bihan, O. Lespinet, A. Mourenko, F. Panizzoli, D. Placko and E. Santander, Procédé de mesure d'épaisseur de paroi d'une aube creuse. 2000, French patent 0008368, CNRS-Snecma Moteurs.

[4] P.G. Ciarlet, Introduction à l'analyse numérique matricielle et à l'optimisation, Masson, Paris, 1994.

[5] K. Hornik, M. Stinchcombe and H. White, Multilayer feedforward networks are universal approximators, *Neural Network*, **2** (1989) 359-366.

[6] A. Barron, Universal approximator bounds for superposition of a sigmoidal function, IEEE trans. on Information Theory, **39** (1993) 930-945.

[7] M.T., Hagan and M. Menhaj, Training feedforward networks with the Marquardt algorithm, IEEE trans. on Neural Networks, **5** (1994) 989-993.

[8] W.S. Sarle, Generalization, *Neural Network FAQ, part 3 of 7*, ftp://ftp.sas.com/pub/neural/FAQ.html.

[9] Guide to the Expression of Uncertainty in Measurement, ISO, Geneva, 1995.

Electromagnetic Nondestructive Evaluation (VIII)
T. Sollier et al. (Eds.)
IOS Press, 2004

Fundamental Study on the NDT Method Based on Electromagnetic Waves

H. Hashizume, S. Kitajima, T. Shibata, Y. Uchigaki, K. Ogura
Department of Quantum Science and Energy Engineering,
Tohoku University, Aramaki, Aoba, Sendai, 980-8579, Japan

Abstract It is considered that the NDT method using electromagnetic waves has strong possibility to detect defects in large structures, or large pipes because the electromagnetic waves propagate inside the pipe up to a long distance. The purpose of this study is to experimentally verify the validity of this method. It is confirmed that the microwave of circular TM_{01}-mode, which is necessary to detect the circumferential crack, is converted from the rectangular TE_{01}-mode. It also becomes clear that the reflected wave carries a lot of information on crack width and depth. Experimental results indicate that the NDT method using the electromagnetic waves has excellent prospect.

1. Introduction

It is clear that using fission reactors for a long term contributes to reducing both electric cost and carbon dioxide emission. In order to maintain structural integrity for long-term usage it is necessary to utilize NDT methods to detect defects existing in the system. The eddy current testing (ECT) with various kinds of advanced EC probes can successfully detect defects occurring in small structures like steam generator tubes [1][2]. When the ECT system is utilized for evaluation of large pipes or large structures, however; exciting coils must get larger or long inspection time is needed. Therefore, it is necessary to develop some system to find quickly the crack existing inside the pipe. For this purpose, an electromagnetic wave is suitable because the wave can propagate inside the pipe up to a long distance. And then by combining the conventional detecting method with this new one, it becomes possible to find the crack quickly and determine its size precisely, which is very attractive.

In a previous study [3], it was shown by numerical analysis that electromagnetic wave of TM_{01}-mode is useful for detecting a circumferential defect. It was also demonstrated that the electromagnetic wave below the cut-off frequency could be available to identify its location, because the wave intensity inside the pipe is damped exponentially.

Based on these numerical predictions, an experimental investigation is carried out here to demonstrate the validity of the method. At first, an experimental system is constructed to generate the circular TM_{01}-mode microwave from the rectangular TE_{01}-mode one emitted from a gun oscillator with controlling the frequency continuously. Then the performance of new method for crack detection is estimated by experiments using a test pipe with an artificial defect.

2. Characteristics of microwave propagating inside the pipe

The principle of defect detection is as follows. If microwaves propagate inside the pipe with a crack, the crack prevents the surface current from flowing, and then a reflected wave occurs. Thus it becomes possible to detect the existing crack by detecting it. In the following, some mathematical background is introduced.

2.1 Guide wavelength and cut-off frequency

Electromagnetic waves inside the pipe can propagate according to guide wavelength λ_g

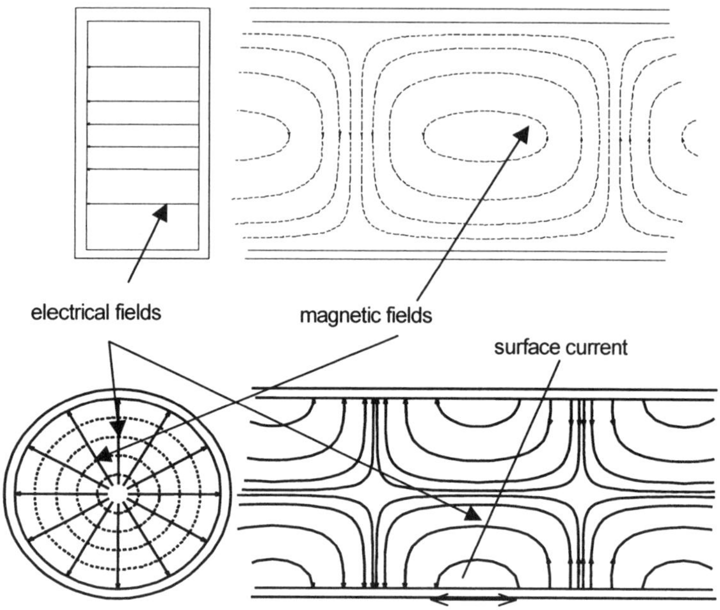

Fig.1 Microwave mode (rectangular TE_{01}-mode and circular TM_{01}-mode)

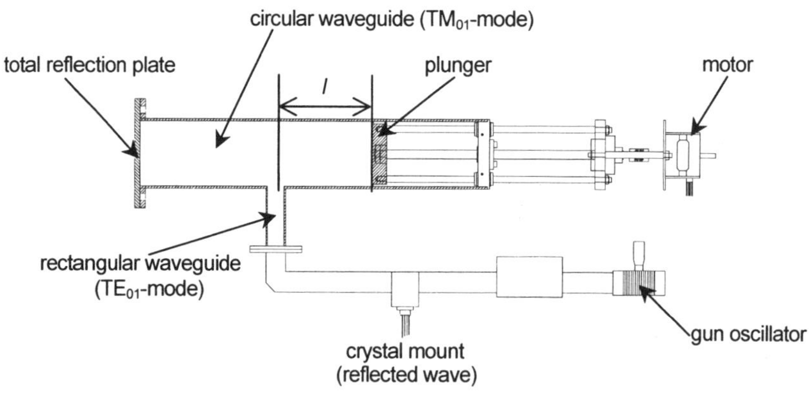

Fig.2 Experimental system (mode converter)

when the frequency is larger than the cut-off frequency of the pipe, f_0. The guide wavelength λ_g is given by the following equation;

$$(1/\lambda_g)^2 = (1/\lambda)^2 - (1/\lambda_c)^2, \quad (1)$$

where λ is the wavelength in free space and λ_c is called the cut-off wavelength. In the case of a circular waveguide λ_c is given by;

$$\lambda_c = 2\pi a/r_0 = c/f_c \text{ (c: velocity of light)}, \quad (2)$$

where "a" indicates a radius of the inspected pipe and "r_0" is some constant determined by the Bessel function (r_0= 2.405 for the circular TM_{01}-mode, r_0= 1.841 for the circular TE_{11}-mode and r_0= 3.054 for the circular TE_{21}-mode).

On the other hand below the cut-off frequency, its electric field propagates decreasingly according to the following equation;

$$E = E_0 \exp(-Kz)\exp(i\omega t). \quad (3)$$

Here "K" shows a wave number inside the circular waveguide and "ω" is angular frequency. Equation (3) means that changing the frequency can control the propagating distance of microwave below f_c. Using these characteristics, it may become possible to identify the crack location by microwave near the cut-off frequency.

2.2 Mode converter

Since a gun oscillator emits TE_{01}-mode microwave, mode-converting system to TM_{01}-mode is needed to perform the experiment. In the TE_{01}-mode the component of the electric field in the longitudinal direction becomes zero and the one of the magnetic field in the same direction exists inside the waveguide, while in the TM_{01}-mode the electric field in the longitudinal direction exists and the magnetic field becomes zero. Figure 1 schematically shows the electric and magnetic field distributions of rectangular TE_{01}-mode and circular TM_{01}-mode. In the circular TM_{01}-mode surface currents flow into the longitudinal direction inside the waveguide by the electrical field in that direction. Then the circumferential defect will disturb the surface current flowing to generate some reflected microwave. By detecting this microwave it becomes possible to detect the crack. The rectangular TE_{01}-mode is converted to the circular TM_{01}-mode by joining the circular waveguide to the rectangular waveguide as shown in Fig. 2. The electrical field of the TE_{01}-mode is converted to the one of the TM_{01}-mode since the direction of both fields is the same. The conditions to resonate or damp the mode are given by the following equations;

$$l = \lambda_g \cdot m/2 \quad \text{(for resonating)}, \quad (4)$$

$$l = \lambda_g \cdot (2 \cdot m + 1)/4 \quad \text{(for damping)}. \quad (5)$$

Here "m" and "l" are arbitrary integers and the length between the center of rectangular waveguide and a plunger shown in Fig. 2, respectively. "λ_g" is calculated from Eq. (1), and its value depends on the mode. In order to resonate the TM_{01}-mode, we must change the frequency and the length l simultaneously to satisfy Eq. (4).

3. Experimental methods and results

At first, the mode converter performance is evaluated and then the feasibility of crack detection is investigated by introducing an artificial crack in a test pipe.

3.1 Evaluation of the mode converter performance

In this experiment we use one mode converter with a perfect reflection plate, which is installed at the end of the converter. Figure 2 shows this experimental system. The

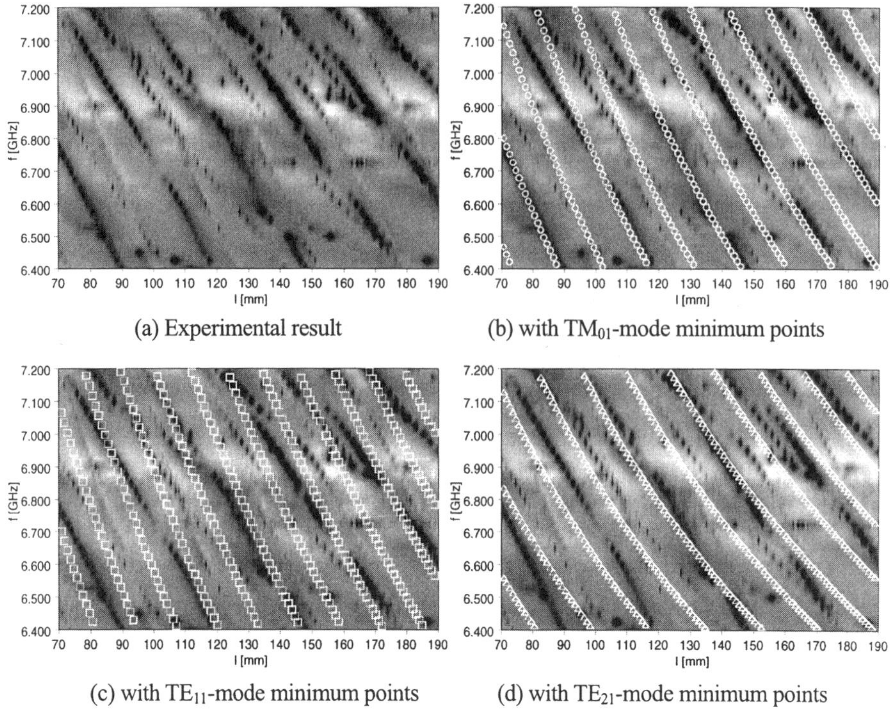

(a) Experimental result

(b) with TM_{01}-mode minimum points

(c) with TE_{11}-mode minimum points

(d) with TE_{21}-mode minimum points

Fig.3 Magnitude of the reflected wave

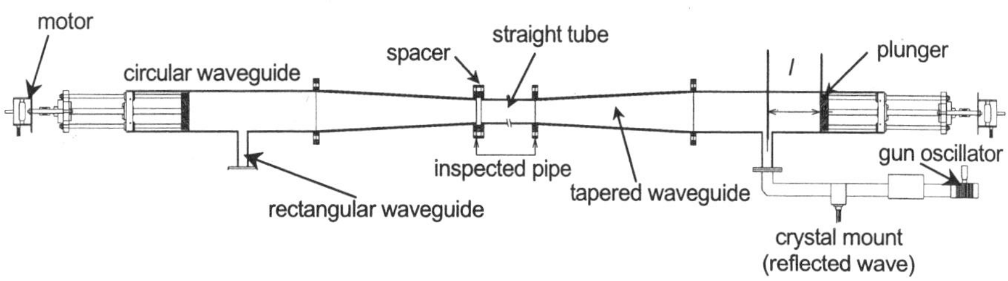

Fig.4 Experimental system (crack detection)

rectangular TE_{01}-mode emitted by the gun oscillator is converted to circular TM_{01}-mode by the mode converter. And the circular TM_{01}-mode reflected by the reflection plate is converted to rectangular TE_{01}-mode by the mode converter again, and is detected as "reflected wave" at the crystal mount installed in front of the gun oscillator. The length l between the center of rectangular waveguide and the plunger, is changed from 70[mm] to 190[mm]. The frequency f used in this experiment is between 6.4[GHz] and 7.2[GHz].

Figure 3 shows the result of this experiment, where the black area corresponds to the minimum intensity of reflected wave, and the white ones to the maximum. The theoretical points, where each mode becomes minimum, are represented by white circles. It can be confirmed which modes are generated inside the mode converters by coincidence of the black area and white circles.

Figure 3(b) shows the comparison of the black area with the minimum points for the circular TM_{01}-mode, which are calculated from Eq. (5). Since most of the white circles correspond with the black area, it is considered that the circular TM_{01}-mode is generated inside the mode converter. However, areas where the white circles disagree with the black area have been observed, which indicates that other modes are generated in the mode converter.

Figure 3(c) shows the comparison with circular TE_{11}-mode. The white circles of the circular TE_{11}-mode appear between the black areas. This result means that microwaves converted to the circular TE_{11}-mode were not so large.

On the other hand, in Fig. 3(d) we can see some coincidence of the white circles of TE_{21}-mode with the black areas especially when l becomes relatively large. From these results it can be said that the circular TM_{01}-mode is generated together with higher order modes like TE_{21}, which means the mode converter will successfully work.

3.2 Crack detection testing

Figure 4 shows the experimental system for the evaluation of the crack detection performance. The diameter of the test tube is 34 [mm], and cut-off frequency is calculated to be 6.755[GHz] for the TM_{01}-mode. The frequency f used in this experiment was chosen between 6.5[GHz] and 7.0[GHz] to include the cut-off frequency. The tapered waveguide is employed to introduce the microwave below the cut-off frequency of the inspected pipe, which enables us to control the wave propagating distance inside the pipe by changing the wave frequency. The length l between the center of rectangular waveguide and the plunger is changed from 90[mm] to 180[mm]. The diameter of spacer installed to simulate the defect is 42[mm] and 38[mm], which are larger than that of the straight tube to simulate cracks of 4[mm] and 2[mm] depth. The spacer thickness is set to be 2[mm] or 1[mm], which is relatively large to investigate experimentally the possibility of proposed method. We also use a thin spacer of 34[mm] diameter so as not to change the pipe length when the reference data without crack is measured.

The performance for the crack detection is evaluated by comparing the reflected wave intensity with and without crack. The experiment is performed with changing both the length l and frequency. In Fig. 5, the black areas suggest that the differences of reflected wave intensity are relatively large. The cut-off frequency of the circular TM_{01}-mode (6.755 [GHz]) is also drawn by the solid line in the figure. We can see large difference in signal at three frequency regions. One is just below the cut-off frequency and the others are above the cut-off frequency. This means that not only the microwaves near the cut-off one but also those above the cut-off carry informations on the crack.

Figure 6 shows the distribution of the signal difference in the reflected wave intensity with and without the crack at frequency, f=6.925 [GHz], 6.825[GHz], and 6.725 [GHz] by changing the length l along dotted lines shown in Fig. 5. In Fig. 6 it can be seen that the

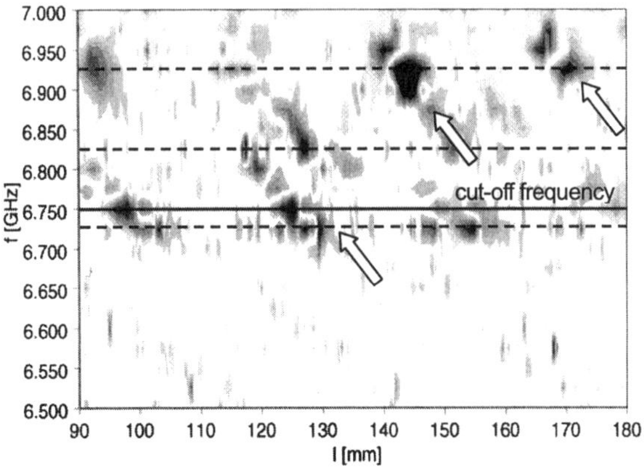

Fig.5 Distribution of differences of reflected wave intensity

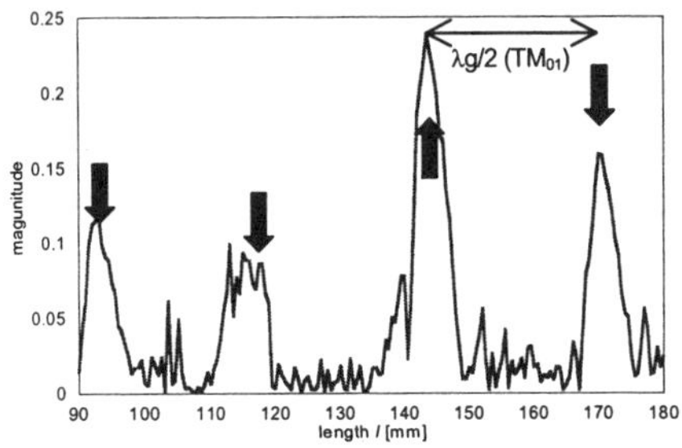

(a) f=6.925[GHz]

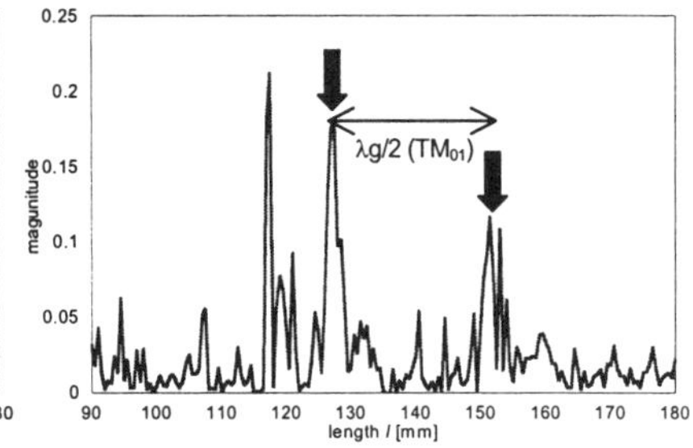

(b) f=6.825[GHz]

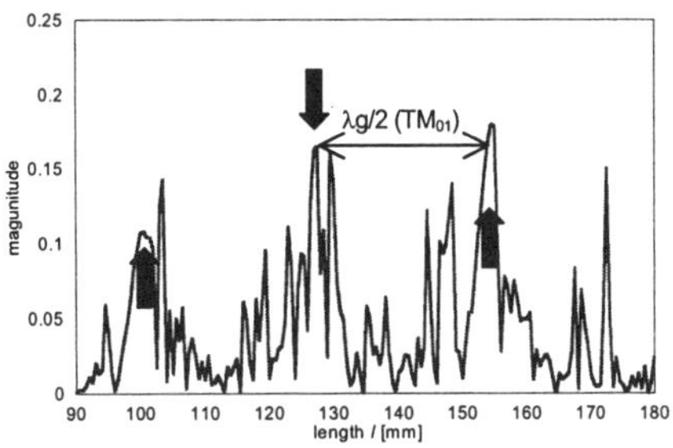

(c) f=6.725[GHz]

Fig.6 Profile of difference of reflected wave intensity

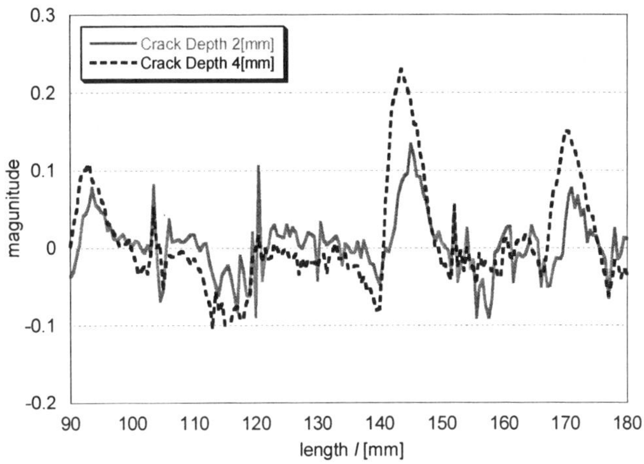

Fig.7 Effect of crack depth on the intensity of the reflected wave at 6.926[GHz]

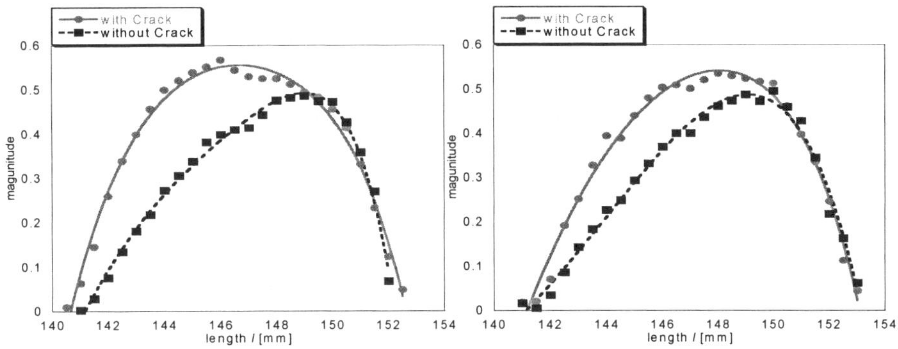

(a) Spacer Thickness 2.0[mm] (b) Spacer Thickness 1.0[mm]

Fig.8 Effect of crack thickness on the intensity of the reflected wave at 6.926[GHz] length between peaks of the reflected wave intensity corresponds with $\lambda_g/2$, the half guide wavelength of circular TM_{01}-mode. These results indicate that the TM_{01}-mode is generated and the microwave near the cut-off frequency is affected by existence of the crack. In Fig. 6(c) where the frequency is smaller than the cut-off one, it seems that there are many regions showing the difference of reflected signals. This result also suggests that the reflected wave below the cut-off frequency carries more information on the crack as indicated by the numerical analysis.

In the following, the size of crack is changed to evaluate its effect on the signal. Figure 7 shows the difference of reflected signal from the reference one in cases of 2[mm] and 4[mm] crack depths at 6.925[GHz]. It can be seen that the reflected wave intensity is strongly affected by the crack depth. Figure 8 shows the distribution of reflected wave intensity both with and without crack at 6.925[GHz] for the spacer thickness of 2[mm] and 1[mm]. The intensity difference clearly appears when the length l varies from 140[mm] to 150[mm] in both cases. These results mean that the signal differences between with and without the crack above the cut-off frequency of TM_{01}-mode are available to detect the crack existence and become smaller by decreasing the spacer width. That is, the microwave of circular TM_{01}-mode above the cut-off frequency f_c brings a lot of information on the crack thickness and width.

4. Conclusion

Through this study, the following conclusions are reached.
1) By the first experiment, it has been confirmed that the microwave of circular TM_{01}-mode was successfully generated inside the mode converter with higher order modes.
2) The characteristic changes in reflected microwave have clearly appeared near the cut-off frequency of the circular TM_{01}-mode. This result means that the reflected microwave has some information on the crack. By combining the TE_{21}-mode with the TM_{01}-mode, there is a possibility that not only the circumferential crack but the axial one also can be detected.
3) The microwave above the cut-off frequency also shows characteristic signal changes due to the crack size. It becomes, therefore, useful to use the transmitted microwave.
4) For future work, the validity of this method should be investigated by using natural cracks and then we should develop some methodology to determine the crack position by using the microwave near the cut-off frequency.

References

[1] P. Y. Joubert, D. Miller, D. Placko, E. Savin, *Multi-Detector Eddy Current Probe for the Non-Destructive Evaluation of Steam Generator Tubes, Designed for an Imaging Approach*, Studies in Applied Electromagnetics and Mechanics, vol. 15, 1999, pp.34-44

[2] Masaaki Kurokawa, Reizou Miyauchi, Koji Enami, *New Eddy Current Probe for NDE of Steam Generator Tubes*, Studies in Applied Electromagnetics and Mechanics, vol. 15, 1999, pp.57-64

[3] Koji Suguwara, Hidetoshi Hashizume, Sumio Kitajima, *Development of NDT method using electromagnetic waves*, JSAEM Studies in Applied Electromagnetics and Mechanics, vol. 10, 2001, pp.313-316.

Electromagnetic Nondestructive Evaluation (VIII)
T. Sollier et al. (Eds.)
IOS Press, 2004

Development and Validation of a 3D Model of a Magneto-Optic / Eddy Current Imager by a Dyadic Green's Functions Formalism

J.-M. DECITRE[(a,c)], D. PREMEL[(b)] and M. LEMISTRE[(c)]

(a) SATIE - ENS de Cachan, 61 av. du Président Wilson, 94235 Cachan Cedex, France
(b) CEA Saclay - DRT/DECS/SISC Bât. 611, 91191 Gif-sur-Yvette, France
(c) ONERA - DMSE, BP 72, 29 av. de la Division Leclerc, 92322 Châtillon Cedex, France

Abstract. This paper deals with the development of a 3D electromagnetic model of a plane wire inducing eddy-currents in a flat flawed conductive media as in the case of a magneto-optic imager. The described model is based on the dyadic Green's functions formalism to solve the forward problem. The response of the flaw is observed via a pick-up coil scanning the surface above the investigated region. Thus, the normal component of the magnetic field is evaluated by the model. Simulated data provided by the model are compared to 3D finite element data and to experimental data. The validation of the proposed model is intently tested before a future work considering the inverse problem. The proposed model is also exploited in order to simulate a countersunk head rivet with a buried crack.

1. Introduction

The electromagnetic techniques by eddy currents have many advantages in NDT for applications in aeronautics. They ensure a good sensitivity for not very thick conductive plates, a total safety and a compact material. We consider here a device of excitation including a conducting sheet traversed by a current, placed just above the target. It creates parallel lines of induced currents. A magnetic field results from the deviation of this induced current lines in the target due to the presence of defects. It can be measured by several methods, in particular by a small coil or by Faraday effect of a garnet film as in the case of the magneto-optical imager of PRi [1]. This last sensor presents interesting advantages. It is possible to examine straightforwardly an area of several cm^2. Compared to an acquisition using classical sensors, the gain of time is about 8.

Models of MOI® using finite-element techniques have already been developed [2,3,4]. The images provided by the MOI® can be better understood and the parameters defining the driving system may be optimized. The purpose of this paper is to give a faster method for 3D modeling, based on integral equations by using a Dyadic Green's function approach. The data provided by this integral model are successively compared to data provided by a finite element model (FEM) and to experimental data.

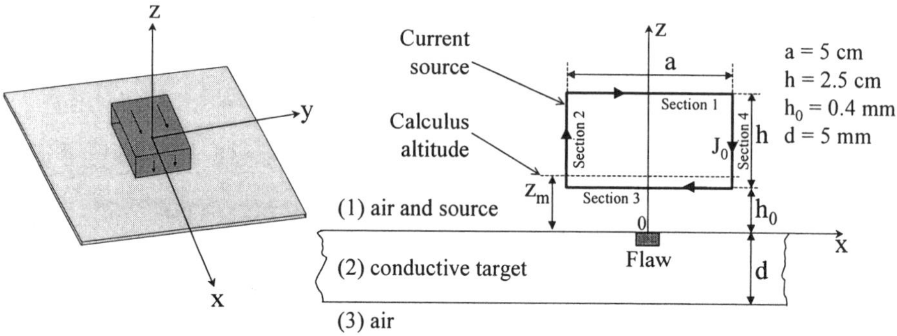

Figure 1: Simplified geometry of induced planes for the model. Two interfaces are considered for the plane target.

2. Geometry of the Excitation Device

The considered excitation device is composed with four copper plates placed as shown in Figure 1 [5]. Such a device is a simplified version of the excitation apparatus enclosed into the Magneto-Optic Imager. In order to compare the results with an EF method, it is necessary to satisfy the condition:

$$\mathrm{div}\vec{J} = 0 \tag{1}$$

The aluminum conductive target features a calibrated notch. Two interfaces are considered for the stratified target.

3. Green's Functions Approach

The method based on the Green's functions is a method of exact resolution of the differential equations of Maxwell. Its principal advantage lies in the speed of calculations. As a consequence, it makes possible to solve the inverse problem faster.

The dyadic Green's approach consists in solving three stages of computations [6][7]. First, the incident electric field is computed inside an investigated region in the sound target. The perturbation of the electric field involving the flaw in the target is evaluated. Finally, the normal component of the magnetic induction can be computed above the surface of the target or the electromotive force (emf) induced in a pick-up coil after the integration of the normal electric field.

3.1 Calculus of the Incident Electrical Field

The current source density, in any point $P'(x', y', z')$, created by the induction foils is given by the following expression:

$$\vec{J}_0(x',y',z') = j_0 \iiint_{\text{section 1}} [\delta(x'-x)\delta(y'-y)\delta(z'-z)] \mathrm{d}x\mathrm{d}y\mathrm{d}z\, \vec{u}_x \\ + j_0 \iiint_{\text{section 2}} [\delta(x'-x)\delta(y'-y)\delta(z'-z)] \mathrm{d}x\mathrm{d}y\mathrm{d}z\, \vec{u}_z \\ - j_0 \iiint_{\text{section 3}} [\delta(x'-x)\delta(y'-y)\delta(z'-z)] \mathrm{d}x\mathrm{d}y\mathrm{d}z\, \vec{u}_x \\ - j_0 \iiint_{\text{section 4}} [\delta(x'-x)\delta(y'-y)\delta(z'-z)] \mathrm{d}x\mathrm{d}y\mathrm{d}z\, \vec{u}_z \tag{2}$$

To calculate the incident magnetic field in this first step, let us consider an isotropic and homogeneous aluminum target. Let us use the dyadic Green's function $\overline{\overline{G}}_{21}^{(ee)}$, which allows to express the electric field $\vec{E}_0$ in any point $P(\vec{r})$ of the target, due to the source [6]:

$$\vec{E}_0(\vec{r}) = i\omega\mu_0 \iiint_{\text{excitation foils}} \overline{\overline{G}}_{21}^{(ee)}(\vec{r},\vec{r}')\vec{J}_0(\vec{r}')\,d^3\vec{r}'(3) \tag{3}$$

3.2 Add of a Defect in the Target

In this second step, let us introduce the normalized contrast function in terms of the relative conductivity:

$$f(\vec{r}) = \frac{\sigma_0 - \sigma_d(\vec{r})}{\sigma_0} \tag{4}$$

Then, the presence of a flaw is translated like a source of fictitious currents:

$$\overrightarrow{J_p}(\vec{r}) = [\sigma_0 - \sigma_d(\vec{r})]\overrightarrow{E_2}(\vec{r}) \\ = \sigma_0 f(\vec{r})\overrightarrow{E_2}(\vec{r}) \tag{5}$$

This current density changes the previous electrical field $\vec{E}_0$. So, let us introduce the dyadic Green's function $\overline{\overline{G}}_{22}^{(ee)}$ to calculate the electric field $\vec{E}_2$ in the target taking into account the fictitious currents [6]. The state equation is then given by:

$$\vec{E}_2(\vec{r}) = i\omega\mu_0\sigma_0 \iiint_{\text{flaw}} \overline{\overline{G}}_{22}^{(ee)}(\vec{r},\vec{r}')f(\vec{r}')\overrightarrow{E_2}(\vec{r}')\,d^3\vec{r}' + \vec{E}_0(\vec{r}) \tag{6}$$

One can note that the internal electric field $\vec{E}_2$ can be computed by the conjugate gradient method.

3.3 Observation

Finally, the magnetic field $\vec{H}_d$ at the location of garnet (i.e. above the surface of the conductive copper plate) due to the disturbance electric field $\vec{E}_2$ in medium 2 is evaluated by:

$$\overrightarrow{H_d}(\vec{r}) = i\omega\mu_0\sigma_0 \iiint_{\text{flaw}} \overline{\overline{G}}_{12}^{(me)}(\vec{r},\vec{r}')f(\vec{r}')\vec{E}_2(\vec{r}')\,d^3\vec{r}' \tag{7}$$

The dyadic Green's function $\overline{\overline{G}}_{12}^{(me)}$ [6] indicates the magnetic field in the air region 1 due to a point source located in the damaged zone.

3.4 Discretization

The discretization of the two coupled equations (5) and (6) is performed by subdividing the supposed damaged zone into a regular mesh of $N = N_x \times N_y \times N_z$ volumetric cells. By using the Galerkin's variant of moment method, we obtain a system of two coupled matrix equations:

$$\begin{cases} \mathbf{y} = \mathbf{\Gamma}_{12}\, \mathbf{F}\, \mathbf{e}_2 \\ \mathbf{e}_2 = \mathbf{\Gamma}_{22}\, \mathbf{F}\, \mathbf{e}_2 + \mathbf{e}_0 \end{cases} \tag{8}$$

where e_0 is the $3N\times1$ incident electric field vector and e_2 is the $3N\times1$ total electric field vector, the number of components of these vectors come from the three components of each $N\times1$ electric field vector E_x, E_y and E_z. The $M\times1$ observation vector y contains M complex values of the normal magnetic field according to a regular grid. Γ_{12} et Γ_{22} are the matrix derived from the Green's functions $\overline{\overline{G}}_{12}$ and $\overline{\overline{G}}_{22}$. Finally, F is a diagonal matrix which contains the values of the contrast function f.

4. Validation with a Finite Element Method

A comparison between the proposed model and a finite element model (FEM) using the Ansys® software is performed to validate the model. A simple parallelepipedic defect is chosen. It is 12 mm long, 10 mm wide and 1 mm deep in a 5 mm deep aluminum plate. The simulated experiments are carried out at a frequency of 10 kHz. It is important to note that only a 40×30×4 regular mesh is used to discretize the damaged zone whereas the finite element formulation in Ansys® requires the discretization of the free space and uses more than 80000 elements to perform the same simulations with the FEM. The inductive plate is fed by a current density equal to 10^8 A/m^2.

Figure 2 and Figure 3 present the magnetic induction at the altitude $z_m = 0.62$ mm obtained with the Green's formalism and the FEM respectively. The mean square differences between the two methods are listen in the Table 1 for each component, without and after a normalization for each component.

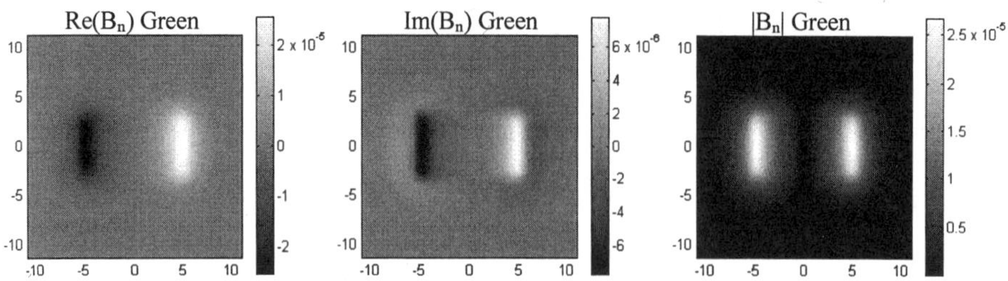

Figure 2 : Normal induction field at 10 kHz obtained with the proposed model (x and y axis are given in mm).

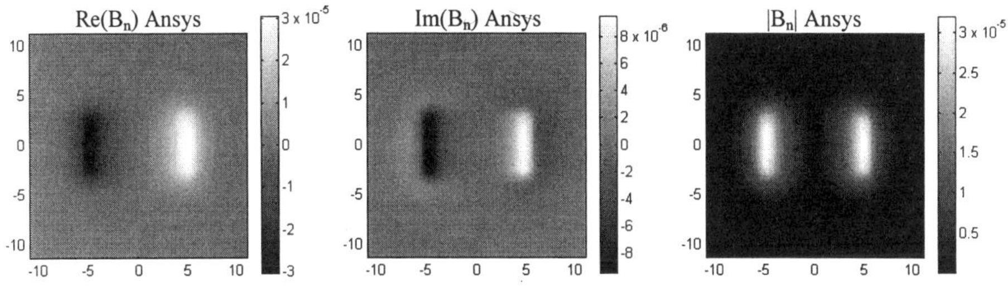

Figure 3: Normal induction field at 10 kHz with the finite element method using Ansys® software (x and y axis are given in mm).

Table 1 : The mean square difference between magnetic induction for the two models.

	Real component	Imaginary component	Modulus
Without normalization	17 %	20 %	17 %
With normalization	5.8 %	5.0 %	3.9 %

Table 2 : The geometrical characteristics of the small coil.

Average radius	0.7 mm
Height	1.25 mm
Number of turns	71

The variation of the results between the two models is not very significant. The difference has several origins. Firstly, the relationship between the greatest segment to be meshed and the smallest in FEM model is unfavorable (of the order of 80). More, the mesh have been limited because of the huge number of elements which is required. Secondly, the side effects due to the non infinite target according to x and y in the FEM can disturb the result in a significant way. Lastly, the height of the inductor is not taken into account in our Green's approach.

In a general way, the variation can be minimized by using a finer grid for both methods. However, the undeniable advantage of the method using the Green's functions lies in the computing time. A saving of time between one and two orders of magnitude is noted. Moreover, the hardware configuration must less important with Green's approach, in particular about the memory.

5. Validation with Experimental Data

This paragraph aims to compare the results of the semi-analytical model with the cartography of the magnetic induction. The used sensor is a small coil. Its axis is normal to the surface of the target and the coil is disposed just above the inferior inductor plate. The electromotive force (emf) on its terminals is measured. The geometrical characteristics of the coil are given in Table 2.

The excitation device is displayed in Figure 4, while the dimensions are similar to those of the Figure 1.

The observation equation (7) provides the normal component of the magnetic field. Thus, it must be consequently modified to obtain the emf of the coil. The emf is computed by the integration of the disturbance electric field on contour Γ of the turn of the pick-up coil:

$$e(\vec{r}) = \int_{\Gamma} \vec{E}_d(\vec{\ell}).\vec{d\ell} \tag{9}$$

and the disturbance electric field $\vec{E}_d$ is obtained after the dyadic Green's function computation $\overline{\overline{G}}_{12}^{(ee)}$. $\vec{E}_d$ is given by the following equation:

$$\vec{E}_d(\vec{r}) = i\omega\mu_0\sigma_0 \iiint_{\text{flaw}} \overline{\overline{G}}_{12}^{(ee)}(\vec{r},\vec{r}\,')\, f(\vec{r}\,')\, \vec{E}_2(\vec{r}\,')\, d^3\vec{r}\,' \tag{10}$$

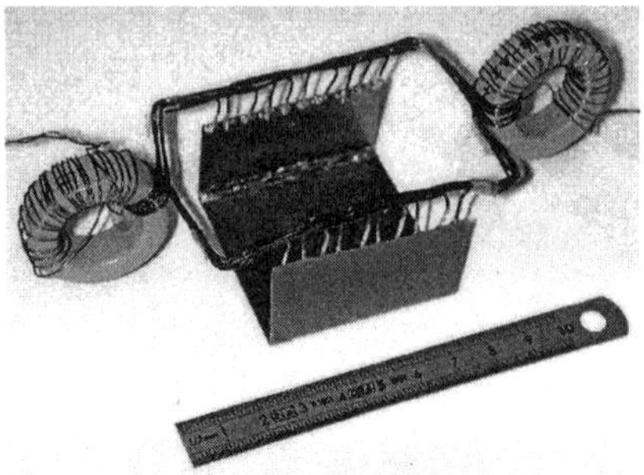

Figure 4 : Excitation device used for the acquisition of the experimental data.

Figures 5 and 6 give the results provided by the proposed model comparing to experimental data considering the pick-up coil.

After normalization, the mean square difference are about 30 %, 17 %, 18 % for the imaginary part, the real part and the absolute value of the emf, respectively. The differences can be attributed, on the one hand, to the latent lack of accurate knowledge of physical parameters, and, on the other hand, to the difference between the form and the geometry of the exciting device.

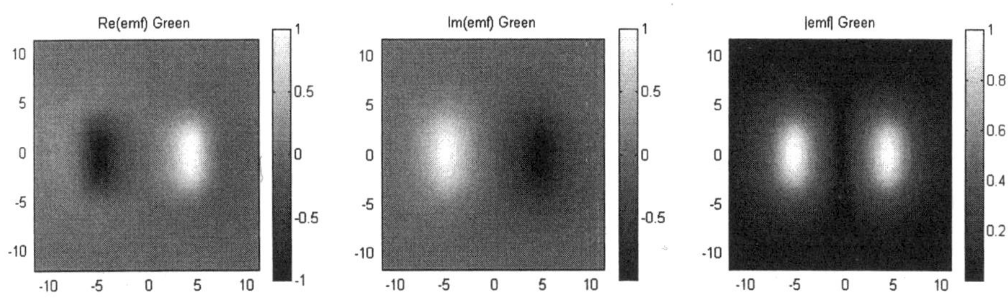

Figure 5 : Emf obtained with the Green's model.

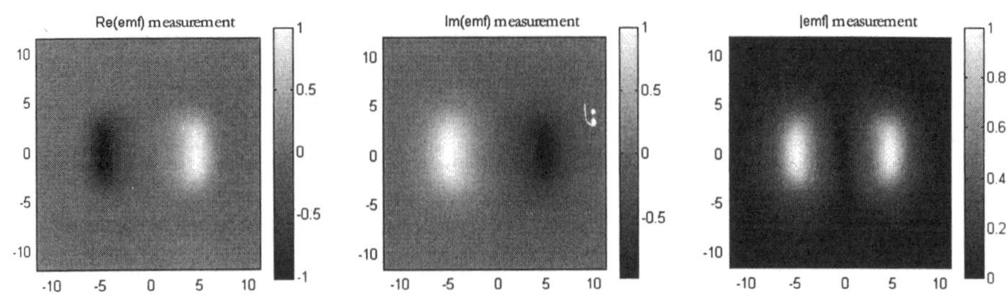

Figure 6: Emf experimental data obtained with a pick-up coil.

6. Use of the Model

The following figures give some results obtained with a rivet with or without crack. The first anomaly zone simulates a 3.5 mm diameter aluminum countersunk head rivet. It is riveted in a 2 mm deep aluminum target. The air-gap between the target and the rivet is 0.4 mm width. The anomaly zone is discretized in 25×25×4 elements. For each voxel, it is assigned, a value between 0 and 1, proportional to the volume of the air and aluminium. The black voxels on the Figure 7a correspond to 100% air (and the contrast function f is equal to 1), and the white voxels correspond to 100% aluminum ($f = 0$). The operating frequency is equal to 10 kHz. The geometry and normal magnetic induction modulus are given on Figure 7.

On Figures 8 and 9, the same rivet is used, but a buried crack is added. It is 0.25 mm below the surface, 6.5 mm length in regard to the rivet axis, 0.6 mm width. Its orientation is the same as the excitation current (Figure 8) or perpendicular to excitation current (Figure 9).

At 10 kHz, the thickness of the skin is about 1.5 mm in the target and the crack is not very visible on Figure 8a. In return, at 5 kHz, the crack effect is more significant, but its spatial resolution is not so good.

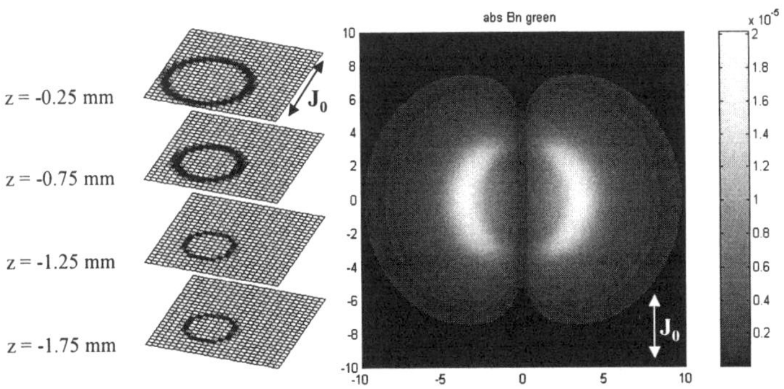

(a) (b)

Figure 7 : Simulation of a rivet without crack. (a) Anomaly zone geometry with 4 layers, (b) normal magnetic induction modulus at 10 kHz.

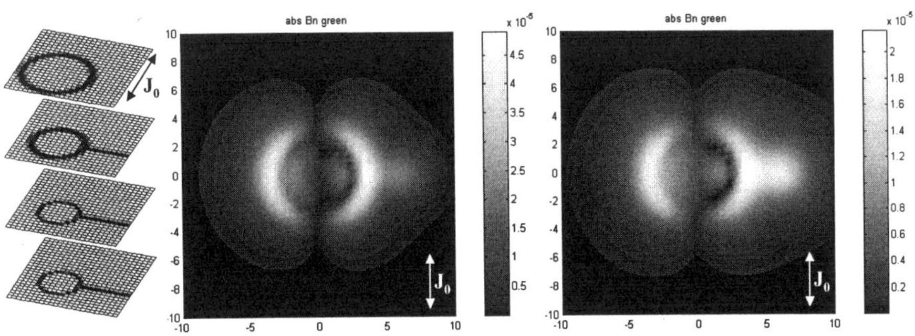

(a) (b) (c)

Figure 8 : Simulation of a rivet with a crack. Crack and excitation currents are parallel. (a) Anomaly zone geometry, (b and c) normal magnetic induction modulus at 10 kHz and at 5 kHz.

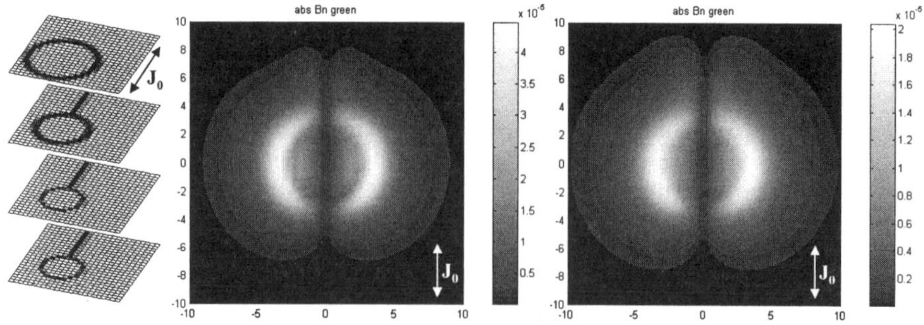

(a) (b) (c)

Figure 9 : Simulation of a rivet with a crack. Crack and excitation currents are perpendicular. (a) Anomaly zone geometry, (b and c) normal magnetic induction modulus at 10 kHz and at 5 kHz.

7. Conclusion and Perspectives

We expect these results to be sufficient to solve the inverse problem using experimental data obtained with a single sensor as a coil. The sensor will be replaced by an magneto-optical device based on a ferrimagnetic garnet. The final goal is to design a quantitative magneto-optical imager.

References

[1] Fitzpatrick G. L., Thome D.K., Skaugset R.L. and Shih W.C.L., "Magneto-optic/Eddy current imaging of subsurface corrosion and fatigue cracks in aging aircraft", in *Review of Progress in Quantitative Nondestructive Evaluation*, eds. D.O. Thompson and D.E. Chimenti, Vol. 15, 1996, pp. 1159-1166.

[2] Chao C., Udpa L., Xuan L., Fitzpatrick G., Thome D. and Shih W., "Parametric studies of magnetic-optic imaging using finite-element models", in *Review of Progress in Quantitative Nondestructive Evaluation*, eds. D.O. Thompson and D.E. Chimenti, Vol. 15, 2000, pp. 1947-1954.

[3] Xuan L., Shanker B., Udpa L., Shih W. and Fitzpatrick G., "Finite element model for MOI applications using A-V formulation", in *Review of Progress in Quantitative Nondestructive Evaluation*, eds. D.O. Thompson and D.E. Chimenti, Vol. 20, 2001, pp. 385-391.

[4] Rubbinacci G. et al., "Numerical simulation of magneto-optic/ Eddy-current imaging", in *Electromagnetic Non-Destructive Evaluation*, Saarbrücken, 2002.

[5] J.M. Decitre, D. Prémel and M. Lemistre, "3D modeling of a magneto-optic imager by a dyadic green's functions approach", Review of Progress in QNDE, 2002.

[6] Chew C.C., *Waves and fields in inhomogeneous media*, 1990.

[7] Denis Prémel and Alexandre Baussard, "Eddy-current evaluation of three-dimensional flaws in flat conductive materials using a Bayesian approach", *Inverse problems*, Vol. 18, pp. 1873-1889 (2002).

Electromagnetic Nondestructive Evaluation (VIII)
T. Sollier et al. (Eds.)
IOS Press, 2004

TEDDY PLUS$_{DIGITAL}$: A new generation of ET Data Acquisition Systems

Belén RIBES, Jerónimo HERNANDEZ, Victor BARCENILLA, Jesús VAZQUEZ
Tecnatom, S.A., Avd. Montes de Oca 1, 28709 San Sebastián de los Reyes, Madrid

Abstract. Tecnatom is developing a new line of eddy current data acquisition systems (ET-DAS). The main innovations of the system consist of a digital reception technique approach, dynamic gain control, real-time signal data processing through DSP's, array probes support as well as electronic embedded probes support and integration with sound, video, UT and mechanical controllers in the same rack. Another advantage of the system is its small size due to the high-tech electronics components employed. With regards to signal processing, several tools have been implemented both for data acquisition and data analysis purposes. In this sense, a complete new enhanced tool for automatic data analysis has been developed. This software module, named *Expert,* together with the current automatic analysis process, makes TEDDY a powerful analysis tool since it can perform a complete double automatic analysis. This paper presents the new module as well as the new line of eddy current data acquisition systems, named TEDDY PLUS$_{DIGITAL}$.

1. Introduction

Many research studies have been performed[1] in the field of automating the eddy current data analysis process during steam generator inspections since these processes result in significant economic and technical benefits.

Since 1989 up to now, Tecnatom has analyzed several methodologies and has made different studies on automatic analysis systems, that go from Computer data Screening techniques (based on level thresholds, phase, shape of the signals, behavior of parameters in different channel combination, mixing of channels, etc), pattern recognition studies (based on parameters like amplitude and phase, Fourier descriptors, moments, energies and autoregressive models) up to expert systems.

In 1990 Tecnatom developed its first automatic analysis software (EVA) that was first used in 1991. The system automatically classifies signals using a rule-based sorting criterion. In the case of Steam Generators tube inspection two independent analyses are required by regulatory bodies, usually called primary and secondary analysis. Currently, the first of these analyses is performed automatically by the computer and the other manually by an analyst. There is no commercial system that can provide two independent automatic analyses, that is why Tecnatom decided to develop a second automatic analysis software. This automatic analysis is based on a classification given by an expert system. The expert system is based on Multiple Logistic Regression, its principles and its implementation are described in the following sections.

2. Multiple Logistic Regression for Automatic Eddy Current Signal Interpretation

Logistic Regression[1] is most frequently used to model the relationship between a binary outcome variable and a set of covariates, and with few modifications it can model the relationship between a polytomous outcome variable and a set of covariates.

In the case of eddy current signal classification, the categories of the polytomous outcome variable Y, coded as 0,1,2,…,n, represent the different types of defect signal types to be classified; while x represents the vector of covariates of length p+1 describing the different parameters evaluated over an eddy current signal (amplitude, phase, Fourier descriptors coefficients, fast Fourier transform coefficients, autoregressive model coefficients, etc).

To understand logistic regression let us start with the simplest problem, the case of a binary outcome variable. The relationship between the outcome (Y) and the independent variable (x), is modeled through a logistic distribution:

$$E(Y/x)=\pi(x)=\frac{e^{\beta_0+\beta_1 x}}{1+e^{\beta_0+\beta_1 x}} \qquad \text{Eq. 1}$$

A logit transformation of $\pi(x)$ is applied, such that the resulting logit, g(x), is linear in its parameters and thus has many of the desirable properties of a linear regression model.

$$g(x)=\ln\left[\frac{\pi(x)}{1-\pi(x)}\right]=\beta_0+\beta_1 x \qquad \text{Eq. 2}$$

In order to find the better estimates $\hat{\beta}$ of the beta coefficients, a likelihood function is build and the maximum likelihood estimators of these parameters are chosen to be those values which maximize this function. The likelihood function has the following expression in the case of a binary variable:

$$l(\beta)=\prod_{i=1}^{n}\pi(x_i)^{y_i}[1-\pi(x_i)]^{1-y_i} \qquad \text{Eq. 3}$$

In the case under study, Y has n categories and x represents a vector of dimension p+1 with $x_0=1$ to account for the constant term, where each x_i is a parameter measured over the signal of interest.

Following equation 2, the logit functions will have the following expressions:

$$g_i(x)=\ln\left(\frac{P(Y=i/x)}{P(Y=0/x)}\right)=\beta_{i0}+\beta_{i1}x_1+........+\beta_{ip}x_p \qquad \text{Eq. 4}$$

And the conditional probability of each outcome category or type of defect:

$$\pi_i(x)=P(Y=j/x)=\frac{e^{g_j(x)}}{\sum_{k=0}^{n-1}e^{g_k(x)}} \qquad \text{Eq. 5}$$

The problem of finding the values of the beta coefficients is then solved by constructing the likelihood function associated and finding the maximum likelihood estimator, $\hat{\beta}$.

3. Expert Product Description

A software package has been built, named EXPERT, based on multiple logistic regression as a classification technique. With the software it is possible to:

- build knowledge bases
- build classification models

- study the validity of a model

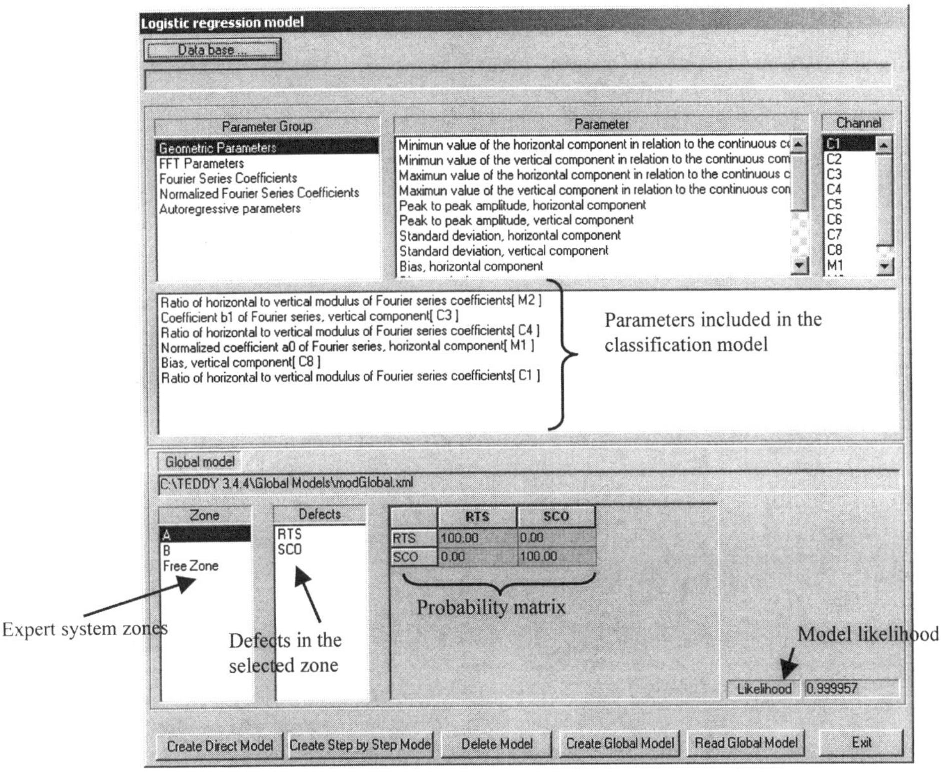

Figure 1 Example of a classification model

- validate a classification model
- classify a signal by a simple query
- use the expert classification in an automatic process.

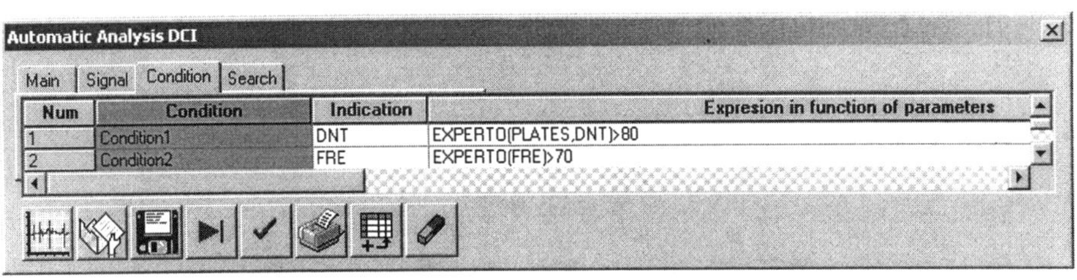

Figure 2: Including the expert classification in a rule

4. Automatic Analysis implemented in TEDDY Plus

Two different ways of classifying automatically a signal have been implemented,

(a) **Unattended automatic analysis**: Its main advantages are the speed, the ease to include new rules and that it is understandable for the operator. The condition for a signal classification can be a combination of:

- A rule based on the analyst experience, i.e.:
  Voltage > 3 volts and 30<phase<130
- The result of a classification tree (pattern recognition techniques), i.e.:
  0.006 miny – 0.1348 fftmfy + 0.015496 fsca0x –1.5 > 0 then fretting on AVB
- Expert based on logistic regression
  Probability of being fretting > 80%

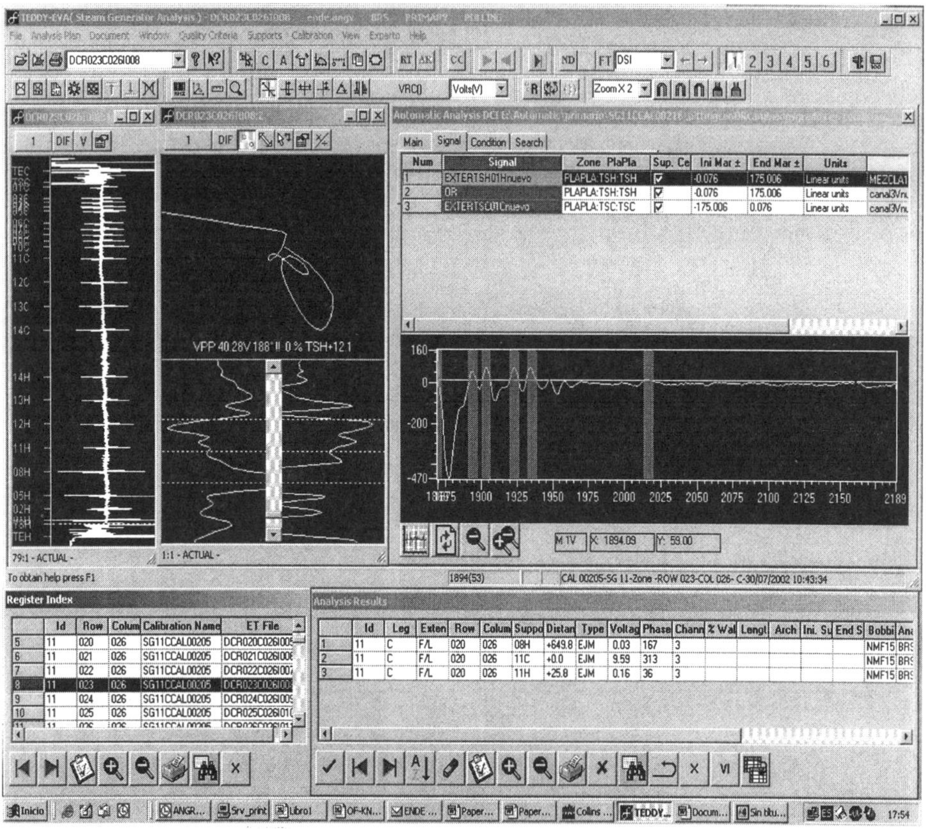

Figure 3: Automatic Analysis running on background

and, (b) **Automatic analysis of signals under analyst request**: which is based on classification models and in informs of the defect with a higher probability

The strategy of the unattended automatic analysis consists on the following steps:

- Automatic calibrations and mixing generation
- Automatic detection of geometrical structures
- Automatic signal analysis based on the part of the tube where the signal is located
- Automatic report and defect sizing
- Transfer of the defect data to a Data Base
- Final report generation

Besides the new automatic classification module developed (EXPERT) already described, other enhancements have been developed for the automatic eddy current signal interpretation:

- An algorithm able to detect noisy tubes, like pilgrim noise. These tubes will always produce lots of false call indications otherwise.
- Also it is possible to combine two or more rules when trying to detect and classify difficult defects, like for example very small defects, producing only one call although several of those rules have found and classified the same defect.
- A trace mode capability in order to follow all the steps performed in the automatic analysis process, from signal filtering, threshold selection, up to rule behavior.

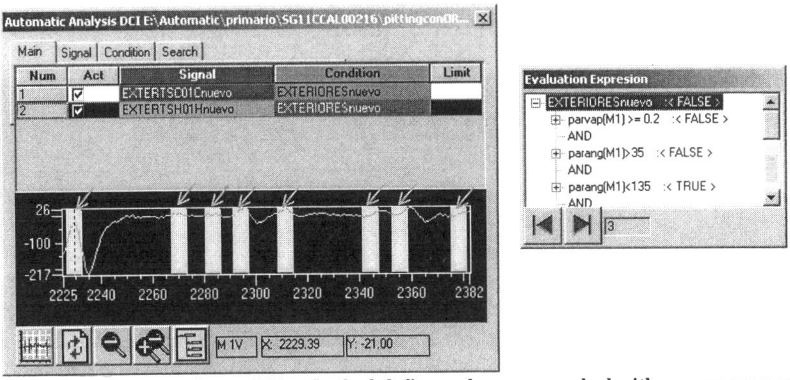

Figure 4: Trace mode capability. In the left figure the areas marked with arrows are possible defect signals. In the right figure the result of the condition evaluated over a signal.

- Since the automatic process runs in background, a new window has been built, showing statistics of the number of tubes analyzed, number of tubes left, time remaining, etc.

The automatic detection of the structures, analysis of all the signals and report for a tube is performed in about 1 second.

5. Application and Results

The application described here is based on signal classification for Steam Generator tubes. In order to build a Steam Generator classification model, the following steps should be followed:

- Steam Generator Model Zone Definition: in this step, the Steam Generator tubes are divided into different areas or zones where the different defects might theoretically appear.
- Defect Assignment: taking the list of possible indications that could be present in the Steam Generator, these are assigned to belong to one (or several) zones.
- Signal Parameterization: Taking as a starting point the results of an inspection performed in the Steam Generator (results data base), the system computes the corresponding parameters data base, named *knowledge base*.
- Building Classification Models: two different strategies are implemented to build the classification models:
  - Direct method: the user decides what parameters (x_i variables) will take part to the model
  - Step by step method: the system selects those parameters that maximise the likelihood function ($l(\beta)$)

The classification models were applied to a D3 type Steam Generator with the following defectology:

CODE	MEANING	SEARCH ZONE
DNT	DENTING	Support Plates
CUD	COPPER	Above Tubesheets+Support Plates
DSI	DISTORTED SUPPORT INDICATION	Support Plates
DSS	DISTORTED SUPPORT SIGNAL	Support Plates
DNI	SUPPORT WITH DENT & INDICATION	Support Plates with Dents
NQI	NON-QUANTIFICABLE INDICATION	Free Span
PIT	PITTING	Free Span
ADR	ABSOLUTE DRIFT INDICATION	Free Span + Ubend
MBM	MANUFACTURING BURNISH MARK	Free Span + Ubend
PVN	PERMEABILITY CHANGE	Free Span + Ubend+ Support Plates
AVB	AVB WEAR	Ubend
SLG	SLUDGE	Above Tubesheets
DTI	DISTORTED TUBESHEET INDICATION	All length Tubesheets
DRI	DISTORTED ROLL INDICATION	Roll Expansion
PLP	POSSIBLE LOOSE PART	Above Tubesheets+Above support plates
LPI	LOOSE PART WITH INDICATION	Above Tubesheets+Above support plates
DTS	DISTORTED TUBESHEET SIGNAL	All length Tubesheets

Up to 6 classification models were built taking into consideration the defectology and the different zones in which they might appear, that is: Free Span, Support plates, Above support plates, Tubesheets, Ubend and above Tubesheets. The different indications were grouped into the different zones, and the classification models were built both, by the direct method and by the step by step method. In the case of the direct method, the parameters (x_i variables) were chosen according to those channels (frequencies) where there is more discrimination among the signals to be classified and to the analysts experience. Both methods (direct and step by step) showed similar results. Using the set of tubes where the sample was extracted, the following results are obtained: the percentage of correct classification is around 95% (correctly classified defects/total number of reported defects) and the number of false calls (no defect/total number of reported defects) is less than 3%.

6. TEDDY PLUS$_{DIGITAL}$: An overview

The ET technique has been used by Tecnatom since 1978. The first developments were oriented to data software analysis, and in 1989 the first inspection was made using ENEAS software (based on UNIX-OS). In the mid 90's we started to develop our first ET acquisition system, named TEDDY, that was used for first time in 1997 and since then new hardware and software applications have been developed. Now we are developing a complete new line of ET-Data Acquisition Systems named TEDDY PLUS$_{DIGITAL}$. In the following points the main innovations of this product are highlighted.

- digital reception technique approach,
- probe array support as well as electronic embedded probes support.
- high modular concept design architecture based on PC-104+.
- dynamic gain control,
- real-time signal data processing through DSP's,
- integration with sound, video, UT and mechanical controllers in the same rack.
- small size due to the high-tech electronics components employed.

As an example of the advantages of using probe arrays, the inspection of the CRDM attachment Weld (J-Weld) is described. In the case of surface inspections the use of probe arrays constitutes a clear competitive advantage against the use of conventional probes, since it enables much faster inspections together with the possibility of inspecting complex geometry surfaces. The array module consists of 8 pairs of probes, Pancake or Plus Point type. With the aid of two multiplexers it is possible to excite the probes sequentially in two modes: impedance and pick-up mode, and at the same time, excite sequentially a pair of coils. So it is possible to detect cracks oriented in any direction by just one scan. In the next two figures the array module over a J-Weld mock-up and a C-Scan representation with TEDDY software are illustrated.

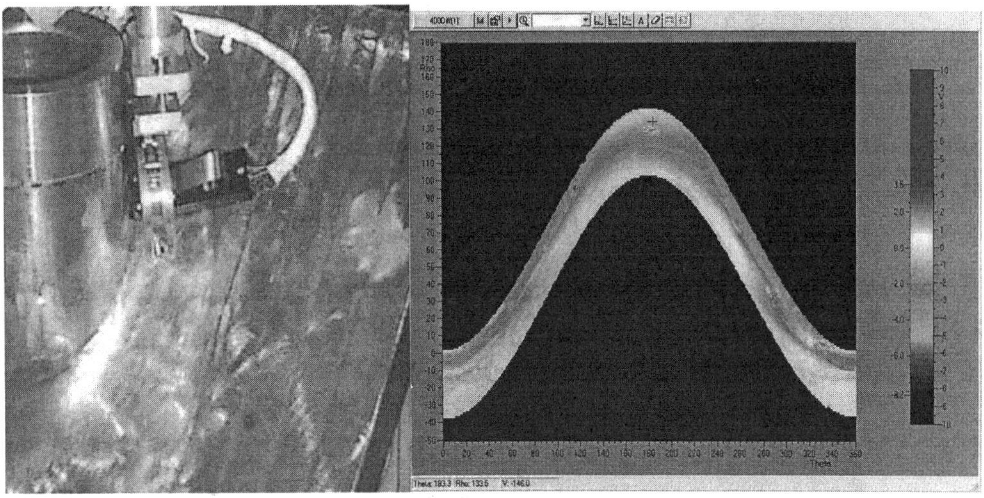

Figure 5. J-Weld Inspection System and C-Scan Representation

7. Conclusions

The use of multiple logistic regression for eddy current signal interpretation has demonstrated its viability.

TEDDY software counts with the possibility of performing a completely double automatic analysis, based on analyst classification rules by one side and an expert system on the other.

The new line of ET-DAS systems turns out to be very promising due to its innovations against its competitors based on a digital reception technique approach as well as its great flexibility, modularity and reduced size.

REFERENCES

[1] Steam Generator Automated Eddy Current Analysis, A Benchmarking Study, EPRI, 1998.

[2] Applied logistic regression. David W. Hosmer, Stanley Lemeshow; John Wiley &Sons, 1989.

Electromagnetic Nondestructive Evaluation (VIII)
T. Sollier et al. (Eds.)
IOS Press, 2004

Recognition of INCONEL weld conductivity variation by means of eddy current testing

Ladislav JANOUSEK, Noritaka YUSA, Zhenmao CHEN, Kenzo MIYA
International Institute of Universality,
Bldg. 8F, 1-4-6 Nezu, Bunkyo-ku, Tokyo 113-0031, Japan

Abstract. The paper deals with the prediction of electromagnetic characteristics variation of an INCONEL welding part based on the eddy current testing (ECT). Two different inverse algorithms are employed to cope with a task of conductivity profile reconstruction across the weld line.

1 Introduction

Alloy 600 is one of the base material used in manufacturing internal components of nuclear reactor vessels. These components contain welds due to their size and configuration. The combination of the chosen material and the weld-induced residual stresses was found to be the cause of the problem leading to stress corrosion cracking (SCC) [1]. Therefore the non-destructive testing (NDT) of INCONEL welded structures in nuclear power plants has been of great interest.

It is well known, that ultrasonic NDT of INCONEL welds is subjected to serious difficulties such as beam skewing, splitting and distortion, unexpected beam propagation directions, ghost echoes, and for heterogeneous media, beam curvature in the weld. These difficulties arise from coarse and oriented grain structure with grain size typically of the order of the ultrasonic wavelength [2]. The eddy current (ECT) and the ultrasonic (UT) techniques are both employed in INCONEL weld inspection, while ECT is applied to the defect detection and UT is applied to defect evaluation [1]. The possibility of defect sizing based on UT inspection is (due to the difficulties mentioned above) limited to the minimal depth of approximately 6 mm [3]. Also deep cracks are not well evaluated with UT because UT beams are scattered at the edges of SCC. Cracks shallower than 6 mm are exactly the promising space for employing the ECT technique. Inversion of ECT signal is becoming very sophisticated [4] and for this reason a great attention has been put toward the ECT inspection of INCONEL welds.

Inspection of INCONEL welds with ECT is subjected to overcome problems connected with a noise caused by the bead surface and the electromagnetic characteristics change of welded parts. To solve a problem of probe optimization and signal inversion, the basic features of the weld should be taken into account in numerical analyses.

Few papers have reported the utilization of ECT techniques to material electromagnetic parameters characterization. Basically they deal with the thickness and conductivity determination of isotropic non-magnetic conducting tubes and foils [5], [6], [7]. A layered metal structure whose conductivity varies continuously with depth, while attention is focused on certain conductivity profiles, has been studied in [8], [9]. Neural networks utilization for quantitative depth profiling with quasi-continuously varying electrical conductivity shapes is presented in [10].

The present paper deals with a recognition of the electromagnetic characteristics perturbation of INCONEL weld from measured ECT signals. In order to tackle this problem that requires estimation of local parameters, unlike the studies above, two inverse algorithms are employed to determine the electrical conductivity and magnetic permeability profiles along the weld line from the measured data.

2 Principle of recognition

When a coil excited with alternating current is placed over an electrically conducting material, eddy currents are induced in the material. Defects, variations in electrical conductivity and/or magnetic permeability, etc. disturb the eddy current flow, and in turn, alter the coil impedance. The fact that the coil (through its impedance) carries information about the material electromagnetic characteristics is the basis of the recognition proposed in this paper.

A self induction pancake coil with outer diameter of 3.2 mm, inner diameter of 1.2 mm, hight of 0.8 mm and number of turns 140 is used for this study. Signals caused by the conductivity and permeability difference between welded part and base material are measured. Calibration of the signals is performed by using INCONEL plate with known parameters and dimensions (calibration plate). The INCONEL welded specimen under study is shown in Fig. 1. The signal is gained by scanning over the target with a constant lift-off across the weld line.

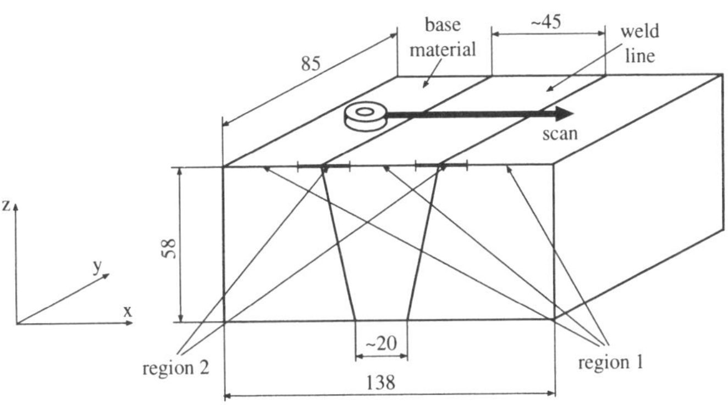

Figure 1: Dimensions of the INCONEL welded specimen

Two assumptions are made to solve an inverse problem of electromagnetic characteristics reconstruction across the weld line:

1) the surface of welded specimen is divided into two regions. The first (Fig. 1 - region 1, part of the base material, middle part of the weld) is quasi stable one, where the signal variation is small. The second, transient one (Fig. 1 - region 2) covers heat affected zone and the boundary of weld, where the signal variation is large.

2) the electromagnetic characteristics are supposed to vary just in a direction perpendicular to the weld line (x direction of the coordinate system shown in Fig. 1).

According to these assumptions, two different prediction methods have been proposed for the two regions separately. The electromagnetic characteristics distributions along the quasi stable regions are estimated based on the relationships derived from an uniform model. It was proved, that such method can deal also with no uniform distribution of the electromagnetic characteristics if their variation is limited. A neural network (NN) approach is utilized to tackle the conductivity and permeability profiles recognition along the transient regions.

3 The uniform model

As mentioned above, the change in electromagnetic characteristics causes the coil impedance variation. Figure 2 shows an example of the dependency of signals on the difference in conductivity and permeability between an isotropic specimen and the INCONEL calibration plate with a frequency of 200 kHz and lift-off of 0.25 mm.

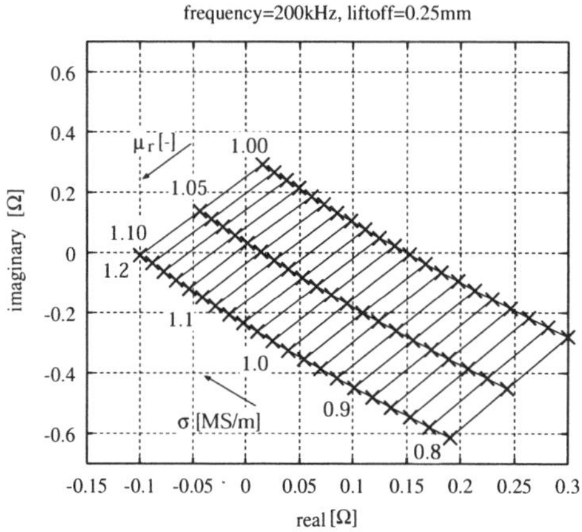

Figure 2: Signals due to the difference in conductivity and relative permeability

From Fig. 2, one can notice that the points representing the signal difference due to the conductivity variation for a constant value of permeability can be approximated by abscissa in this limited range for the given frequency and lift-off. The same observation is valid for the change in permeability with an unchanged conductivity value. The space of frequency ranging from 200 kHz to 500 kHz and the lift-off change from 0.2 mm to 1 mm was calculated and the previous statements hold in this space. General equation of an abscissa in two dimensional x-y plain can be written:

$$y = x \cdot k + q \tag{1}$$

while in the complex plain x component represents the real part Re of the signal, y component represents the imaginary part Im of the signal and k, q are the coefficients of abscissa. The coefficients of the conductivity abscissa depend on the conductivity and can be approximated by its quadratic function. In the similar way, the coefficients of the permeability abscissa are approximated as quadratic function of permeability. These transformations are quite simple. Therefore, an inverse problem of the material electromagnetic characteristics calculation (if they vary in the ranges under consideration) from the measured signal can be easily solved based on the empirical relationships.

The proposed uniform model has been used for the electromagnetic characteristics recognition in the quasi stable regions of welded specimen, where the characteristics vary quasi smoothly in small ranges. Since the coil signal is an integral value over a space given by the coil dimensions, lift-off and frequency, the results of the signal inversion based on the uniform model are average values of conductivity and permeability covering this space. For this reason, a small coil, minimal lift-off and high frequency have been used for the estimation, to focus on as minimal space resolution as possible.

Simulations were carried out to validate the proposed inverse procedure. Figure 3 shows the results of inversion from a simulated signal taking the small variation in conductivity and permeability into account. The conductivity distribution along the scanning line was considered based on the authors' experiences from the experimental and numerical analyses. At the same time the variation in permeability was supposed, although it seems, that the weld is nonmagnetic. The results show that within a limited range of the electromagnetic characteristics variation proposed inverse method based on the uniform model works very well.

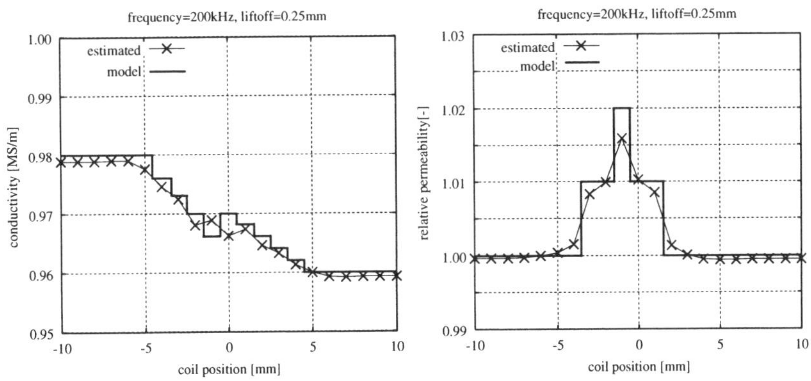

Figure 3: Reconstructed profiles of conductivity and relative permeability based on the uniform model

4 Neural Network approach

An NN structure [11] used in the reconstruction of conductivity and permeability profiles in the transient regions is shown in Fig. 4. The structure is a feed-forward network with three node layers - input, hidden and output layer. Input nodes with linear activation function simply copy the input value at their output. The hidden and output nodes possess nonlinear hyperbolic tangent activation function (f_1, f_2).

The bias of the map is introduced as a constant input source node. This architecture contains additional direct forward input-output connections, which account for mapping linearities. The mapping relationship modeled by the network is governed as:

$$y^T = f_2\left(\left[x^T, f_1\left(x^T \cdot \mathbf{W_{ih}}\right)\right] \cdot \left[\begin{array}{c}\mathbf{W_{io}} \\ \mathbf{W_{ho}}\end{array}\right]\right), \tag{2}$$

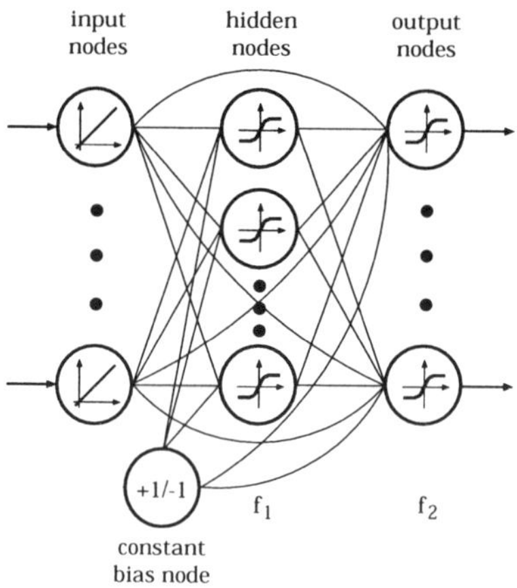

Figure 4: The architecture of neural network

where x^T and y^T are input and output vectors, respectively, $\mathbf{W}_{io}$ is the weight matrix of the input-output connection, $\mathbf{W}_{ih}$ is the weight matrix of the input-hidden connection, and $\mathbf{W}_{ho}$ is the matrix of hidden-output connection. The role of the training is to modify iteratively the variable weights of the network. In this case, only the hidden-output and input-output weights are modified, while random values are given to the rest and kept constant. This guarantees independent learning for each output. Another feature of the used algorithm is that the training starts with only one hidden node, and for each epoch a new hidden node is created, while the new input-hidden connection receives random weights. The weight coefficients are solved by a least-squares minimization, based on the singular value decomposition, which is strictly related to the regularization approach. The continuous node creation and increasing-resolution tuning of the map is the most important quality of this algorithm.

Scanning data, both the real and imaginary part of the signal, are taken as inputs to the network x^T. The outputs of the NN are binary data. Seven output nodes carry coded information about the electromagnetic characteristics at one scanning point. The conductivity variation at this point is transformed into a four bit binary number covering a range from 0.9 to 1.1 MS/m. Similarly, three bit binary coding is used for the relative permeability variation in a range 1.0 - 1.06 for one scanning point. The limits of the electromagnetic characteristics variations were adjusted according to the results of recognition along the quasi stable regions based on the uniform model.

The database of the simulated signals and the most probable conductivity and permeability profiles was used for the network training. To decrease a degree of freedom from the characteristics variations point of view, some preliminary study had been carried out by comparing the measured and simulated signals. On the basis of the study, the most probable profiles of conductivity and permeability were used for numerical modeling. Further reduction of the characteristics variation space came from the results of reconstruction in the quasi stable regions.

5 Results of recognition

A scanning across the weld line shown in Fig. 1 was performed, and the coil signals for three different frequencies of 200 kHz, 300 kHz, 400 kHz and a lift-off of 0.25 mm were stored for further analyses. The length of scanning line is 90 mm, and signals were scanned with 1 mm pitch. Figure 5 shows the profiles of the real and imaginary parts of the coil signal for the frequency of 200 kHz.

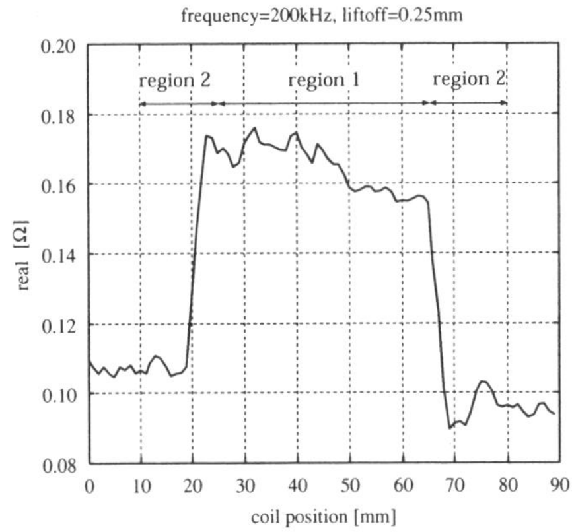

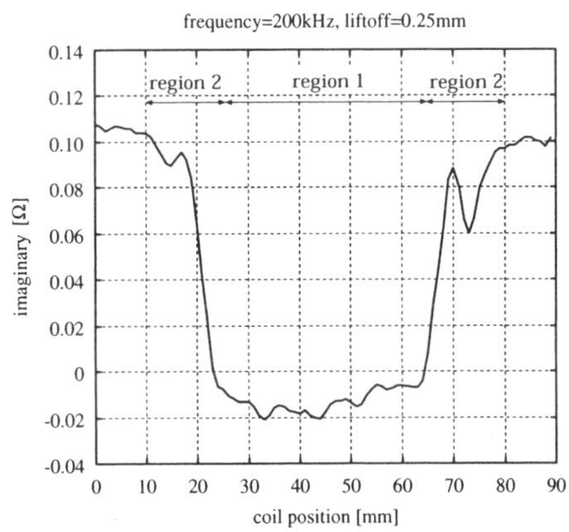

Figure 5: The profile of real and imaginary part of the coil signal across the weld line

The scanning line is divided into two regions based on the strategy described in the section 2 (Fig. 5): the quasi stable region - middle part of the weld (region 1) and the beginning (0-10mm) and the end (80-90mm) of scanning line, and the transient region (region 2).

The reconstructed profile of conductivity across the weld line is shown in Fig. 6. The electromagnetic characteristics in the quasi stable regions were estimated based on the uniform model for all the three frequencies. The plot clearly shows that the characteristics depend also on the skin depth, because the curves for different frequencies are shifted. This is mainly caused by a mechanical treatment of the specimen surface. The weld has approximately 7-8%

lower conductivity than that of the base material. The variation of permeability between the quasi stable regions of base material and weld is on the level of error of the method and can be neglected. Therefore, one can conclude that the permeability does not change significantly in the weld region of INCONEL steel.

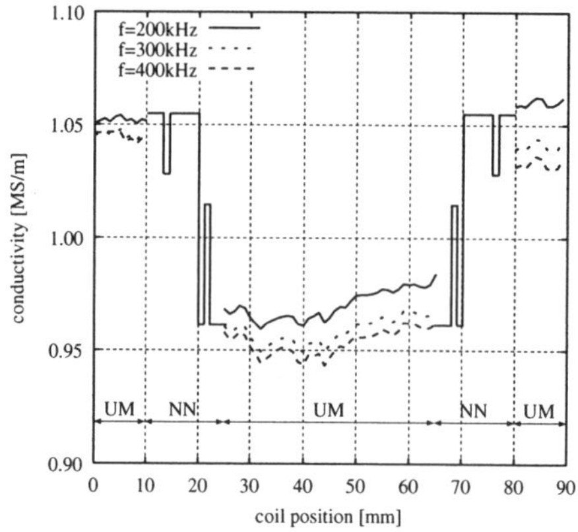

Figure 6: The reconstructed profile of conductivity

In the transient regions, the electromagnetic characteristics were estimated by using the neural network (Fig. 6 - NN) for the frequency 200 kHz. The change of permeability can be also neglected. The conductivity changes sharply at the beginning of weld line.

To prove the results of NN reconstruction, the coil signal along the transient region was simulated using the reconstructed characteristics profiles. The numerical result coincides with measured one in both shape and values as it can be seen in Fig. 7.

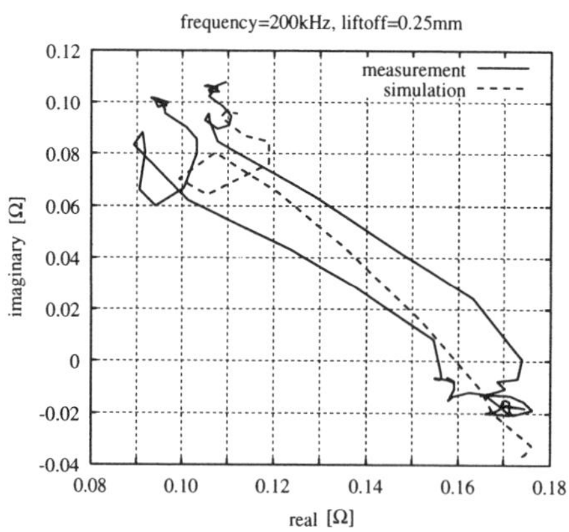

Figure 7: Comparison between the measured and simulated signal along the transient region

6 Conclusion

The profiles of electromagnetic parameters across the INCONEL weld were estimated by means of the eddy current testing. Two independent inverse techniques were employed to reconstruct the characteristics of INCONEL weld along two regions. It was found out that the value of weld conductivity is about 7-8% smaller than that of base material in this case, and changes rapidly at the beginning of weld line. The variation of permeability in the weld region and also heat affected zone is negligible.

Further work will be focused on the study of more welded specimens because the range of INCONEL weld conductivity variation depends on the particular conditions of welding.

References

[1] R.M. Horn, G.M. Gordon, F.P. Ford, R.L. Cowan: Experience and assessment of stress corrosion cracking in L-grade stainless steel BWR internals. In: Nuclear Engineering and Design 174, 1997, pp. 313-325.

[2] W. Cheng, M.Shiwa, H. Yoneyama, H. Huang, T. Takagi and T. Uchimoto: Ultrasonic and Eddy Current Testing of Defects in Inconel Welding Metals. In: Proceedings of 12th MAGMA conference, Oita, Japan, 27-28 of March 2003, pp. 187-190.

[3] J. P. Lareau: Phased Array Imaging First Use Qualification Effort: BWR Feedwater Nozzle Inner Radius Inspection from Vessel OD for a US Nuclear Power Plant. In: NDT.net - May 2002, Vol. 7, No. 5.

[4] B.P.C. Rao, B. Raj, T. Jayakumar, P. Kalyanasundaram: An artificial neural network for eddy current testing of austenitic stainless steel welds. In: NDT&E International,Vol. 35, 2002, pp. 393-398.

[5] Ananth Sethuraman and James H. Rose: Rapid Inversion of Eddy Current Data for Conductivity and Thickness of Metal Coatings. In: Journal of Nondestructive Evaluation, Vol. 14, No. 1, 1995, pp. 39-46.

[6] Benoit de Halleux, Bruno de Limburger Strum and Andrei Pitchblendes: Eddy current measurement of the wall thickness and conductivity of circular non-magnetic conductive tubes. In: NDT&E International, Vol. 29, No. 2, 1996, pp. 103-109.

[7] Benoit de Halleux and Andrei Ptchelintsev: Precision measurement of thickness and conductivity of circular and plane products with eddy current metrology. In: INSIGHT, Vol. 39, No. 12, December 1997, pp. 854-857.

[8] T.P. Theodoulidis, T.D. Tsiboukis and E.E. Kriezis: Analytical Solution in Eddy Current Testing of Layered Metals with Continuous Conductivity Profiles. In: IEEE Transactions on Magnetics, Vol. 31, No. 3, May 1995, pp. 2254-2259.

[9] Erol Uzal, John C. Moulder, Sreeparma Mitra and James H. Rose: Determining the conductivity and thickness of continuously varying layers on metals using eddy currents. In: Review of Progress in Quantitative Nondestructive Evaluation, Vol. 12, 1993, pp. 251-258.

[10] C. Glorieux, J. Moulder, J. Basart and J. Thoen: The determination of electrical conductivity profiles using neural network inversion of multi-frequency eddy-current data. In: J. Phys. D: Appl. Phys.,Vol. 32, 1999, pp. 616-622.

[11] Radu C. Popa and Kenzo Miya: Approximate Inverse Mapping in ECT, Based on Aperture Shifting and Neural Network Regression. In: Journal of Nondestructive Evaluation, Vol. 17, No. 4, 1998, pp. 209-221

Author Index